The Lever of Riches

THE LEVER OF RICHES

Technological Creativity and Economic Progress

JOEL MOKYR

OXFORD UNIVERSITY PRESS
New York Oxford

Oxford University Press

Oxford New York Toronto
Delhi Bombay Calcutta Madras Karachi
Petaling Jaya Singapore Hong Kong Tokyo
Nairobi Dar es Salaam Cape Town
Melbourne Auckland

and associated companies in
Berlin Ibadan

First published in 1990 by Oxford University Press, Inc.,
200 Madison Avenue, New York, New York 10016

First issued as an Oxford University Press paperback, 1992

Library of Congress Cataloging-in-Publication Data
Mokyr, Joel.
The lever of riches:
technological creativity and economic progress
Joel Mokyr.
p. cm. Includes bibliographical references.
ISBN 0-19-506113-6
ISBN 0-19-507477-7 (pbk.)
1. Technological innovations—Economic aspects—History.
2. Economic development—History. I. Title.
HC79.T4M648 1990 338'.064—dc20 89-28298

4 6 8 10 9 7 5 3

Printed in the United States of America

To the Memory of My Mother
Who did it all by Herself

Preface

This is a book about technological creativity. Why would an economic historian write such a book? Technology is about how to make things and services that are useful and enjoyable, that is, it is about production. The difference between rich nations and poor nations is not, as Ernest Hemingway said, that the rich have more money than the poor, but that rich nations produce more goods and services. One reason they can do so is because their technology is better; that is, their ability to control and manipulate nature and people for productive ends is superior. If the West is on the whole comfortable, even opulent, compared to the appalling poverty still rampant in most of Asia and Africa, it is in large part thanks to its technology. As one author has put it, we have to "give credit to just those dull, everyday, pragmatic, honest betterments in simple technology . . . that were taking place in Europe . . . doses of which would wash away much of the misery in the world today" (Jones, 1981, p. 69). Western technological superiority has deep historical roots, and can only be understood—if at all—by an analysis that is willing to look back centuries, even millennia. To be sure, technology cannot take all the credit: the development of law, trade, administration, and institutions were all part of the story. Yet, as I shall try to show, technological creativity was at the very base of the rise of the West. It was the lever of its riches.

How does technology change? Like science and art, it changes through human creativity, that rare and mysterious phenomenon in which a human being arrives at an insight or act that has never been accomplished before. Of course, technological creativity is quite different from artistic or scientific creativity. It tends to be more down-to-earth, with such mundane characteristics as dexterity and greed at the center of the act. Yet it does share with the arts and the sciences its occasional dependence on inspiration, luck, serendipity, genius, and the unexplained drive of people to go somewhere where none has gone before. Although some portion of invention comes today from the cold and calculating minds of Research and Development engineers in white labcoats worn over three-piece-suits, much of the technological creativity that made our economic world what it is came from different sources. I will make an effort to describe and then explain this creativity. However, it is a task that can never be carried out completely. At some level we stand in uncomprehending awe before the miracle of human genius. We can no more "explain" the breakthroughs inside the minds of a Montgolfier or a Westinghouse than we can explain what went on inside the head of Beethoven when he wrote the *Eroica.* Economists and historians alike realize that there is a deep difference between *homo economicus* and *homo creativus.* One makes the most of what nature permits him to have. The other rebels against nature's dictates. Technological creativity, like all creativity, is an act of rebellion. Without it, we would all still live nasty and short lives of toil, drudgery, and discomfort.

It is always a pleasant task to thank the many friends and colleagues who have made this book possible, and are responsible in some way for many of the arguments and opinions expressed in it. As my name appears on the title page and theirs do not, however, I alone shall carry the blame for all errors and misstatements and suffer the wrath of the wronged and misquoted.

The project began many years ago when my former colleague F. Michael Scherer walked into my office at the Northwestern University Economics Department and proposed a "small monograph" on the history of technological progress that, he assured me, would take me at most a few months over the summer to write. I did not buy the Brooklyn Bridge from him that afternoon, but I did make a promise that would result in many long nights in libraries and in front of my wordprocessor. Mike has thus sown the seed for this book, and while it was I who had to carry it to term, he has been supportive and helpful throughout the project. I can only hope that the readers will not hold anything in this book against him.

Intellectually, the inspiration for this book came from a variety of sources, many of which I cannot recall. Teaching economic history for fourteen years means that one inevitably absorbs large amounts of information that cannot possibly be traced to their sources. A number of books, however, stick out as I find myself going back to them again and again for insight, inspiration, and information. They are David Landes' *Unbound Prometheus,* Lynn White's *Medieval Religion and Technology,* Eric Jones's *European Miracle,* Abbott Payson Usher's *History of Mechanical Inventions,* and Donald Cardwell's *Turning Points in Western Technology.*

Of my Northwestern colleagues, I would like to mention those who have not only read the manuscript in its numerous half-baked versions and gave me good advice and encouragement, but also provided the comradeship and emotional support few can do without. They are Louis Cain, Charles Calomiris, Karl de Schweinitz, Jack

Goldstone, David Hull, Jonathan R. T. Hughes, and Sarah Maza. Many of my graduate students, upon whom successive versions of the manuscript were inflicted, made valuable contributions. Among them, I am particularly grateful to Katherine Anderson, Avner Greif, Paul Huck, Lynne Kiesling, John Nye, Gabriel Sensenbrenner, Dan Shiman, Richard Szostak, and Martha Williams. I have come to realize the profound truth in Rabbi Akiva's famous dictum, that he learned much from his teachers, more from his colleagues, and most from his students.

Outside Northwestern, my greatest debt as always is to Cormac Ó Gráda, whose almost legendary patience and wisdom have been an indispensable resource to me for many years now. The manuscript was read by an embarrassingly long list of other friends, who found an equally embarrassing list of errors and omissions, and made many invalid complaints, the most vociferous of which was that I did not write the book they would have if they had written it. This list contains Robert C. Allen, William Baumol, Reuven Brenner, Julia Burns, Paul A. David, Jan De Vries, Stefano Fenoaltea, George Grantham, C. Knick Harley, Dan Headrick, Eric Jones, William McNeill, Donald McCloskey, William N. Parker, Richard Szostak, Andrew Watson, K. D. White, and Bing Wong. A conference of the All University of California group in Economic History in October 1988 discussed an earlier version of this book and provided me with many good ideas, not all of which I could carry out.

My research assistant Erik Zehender loyally assisted me in plundering the shelves of the Northwestern Libraries for three years, as well as provide the Interlibrary Loan office with a medium run increase in the demand for their services. Mrs. Barbara Karni edited the manuscript with her usual thoroughness and competence. Mr. Herbert Addison at Oxford University Press contributed generously from his knowledge of books and publishing to this book in its final stages. The secretaries at the Northwestern department of economics, Mrs. Angie Campbell, Mrs. Ann Roth, and Mrs. Florence Stein, have given me their help and support over many years. A special debt is acknowledged to Mr. Jack Repcheck of Princeton University Press.

My wife Margalit was always at my side when I needed her. My daughters Naama and Betsy were there even at times when I did not.

Evanston, IL J.M.
January 1990

Contents

PART ONE

ECONOMIC GROWTH AND TECHNOLOGICAL PROGRESS

CHAPTER ONE

Introduction

Past economic growth is crucial to the material aspects of our existence: the best predictor of the living standard that a newborn baby can expect to enjoy is the accident of where he or she is born. There is very little in common between the quality of life that can be expected by an average person born in, say, rural Cameroon or urban Java, and one born in Greenwich, Connecticut or Oslo, Norway. The difference is captured by a somewhat artificial statistic concocted by economists known as national income or gross national product per capita. The current level of this magnitude is determined by its past. In economics, history is destiny.

To say, then, that a country is rich is to say that it experienced economic growth in the past. While this statement describes everything, it explains nothing. The causes of economic growth—why some societies grew rich and others did not—have been pondered by economists, sociologists, historians, and philosophers for centuries. This book is another attempt to struggle with this issue of issues. It focuses on what I believe to have been one key ingredient of economic growth: technological creativity. The book puts forth an argument about the causes of economic progress, about rising living standards, about improving nutrition, clothing, housing, and health, and about reducing toil, drudgery, famine, and disease. Technological progress has been one of the most potent forces in history in that it has provided society with what economists call a "free lunch," that is, an increase in output that is not commensurate with the increase in effort and cost necessary to bring it about.

This view of technological change is inconsistent with one of the most pervasive half-truths that economists teach their students, the hackneyed aphorism that there is no such thing as a free lunch. It is the purpose of this book to highlight the greatest counterexample to this statement.[1] Economic history is full of examples of free lunches,

1. For a similar statement, rare among economists, see Kamien and Schwartz (1982, p. 216). Kamien and Schwartz, too, regard technological change as a "trick" that makes it possible to avoid making a choice when faced with Samuelson's famous query "which

as well as (more frequently) very cheap lunches. At the same time, there are endless instances of very expensive meals that ended up inedible and in some cases lethal. To phrase it differently, technological change is primarily the study of what economists call outward shifts of the production possibility frontier, that is, increases in the productive potential of the economy. Through most of history, however, societies have not been *on* the frontier, where they fully exploited their resources, but at a point inside it, where waste and inefficiency meant that living standards were lower than they would have been if the resources available had been deployed efficiently. War, discrimination, unemployment, superstition, barriers to trade and economic freedom, resource mismanagement, and many other instances of human inefficiency meant that only a fraction of the resource potential available was utilized to produce goods that conveyed economic utility. By reducing waste, economies could increase their living standards without increasing the amounts of labor and other resources needed. Although such an increase might well be regarded as some kind of free lunch, it is not the kind I shall be concerned with. Instead, I shall focus on shifts *of* the production frontier itself, that is, increases in the productive potential of the economy, as they are most frequently identified with economic growth.

Not all economic growth is necessarily related to technology. Roughly speaking, economic growth can occur as the result of four distinct processes:

1. Investment (increases in the capital stock). The productivity of labor, and with it the average standard of living, depends on the quantities and qualities of the equipment and tools an average worker has at his or her disposal (known among economists as the capital–labor ratio). When capital accumulates at a rate faster than the growth of the labor force, so that each worker has increasingly more capital to work with, economic growth will take place; that is, output per capita will increase. We can call this kind of phenomenon *Solovian growth,* in honor of Robert Solow, who laid the foundation of the modern theory of economic growth. At first glance, this type of growth involves no free lunch. Investment is made possible by saving; to save by definition is to refrain from consumption in the present in order to consume more in the future. Thus, all future benefits are paid for by a willingness to abstain in the present, which is in itself undesirable and thus costly.

one," and have "both." Their analysis is cast largely in terms of a modern market in which research and development is carried out in a systematic way. McCloskey (1981, p. 117) points out that the Industrial Revolution must be "a bitter disappointment to the scientists of scarcity" and refers to the no-free-lunch dogma as "a mildly comical jargon."

2. Commercial expansion. It is the standard fare of intermediate courses in microeconomics to illustrate that an increase in the exchange of goods, services, labor, or capital can be beneficial to all partners involved. Abstracting from the cost of transacting, the creation of commerce and voluntary exchange between two previously disjoint units—be they individuals, villages, regions, countries, or continents—leads to increases in income for both partners. These gains are known as the gains from trade and represent a good example of a free lunch. Trade, as Adam Smith pointed out in 1776, leads to a growth in the Wealth of Nations. Smith's mechanism of growth was based on the idea that a finer division of labor leads to productivity growth through specialization and the adaptation of skills to tasks. Economic growth caused by an increase in trade may be termed *Smithian growth* (following Parker, 1984).[2] Trade is created by a decline in transactions costs (that is, the costs involved in bartering or buying and selling goods and services on the market), or the improved assignment and enforcement of property rights. It is, however, not the kind of free lunch I shall be much concerned with in this book.

3. Scale or size effects. It is sometimes maintained that population growth itself can lead to per capita income growth (e.g., Simon, 1977; Boserup, 1981). Clearly, if the division of labor increases prosperity, then for very small populations growth in numbers alone would make specialization possible and lead to gains in output. Moreover, at least up to some point there are fixed costs and indivisibilities, such as roads, schools, property-rights enforcement agencies, and so on, that can be deployed effectively only for relatively large populations (North, 1981). When an increase in the scale of the economy by itself leads to growth in the economy's per capita productive potential through the mere multiplication of numbers, it will appear to individuals as a free lunch. A continuous growth in population, however, will increase the pressure of population on other resources that do not grow (such as land and other natural resources) or grow more slowly, and the economy will move from a regime of increasing to a regime of diminishing returns. When this crowding effect begins to be felt, further population growth will lead to intensification of production, which causes average income to decline. It is still possible for other factors to offset diminishing returns, so that population

2. The term is slightly misleading because Adam Smith emphasized the gains from trade that derived from the division of labor, specialization, and the resulting productivity gains. The standard gains from trade model, developed by David Ricardo, is based on comparative advantage and does not depend on the Smithian notions of specialization. Adam Smith emphasized demand as the limit to specialization, while Ricardo's model holds independent of the size of the market.

growth could be *accompanied* by economic growth. But it would be incorrect to infer in such a case that population growth *causes* economic growth.

4. Increases in the stock of human knowledge, which includes technological progress proper, as well as changes in institutions. Again following Parker, I shall refer to this type of process as *Schumpeterian growth,* in honor of Joseph A. Schumpeter, whom we will encounter repeatedly later in this book. Parker (1984, p. 191) defined Schumpeterian growth as "capitalist expansion deriving from continuous, though fluctuating, technological change and innovation, financed by the extension of credit." This book is indeed about little else but technological change and innovation. By technological progress I mean any change in the application of information to the production process in such a way as to increase efficiency, resulting either in the production of a given output with fewer resources (i.e., lower costs), or the production of better or new products. Unlike Parker, however, I believe we should not confine the idea of Schumpeterian growth to capitalist expansion financed by credit alone. Technological progress predated capitalism and credit by many centuries, and may well outlive capitalism by at least as long.

The choice of the words "application of information" is deliberate: much growth, as we shall see, is derived from the deployment of previously available information rather than the generation of altogether new knowledge (Rosenberg, 1982, p. 143). When all is said and done, from the point of view of economic growth it does not really matter whether income grows because of the application of *entirely new* information to production even if a proper definition of what "new" exactly means here could be agreed upon, or the diffusion of *existing* information to new users.

The historical record of technological change is uneven and spasmodic. Some brief spans in the history of a particular nation—such as Britain between 1760 and 1800 or the United States after 1945—are enormously rich in technological change. These peaks are often followed by periods during which technological progress peters out. Why does that happen? Although economists, sociologists, and historians have written extensively about this question, they have found its explanation elusive. One economic historian (Thomson, 1984, p. 243) has quipped that "technical change is like God. It is much discussed, worshipped by some, rejected by others, but little understood." There are good reasons for this lack of understanding. The diversity of technological history is such that almost any point can be contradicted with a counterexample. Picking up empirical regularities in this massive amount of qualitative and often uncertain and incomplete information is hazardous. Yet without it the painstaking work of technological historians seems pointless, and the role of technology in the history of our economies will remain incomprehensible.

When the resource basis of an economy expands, it can do one of two things: it can enjoy higher living standards, or it can, in H. G. Wells's famous phrase, "spend the gifts of nature on the mere insensate multiplication of common life." In recent history, economic growth has occurred *despite* population growth. Before that, as Malthus and the classical political economists never tired of pointing out, the growth of population relentlessly devoured the fruits of productivity growth, and living standards, as far as we can measure them, changed little in the very long run. It might, therefore, make sense, for some purposes, to define growth as an increase in total income rather than income per capita. This is the approach taken by Jones (1988).

The literature produced by modern economists on technological change is vast.[3] All work on economic growth recognizes the existence of a "residual," a part of economic growth that cannot be explained by more capital or more labor, and that thus must to some extent be regarded as a free lunch. Technological change seems a natural candidate to explain this residual and has sometimes been equated with it forthwith. This literature has not, however, been very successful in explaining why some societies are technologically more creative than others. It is rarely informed by economic history, confining itself mostly to the post-1945 period. It is also rarely informed by technological history. When a historian of technology, such as Abbott P. Usher, is cited, it is more for his interesting but speculative application of Gestalt psychology to invention than for his enormous knowledge of how machines actually evolved over time (see, for instance, Thirtle and Ruttan, 1987, pp. 2–5). Economists typically approach the explanation of technological change by considering the relationship between demand and supply variables, research and development, and productivity growth. In so doing, they implicitly treat technology as an input—albeit one with peculiar features—that is produced and sold in the market for research and development. Such a market analysis may or may not be a useful description of the post-1945 period (Jewkes, Sawers, and Stillerman, 1969; Langrish et al., 1972). It is clear, however, that for an explanation of the diffusion of wind power in medieval Europe, the adoption of iron casting in China during the Han dynasty, or the adoption of intensive husbandry in seventeenth-century Britain, such a framework is wholly inappropriate. Technological change throughout most of history can

3. For recent surveys, see Thirtle and Ruttan (1987); Baldwin and Scott (1987); Coombs et al. (1987); Wyatt (1986).

hardly be regarded as the consequence of an orderly process of research and development. It possessed few elements of planning and precise cost-benefit calculation. How, then, to explain it?

Once the economist ventures outside the safe realm of traditional microeconomics and agrees to consider extraeconomic factors, he or she often discovers that events are hopelessly overdetermined, that is, there are many plausible explanations for every phenomenon. Theories of technological change based on geographic, political, religious, military, and scientific factors are typically easy to concoct and hard to reject. Many explanations make sense. But are they likely to be correct? Posing the question in this way may not be very useful; it might be better to ask whether such explanations are persuasive. Can we amass enough evidence to show that a particular theory is supported by facts and not just logic? In what follows, I shall try to pursue such a methodology.

By focusing on Schumpeterian growth I am not downgrading other forms of economic growth. Technological change unaccompanied by other forms of growth is rare. The four forms of growth reinforce each other in many complex ways. For example, a widely held view of technological change maintains that much technological change is embodied in new capital goods, so that in the absence of capital accumulation, technological change would be slow. To the extent that this view is correct, Solovian growth and Schumpeterian growth go hand in hand. Schumpeterian growth can also lead to Smithian growth, as it did, for example, when technical advances in shipping led to increased gains from trade by lowering transportation costs. To keep matters relatively simple, I shall focus here on technological change proper, dealing with other forms of economic growth only insofar as they touch upon technological change directly.

The study of technological change inevitably must move between the aggregate and the individual levels of analysis. Economic growth itself is by definition an aggregate process, the processes of invention and adoption usually being carried out by small units (individuals or enterprises). The economic historian is therefore directed to the macrofoundations of technological creativity, that is to say, what kind of social environment makes individuals innovative, what kind of stimuli, incentives, and institutions create an economy that encourages technological creativity? Technological creativity is analyzed largely as a social, rather than an individual, phenomenon. In other words, the question I am interested in here is not why some individuals are more creative than others, but why were and are there societies that have more creative individuals in them than others? It is that question that lies at the foundation of the issue of issues, why does economic growth (at least of the Schumpeterian type) occur in some societies and not in others.

It may turn out, as Heertje (1983, p. 46) has put it, that technological change cannot be explained. By that he means, I think, that standard economic theory which deals, after all, with rational choices subject to known constraints, faces a dilemma in dealing with technological creativity. Technological change involves an attack by an individual on a constraint that everyone else takes as given. The methodology of economics is largely based on the idea that economic agents try to do as well as they can given their constraints, but by definition can do nothing to change these constraints themselves. Therefore, much of the research in the economics of technological change has been sidetracked to secondary questions, such as whether a particular technology was relatively labor- or capital saving, or what the effect of cyclical variation in demand was on the rate of patenting.

One distinction between different types of technological progress was made by Francis Bacon. In his view, inventions came in two categories, those that depended on the overall state of knowledge and thus could be made only when the scientific and informational background became available, and those that were purely empirical and that could therefore have been made at any time in recorded history. In practice, this distinction is not always easy to make. Some apparently empirical inventions did in fact depend on subtle changes in the understanding and assumptions of the inventor about his or her physical environment, or changes in the availability of a crucial material or component that made the invention practical. Thus, as Cardwell (1968) points out, Bacon was mistaken when he asserted that moveable-type printing could well have been invented by the ancient Greeks, because it depended on medieval advances in metallurgical technology. Yet it is clear that in many cases there is indeed no good reason why an invention was made at a particular point in time and not centuries earlier. It is tempting to argue that lack of perceived need or demand was responsible, but it is also possible that a particular innovation simply did not occur to anyone before. Some medieval inventions of great usefulness and simplicity, such as the wheelbarrow and the stirrup, must be classified in this category.

There is a rich tradition among economists interested in technological change distinguishing between invention and innovation. Schumpeter pointed out that invention does not imply innovation, and that it was innovation that provided capitalism with its dynamic elements. Invention, by that logic, is in itself of limited interest to economists. Ruttan (1971, p. 83) has proposed more or less to abandon the concept of invention altogether and regard it merely as an "institutionally defined subset of technical innovation." Again, these distinctions are not always helpful. During the implementation stages, inventions were usually improved, debugged, and modified in ways that qualify the smaller changes themselves as inventions. The diffusion of innovations to other economies, too, often required adaptation to local conditions, and has in most cases implied further productivity gains as a result of learning by doing.

Before we rashly decide that inventions should not be the focus of the study of the economic history of technological change, it may be useful to reflect further upon the issue. An invention should be defined as an increment in the set of the total technological knowledge of a given society, which is the union of all sets of individual technological knowledge.[4] It is possible to argue that this set is itself not a meaningful concept, but without it, most cultural and intellectual history makes little sense. More important, it could be objected that this set is irrelevant. New knowledge that is not applied makes no difference to economic welfare, whereas the bulk of Schumpeterian economic growth comes from the application of *old* knowledge. To state it in somewhat different terms, at any moment there is a large gap between *average-* and *best-*practice technology; reducing this gap by disseminating the techniques used by producers at the cutting edge of knowledge is technological progress without invention. The diffusion of technology from advanced regions or countries to backward ones also represents technological progress without invention. But inferring from these observations that invention should therefore not be at the center of any discussion of the history of technology in economic growth would be unwarranted. Any discussion of the gap between average- and best-practice techniques makes little sense unless we have some notion of where the best-practice technique came from in the first place. Without further increments in knowledge, technological diffusion and the closing of the gap between practices will run into diminishing returns and eventually exhaust itself. It is thus incumbent on us to examine the cases in which this exhaustion has not occurred and to ask why.

To put it differently, a basic premise of this book is that invention and innovation are complements. In the short run, this complementarity is not perfect; it is indeed possible to have one without the other. But in the long run, technologically creative societies must be both inventive and innovative. Without invention, innovation will

4. Strictly speaking, the society over which this set is defined should be all of humanity. In the case of largely disjoint parts of humanity it seems to make sense to restrict this set somewhat so that it is possible to speak of independent inventions.

eventually slow down and grind to a halt, and the stationary state will obtain. Without innovation, inventors will lack focus and have little economic incentive to pursue new ideas. I shall argue in Part III that this complementarity is one of the reasons so few societies have been technologically creative. Many diverse conditions have to be satisfied simultaneously. Invention depends on factors that determine individual behavior, as the inventor is ultimately alone in his or her attempt to make something work. Innovation, on the other hand, requires interaction with other individuals, depends on institutions and markets, and is thus largely social and economic in nature.

No single answer to the question of why technological creativity occurs in some societies and not in others will ever satisfy everyone. "A single comprehensive general law of evolution," wrote Gerschenkron (1967, p. 448), "lies on the other side of the line that separates serious research from fanciful superficialities." Yet clearly we can say more than simply that "everything mattered" or identify the salient differences between creative and uncreative societies, respectively, and draw a causal relationship accordingly. Correlation does not prove causation. Ireland's Catholicism, central Africa's climate, and Southeast Asia's dependence on rice have all been falsely identified as causal factors in the lack of technological creativity in these societies. Economic analysis can help us to identify some factors as important, to cast doubt on others, and to design tests to assess those factors that are ambiguous from a theoretical point of view.

To see how and why technological creativity occurs, we thus must distinguish two basic components in the invention-innovation sequence. The first component is that technical problems involve a struggle between mind and matter, that is, they involve, control of the physical environment. Nature, it is said, yields her secrets in a niggardly manner. Teasing these secrets out of her and then manipulating them for material benefit is the essence of any technological breakthrough. The outcome is determined by the daring and ingenuity of the inventor, the limitations of the materials and tools at his or her disposal, and the resistance of the laws of nature to bending to the inventor's will. The other component is social. For a new technique to be implemented, the innovator has to interact with a human environment comprised of competitors, customers, suppliers, the authorities, neighbors, possibly the priest. For a society to be technologically creative, three conditions have to be satisfied. First, there has to be a cadre of ingenious and resourceful innovators who are both willing and able to challenge their physical environment for their own improvement. Innovation of any kind is unlikely in a society that is malnourished, superstitious, or extremely traditional. Second, economic and social institutions have to encourage potential innovators by presenting them with the right incentive structure. In part, such incentives are economic; technological creativity is more likely if an innovator can expect to become rich. Noneconomic incentives can matter too, however. A society can reward successful innovators by awarding them medals, Nobel prizes, or intangible symbols of prestige. Third, innovation requires diversity and tolerance. In every society, there are stabilizing forces that protect the status quo. Some of these forces protect entrenched vested interests that might incur losses if innovations were introduced, others are simply don't-rock-the-boat kinds of forces. Technological creativity needs to overcome these forces.

An essay on the economic history of technological change inevitably contains dates, names, and places. By its nature, the tale of technological creativity requires citing who first came up with an idea and who made the critical revisions and improvements necessary for the idea to work. Yet in the past decades, economic historians have not practiced this type of history. As David (1987) asks, does technology not simply accumulate continuously from the incremental, almost imperceptible, changes brought about by a large number of anonymous people? Some historians insist that almost all invention consists of this "technological drift" (as Jones, 1981, p. 68, has termed it), consisting mostly of anonymous, small, incremental improvements (Rosenberg, 1982, pp. 62–70). As a reaction to heroic theories of invention in which all improvements are attributed to individual geniuses, the drift theory has been justly influential. But is it not possible to go too far in the other direction and give too little credit to major inventions made by a vital few? Some inventions, such as the printing press, the windmill, and the gravity-driven clock, contradict the generality of the gradualist model of technological progress. There have been, and probably always will be, large and discrete changes in technology that sweep the world off its feet and make it rush in to acquire and imitate the novelty. Modern research has shown, to be sure, that most cost savings are achieved through small, invisible, cumulative improvements. But improvements in what? Virtually every major invention was followed by a learning process, during which the production costs using the new technique declined; but for these costs to fall, the novelty had to be invented in the first place. An adult weighing 150 lbs has acquired about 95 percent of his or her weight since birth—does that mean that conception is not important?

In discussing the distinction between minor inventions, whose cumulative impact is decisive in productivity growth, and major technological breakthroughs, it may be useful to draw an analogy between the history of technology and the modern theory of evolution, a comparison that I shall pursue in Chapter 11. Some biologists distinguish between micromutations, which are small changes in an existing species and which gradually alter its features, and macromutations, which create new species. The distinction between the two could provide a useful analogy for our purposes. I define *microinventions* as the small, incremental steps that improve, adapt, and streamline existing techniques already in use, reducing costs, improving form and function, increasing durability, and reducing energy and raw material requirements. *Macroinventions,* on the other hand, are those inventions in which a radical new idea, without clear precedent, emerges more or less ab nihilo. In terms of sheer numbers, Microinventions are far more frequent and account for most gains in productivity. Macroinventions, however, are equally crucial in technological history.

The essential feature of technological progress is that the macroinventions and microinventions are not substitutes but complements. Without subsequent microinventions, most macroinventions would end up as curiosa in musea or sketchbooks. Indeed, in some historical instances the person who came up with the improvement that clinched the case receives more credit than the inventor responsible for the original breakthrough, as is the case of the steam engine, the pneumatic tire, and the bicycle. But without novel and radical departures, the continuous process of improving and refining existing techniques would run into diminishing returns and eventually peter out. Microinventions are more or less understandable with the help of standard economic concepts. They result from search and inventive effort, and respond to prices and incentives. Learning by doing and learning by using increase economic efficiency and are correlated with economic variables such as output and employment. Macroinventions, on the other hand, do not seem to obey obvious laws, do not necessarily respond to incentives, and defy most attempts to relate them to exogenous economic variables. Many of them resulted from strokes of genius, luck, or serendipity. Technological history, therefore, retains an unexplained component that defies explanation in purely economic terms. In other words, luck and inspiration mattered, and thus individuals made a difference. Scholars who cast doubt on the importance of individuals often rely on a dispensability axiom: if an invention had not been made by *X* it would have been made by *Y*, a conclusion typically inferred from the large number of simultaneous discoveries. Although this regularity holds true for some inventions, including the telephone and the incandescent lightbulb, it is not applicable to scores of other important inventions.

If there is any area in which a deterministic view that outcomes are shaped inexorably by forces stronger than individuals—be they supply and demand or the class struggle—is oversimplified, it is in the economic history of technology. Asking whether the major breakthroughs are more important than the marginal improvements is like asking whether generals or privates win a battle. Just as in military history we employ shorthand such as "Napoleon defeated the Prussians at Jena in 1806," we may say that a particular invention occurred at this or that time. Such a statement does not imply that individual inventors are credited with all the productivity gains from their invention any more than the statement that Napoleon defeated the Prussians implies that he single-handedly defeated an entire army. It is useful to organize the narrative around the discrete event.

The distinction between micro- and macroinventions is useful because, as historians of technology emphasize, the word *first* is hazardous in this literature. Many technological breakthroughs had a history that began before the event generally regarded as "the invention," and almost all macroinventions required subsequent improvements to make them operational. Yet in a large number of cases, one or two identifiable events *were* crucial. Without such breakthroughs, technological progress would eventually fizzle out. Usher (1954, p. 64) emphasized the importance of "the art of insight," and although he insisted that no single individual was uniquely necessary, he also recognized that individual qualities and differences were at the center of the process.

This study will necessarily draw from a small sample and be heavily biased toward Western economic history. Before I can turn in Part III to an analysis of why some societies were creative and others were not, it is imperative to survey the facts. Too much writing about the economic aspects of technological progress takes place in an historical vacuum, or creates factual artifacts in which events and facts convenient to the author's interpretation are plucked selectively from the record. A more comprehensive knowledge of the main currents of technological history restrains the urge to make generalizations. In Part II, therefore, I summarize the main technological developments of the past 25 centuries. Such an undertaking seems absurdly ambitious: the five-volume *History of Technology* edited by Charles Singer and others and published in the late 1950s contained 4,000 pages of text and yet is widely considered incomplete. My intention is to survey only the most important innovations in a limited number of areas. I leave out the prehistoric and very early eras, as well as the post-1914 period. Even so, between about 500 B.C. and 1914 there is a record so stupefyingly abundant in facts and evidence, that I can only scratch the surface of a deep and rich seam. By largely ignoring civil engineering, architectural, medical, and military technology, and focusing only on those developments that demonstrably affected living standards, I have kept the survey within manageable proportions.

Following the historical survey, I analyze the differences between creative and uncreative economies. In Chapter 7 I lay out a series of explanations ranging from nutrition to religion, from institutions to values and *mentalités.* To examine the importance of those factors, in Chapters 8, 9, and 10 I apply them to three comparative studies—classical society versus the medieval West, China versus Europe in the post-1400 period, and the Industrial Revolution in Britain and the rest of Europe and the eventual decline of Britain as a leading technological society—that juxtapose societies that were technologically dynamic with others that were not.

Part IV looks at the dynamics of technological change. Specifically, it asks whether technological change occurs in leaps and bounds, or whether it takes place gradually and continuously. To answer this question, I set up an analogy between technological progress and biological evolution, and try to apply the concept of punctuated equilibrium to the analysis of technology.

The history of technological progress is open to accusations of "Whiggishness." It tends to be a tale of progress, of improvement, of the irrepressible advance toward a better and richer age. Given the difference between the standard of living in the West today and three centuries ago, I find criticisms of Whiggishness easy to ward off. E. H. Carr (1961) tells the tale of Czar Nicholas I, who was reputed to have issued an order banning the word *progress* from his realm, and adds acerbically that historians of the West have belatedly come to agree with him. It seems hardly open to dispute that what has happened to people's ability to manipulate the laws of nature in the service of economic ends is unidirectional and deserves the word progress. Without assessing the wider repercussions of changing technology on noneconomic intangibles, such as liberty and brotherly love, an economist's judgment of history, colored by the evidence of the perpetual struggle against poverty and drudgery, finds technological change worthy of the word progress. Of course, if technological change eventually leads to the physical destruction of our planet, survivors may no longer wish to use the word progress in their descriptions of technological history. Until then, however, I feel justified in using the term, not in the teleological sense of leading to a clearly defined goal, but in the more limited sense of direction. If the same bundle of goods can be produced at lower cost, provided that these costs are accurately measured to include social costs such as environmental damage—the term progress is suitable.

Yet the central message of this book is not unequivocally optimistic. History provides us with relatively few examples of societies that were technologically progressive. Our own world is exceptional, though not unique, in this regard. By and large, the forces opposing technological progress have been stronger than those striving for changes. The study of technological progress is therefore a study of exceptionalism, of cases in which as a result of rare circumstances, the normal tendency of societies to slide toward stasis and equilibrium was broken. The unprecedented prosperity enjoyed today by a substantial proportion of humanity stems from accidental factors to a degree greater than is commonly supposed. Moreover, technological progress is like a fragile and vulnerable plant, whose flourishing is not only dependent on the appropriate surroundings and climate, but whose life is almost always short. It is highly sensitive to the social and economic environment and can easily be arrested by relatively small external changes. If there is a lesson to be learned from the history of technology it is that Schumpeterian growth, like the other forms of economic growth, cannot and should not be taken for granted.

PART TWO

NARRATIVE

CHAPTER TWO

Classical Antiquity

Until recently, the consensus on classical civilizations (Greek, Hellenistic, and Roman) was that these societies were not very successful technologically (Finley, 1965, 1973; Hodges, 1970; Lee, 1973). As some recent critics have pointed out, such a judgment is overly harsh (K. D. White, 1984). First, some significant technological breakthroughs were made during classical times, and their extent is likely to be underrated by historians because of the scant literary and archaeological evidence remaining. Second, the notion that science could be applied to concrete objectives rather than be admired for its own sake was certainly present, and was particularly well developed among the Hellenistic mechanicians. Third, as Finley (1973, p. 147) emphasizes, in some sense the harsh judgment of classical societies imposes our own value system on a society that had no interest in growth. "As long as an acceptable life-style could be maintained, however that was defined, other values held stage." In the areas that mattered most to *them,* the Greeks and Romans achieved huge successes. Cipolla (1980, p. 168) adds that because our own society has a civilization that is mechanistic, we tend to identify technology to a large extent with mechanics, whereas classical civilization was oriented toward other forms of technology. Some of the important achievements of classical technology were in those aspects of technology that were nonphysical in nature: coinage, alphabetization, stenography, and geometry were part of the information-processing sphere rather than the physical production sphere of the economy. Even when their achievements were in the physical sphere, they were mostly in construction and architecture, rather than in mechanical devices. Nonetheless, the judgment reflects our instinctive disappointment with a civilization that celebrated such triumphs in literature, science, mathematics, medicine, and political organization.[1] Even in nonmechani-

1. In contrast, the early Iron Age (1100 B.C. to 500 B.C.) saw, in addition to the development of the use of iron, the development of most carpenters' tools (lathes, saws, pegs), shears, scythes, axes, picks, and shovels; the invention of the rotary quern (used to grind flour and ores); and improvements in the construction of ships and wagons.

cal aspects of technology, such as chemistry and farming, the record of the classical world seems to fall short of what appears to us to have been its potential.

What technological progress there was in the classical world, especially in Roman times, served the public, rather than the private, sector. Roman leaders acquired popularity and political power by carrying out successful public works. The history of the Roman Empire, especially, reveals the importance of men such as Agrippa and Apollodorus, who helped their respective masters (Augustus and Trajan) carry out vast public works. The Rome of 100 A.D. had better paved streets, sewage disposal, water supply, and fire protection than the capitals of civilized Europe in 1800. Most agriculture, manufacturing, and services were carried out by the private sector, however, and achievements there were few and slow. The major areas for which the Greeks and Romans became famous were in civil and hydraulic engineering and architecture. Water conduits, built for both the supply of fresh water and for drainage, date from early classical Greece.[2] The Roman Empire, which commanded enormous resources, brought public-good engineering to great heights, although most of its engineering feats, including roads and aqueducts, utilized existing technology. The supply of Rome with water was begun by Appius Claudius in 312 B.C., and the system reached unprecedented complexity by the first and second centuries A.D.[3] Sewage and garbage disposal were also highly developed. A system of central heating was developed for use in both homes and bathhouses. In describing the urban bias in Roman technology, Hodges (1970, p. 197) concludes that "the Roman city was more interesting for its scale than for the novelty of its design."

Equally important in Roman engineering was the infrastructure of land transport: the roads and bridges built by the Romans are justly admired as one of their greatest achievements. Much of the success here was due to their discovery of cement masonry, which Forbes (1958b, p. 73) has called "the only great discovery that can be as-

2. The first such large project, the Samos aqueduct, was built by Eupalinus of Megara, around 600 B.C. It brought water from a lake to the town along a tunnel about a mile in length and 8 ft. in diameter.

3. By the time of the water superintendent Frontinus, writing between 97 and 104 A.D., many houses in Rome had hot and cold running water. The use of lead pipes was an ingenious solution to the short supply of construction material, but may have had severe side effects on the population's health. Interestingly enough, Roman writers were aware of the problems of lead poisoning, but their warnings were not heeded, and it was not until the time of Benjamin Franklin that lead poisoning was again identified as a hazard (K. D. White, 1984, pp. 164–65).

cribed to the Romans."[4] The economic significance of Roman roads should not be exaggerated. The roads that survived deep into the Middle Ages were those that were not used, whereas most of the heavily traveled and poorly maintained roads of Gaul became unusable. Roman roads were built for military purposes and their use by the general public for the purpose of trade was incidental (Leighton, 1972, pp. 48–60). Late imperial governments imposed severe limitations on the weight that could be hauled, and without the complementary technology of horse harnessing and -shodding and wagon building their economic impact was probably limited to high-value, low-weight cargoes. Roman roads were steeply graded, which did not matter much for the movement of infantry and light loads, but proved too cumbersome for commercial loads.[5] In bridges and aqueducts, the Romans used the revolutionary technique of weight-supporting arches or pillars made of cement. Some of their aqueducts, such as the famous Pont du Gard near Nîmes, have survived. Others, such as the wooden trestle bridge that Julius Caesar's troops built on the Rhine in 10 days (in 55 B.C.), are known to us only from documents.

Another area in which technical ingenuity improved efficiency in the public sector was in the construction of war machines. Although military technology is not of major concern here, it is worth noting that Greek and Roman military technology provides one of the few areas of a successful collaboration between scientists and technology.[6] Strangely, the Romans did not improve much on Greek and Hellenistic military machines, although they used them extensively and built bigger and more powerful versions.

In what we would today call *machines,* the contribution of the classical world, especially Hellenistic civilization, was in recognizing fully the importance of the elements of machines, such as the lever, the wedge and the screw, as well as the elements of motion transmission, such as the ratchet, the pulley, the gear, and the cam. Yet these insights were applied mostly to war machines and clever gadgets that

4. This credit is not entirely deserved, since cement was used in Asia Minor before it was used in Rome. But the Romans vastly improved its use, and employed such quality control that they may well be thought of as the originators of the technique. Specific to Roman construction was the use of waterproof cement, known today as *pozzolana,* made from the volcanic dust found near Naples.

5. The cost of transporting goods over land remained more than 20 times higher than the cost of carrying them by sea (Greene, 1986, p. 40).

6. The story of Archimedes helping to build war machines in a futile attempt to defend his native Syracuse against Roman armies is well known (Hacker, 1968). Other engineers who made major contributions to the construction of catapults are Ctesibius (third century B.C.) and Philo of Byzantium (about 180 B.C.).

were admired for their own sake but rarely put to useful purposes. Many of these ideas lay dormant for millennia. Perhaps the most brilliant inventor and engineer of antiquity whose works have come down to us is Hero of Alexandria, whose dates are somewhere in the late first century A.D. (Landels, 1978, p. 201). Among the devices credited to Hero are the *aeolipile,* a working steam engine used to open temple doors; a coin-operated vending machine (for holy water in the temple); and the *dioptra,* an instrument similar to the modern transit used in surveying and construction, combining the *theodolite* (used for measuring angles) with the level. Most of Hero's inventions served at best recreational purposes. The same can be said of Ctesibius, who lived in the third century B.C. and is called by some the Edison of Alexandria. Ctesibius reportedly invented the hydraulic organ, metal springs, the water clock, and the force pump.

A recent discovery that reveals the technical genius of Hellenism is the famous Antikythera mechanism, found as cargo on a sunken ship near Crete. It is a geared astronomical computing machine of astonishing complexity, built in the first century B.C. Derek Price, who reconstructed the mechanism, admonished historians to "completely rethink our attitudes toward ancient Greek technology. Men who could have built this could have built almost any mechanical device they wanted to" (Price, 1975, p. 48). This statement seems to push revisionism too far. What the mechanism proves is that Hellenistic civilization had mastered the use of gears and applied geometry beyond what was previously believed possible, and that their astrolabes (invented in the second century B.C.) were mechanically sophisticated. The Antikythera mechanism demonstrates that classical civilization had the ability to build sophisticated astronomical instruments and could construct far more complex geared mechanisms than had hitherto been thought. Yet the mechanism was, after all, a gadget used to reproduce the motions of the moon, the sun, and the planets for scientific and probably astrological purposes. It served, as far as we can tell, no direct economic purpose and in the view of one classical scholar, the special branch of machine design that made the Antikythera mechanism was not representative of the mainstream of invention of the period, nor did it have much impact on it (Brumbaugh, 1966, p. 98).[7] What the discovery shows is that classical civilization had the intellectual potential to create complicated technical devices. The question remains why so little of this potential was re-

7. Some scholars, such as Price and Cardwell, have argued that astronomical instruments eventually led to the incidental invention of clocks and permitted the measurements of time as a by-product. This theory is effectively dismissed by Landes (1983, pp. 54–57).

alized and translated into economic progress. I shall return to the issue of technology in the Graeco-Roman world in Part III.

One of the areas in which Hellenistic and Roman civilization made lasting contributions was water lifting and pumping. Pumps of various designs were in wide use in antiquity, for irrigation, mine drainage, fire fighting, and bilging water from ships. Force pumps, known and used by the Romans, had the disadvantage that they had to be immersed in water, thus making them difficult to operate and install. But the obvious complement to the force pump, the suction pump, was not invented until the fifteenth century (Oleson, 1984; Landels, 1978, ch. 3). Water-lifting devices led to some advances in mechanics, such as the development of power transmission (gears, cams, and chains). Yet identifiable positive externalities from water lifting to other industries are rare, and some important mechanical insights, such as the crank and the fly wheel, were missed.

In the private sectors, such as agriculture, textiles, and the use of power and materials, the progress achieved between 500 B.C. and 500 A.D. was modest. New ideas were not altogether absent, but their diffusion and application were sporadic and slow. From a purely economic point of view the discovery of the principle of the lever, attributed to Archimedes, was the most important technological breakthrough. The lever was combined with the principle of the helix to produce the screw, and was used for gearing, attaching, and pressing. What connects these techniques is the recognition by ancient engineers that a small force operating over a distance can exert greater force as the distance gets smaller. The wine press, an example of this principle, was first mentioned in about 70 A.D. by Pliny, who believed it was a Greek invention made in the previous century. Another invention, the compound pulley, allowed cranes to lift large weights. What part of these innovations was directly owed to Archimedes is not clear; it is likely that here, as in many other instances, theory followed practice rather than vice versa.[8]

In metallurgy and mining the main advances were the use of scoopwheels and Archimedes screws. The Greeks made some improvements in separating extracted minerals from waste, but by and large there were few changes of substance after 300 B.C. Greek and Roman ironmaking was slow and produced an uneven, and thus poor, product. The ore was heated in a bloomery, which burnt the carbon and left a slag of contaminating minerals that had to be removed to produce what we could call today wrought (i.e., carbon-free) iron. The binding constraint on metallurgical technology was the inability

8. We can find antecedents for the use of pulleys in earlier periods. It seems that the Assyrians already possessed a pulley type of device by the eighth century B.C.

of ancient blacksmiths to produce iron at temperatures high enough for casting. The spongy and pasty blooms produced by ancient blacksmiths needed frequent hammering and further heating in order to yield a metal that was serviceable at best. The best quality steel, "seric iron" (or wootz steel), was imported from India, although the bloomeries of the West were capable of producing some steel of low quality (Barraclough, 1984, Vol. I, p. 19). Bellows were probably in use in furnaces by the fourth century A.D., but cast iron was unknown because sufficiently high temperatures could not be generated. Here the classical and early medieval world lagged behind China, where the art of casting iron had been known since the third century B.C. As far as we can determine, the Greeks and Romans made little progress in metallurgy, despite their heavy use of iron. The evidence for progress is weak and tends to come mostly from eastern Europe and Britain, hardly typical of the Mediterranean world ruled by Rome (Tylecote, 1976, p. 53). The best that can be said of the Roman Empire is that it disseminated best-practice techniques, and possibly saw the adoption of somewhat larger furnaces and a few other marginal improvements.

In shipping, too, advances were modest. Shipping was crucial to the economies of the Mediterranean in classical times, when prosperity depended primarily on commerce, that is, on the gains from Smithian growth. This growth was made possible through a complex network of specialization and interregional trade supported by colonization and later by Roman laws and political control. Classical cargo ships were equipped with a single square mainsail, sometimes supplemented with smaller topsails. The myth that ships in classical times could not maneuver against the wind has by now been satisfactorily laid to rest (Casson, 1971, pp. 273–74; K.D. White, 1984, p. 144). Even square-rigged vessels could beat against the wind, though they were designed primarily to sail with the wind and could probably do no better than "one point into the wind," that is, an angle of 79 degrees against the wind. The best way to maneuver against the wind is to use fore-and-aft rigging (in which the sail is set parallel to the keel). For many years historians were convinced that square rigs were the only rigging the Greeks and Romans knew. Casson has argued that classical civilization knew fore-and-aft rigging in at least three forms: the sprit sail, the triangular lateen, and the quadrilateral lateen. The evidence for the existence of these sails is persuasive, but it is also clear that large merchant ships could not use these riggings and that the primitive square sails remained prevalent. The reasons proposed for the delay of widespread fore-and-aft-rigging include the lack of suitable trees for the larger masts, the slower speed of fore-and-aft rigged ships with the wind, and the difficulty

Figure 1. Roman square-rigged merchant vessel, scratched into the wall of a Pompeii house.
Source: Maiuri 1958: Maiuri A., "Navalia pompeiana," *Rendiconti della Accademia di Archeologia di Napoli* 33, 7–34.

of shifting the sail during tacking (the zig-zag motion that allows sailing ships to move against the wind). Steering by means of oars was reported to be difficult, although one authority (Casson, 1971, p. 224) has contested this, and the Romans achieved some improvement here using a pivot device that made maneuvering a ship somewhat easier.

In agriculture advances were mostly local. The tools and techniques used in Roman times showed great variety in adapting the basic primitive techniques of Mediterranean farming to local conditions. Labor saving inventions were few. K. D. White (1984, p. 58) has concluded that "technical development in Roman farming was common while innovation remains rare." We know that the Romans tried to breed better cattle for the sake of fertilizer, but the problem of how to feed the livestock was not solved satisfactorily. The Roman successes in hydraulic engineering found a few applications in drainage and irrigation, but their contributions pale beside the huge hydraulic irrigation works of Egypt and Mesopotamia two millennia earlier.

Whatever change occurred in agricultural implements came from

Figure 2. Reconstruction of a harvesting machine, known as a *vallus*, used in Europe in late Roman times.
Source: Reconstruction of harvesting machine type I *(vallus)*. H. Cüppers, from K. D. White, *Greek and Roman Technology*, Cornell University Press.

outside the Mediterranean perimeter. The Gauls and other Celtic people improved harvesting equipment, and there are a few famous but controversial accounts of primitive harvesting and threshing machines. There is, however, no evidence that these devices were widely used or had a major impact on productivity.[9] Wooden barrels, one of the most important practical inventions of this time, were unknown to the Greeks and were to the Roman world "a gift of the northern peoples" (Forbes, 1956b, p. 136). Around 370 A.D. an anonymous Roman writer noted that "although the barbarian peoples derive no power from eloquence and no illustrious rank from office, yet they are by no means considered strangers to mechanical inventiveness where nature comes to their assistance" (cited in De Camp, 1960, p. 272). Celtic artisans are also credited with the invention of enameling, the spoked wheel, soap, improved agricultural implements, and advanced ironworking techniques. Archaeological evidence shows that Celtic cartwrights used hardwood pegs in the hub-box between the hub and the axle, which helped rotate the wheels in a fashion similar to that of ball bearings (Cunliffe, 1979, p. 118).

One of the most famous inventions made during the Roman Empire was the waterwheel. The *noria,* a water-lifting device driven by the stream itself, was first described by Philo of Byzantium in the third century B.C. The *noria* is not really a waterwheel, and the authenticity of the relevant passage has been questioned. The Roman

9. The harvesting machine developed in Gaul known as *vallus* was essentially a reaping oxcart with toothlike cutting devices in front to cut the grain. It was last mentioned by Palladius in the fifth century A.D., and its subsequent disappearance suggests that it may not have worked very well (K. D. White, 1967; 1969). The classical period also saw the invention of the Roman sickle and the British long scythe (K. D. White, 1984, pp. 49–50). Yet Roman scythes were so rare that their dating was long doubted (Lynn White, 1972, p. 149).

engineer Vitruvius, who lived in the second half of the first century B.C., describes a waterwheel in which gears were used to transmit power from the turning wheel to the grindstones. Even earlier references are known, but they do not describe the machinery in any detail.[10] Until the fourth or fifth century A.D., waterwheels were apparently used only for flour milling, and there too their use was not widespread. Perhaps the reason was that Vitruvian mills were undershot mills, which were not very efficient, though generally cheaper to build and more suitable to the Mediterranean topography. More efficient overshot wheels, which used gravity rather than the impulse of the movement of water, appear in the West possibly as early as the third century (Reynolds, 1983, p. 26), and can be dated with certainty to Athens in the middle of the fifth century A.D. (Blaine, 1976, p. 166).

In assessing classical technology, then, it is important not to overemphasize the lack of achievement. The sources are biased, because many tools and devices made of wood or leather have not survived. Most classic writers did not bother with technological matters, which may be significant in itself.[11] Can archaeology shed more light on the issue? Archaeological evidence has demonstrated recently that waterwheels were more widespread than literary sources had suggested (Wikander, 1985). The famous Lake Nemi ships, exposed in 1929 when the lake was drained, contained a pair of turntables akin to our modern concept of ball- and roller bearings, an idea subsequently not heard of until the sixteenth century.[12] Price has argued that what we do for ancient civilization is tantamount to reconstructing modern society on the basis of a few buildings and paintings.

10. Antipater of Thessalonika wrote the following epigram in 85 B.C.:

> Cease from grinding, ye women who toil at the mill
> sleep late even if the crowing cocks announce the dawn
> For Demeter has ordered the (water) nymphs
> to perform the work of your hands
> and they, leaping down on the top of the wheel,
> turn its axle which with its revolving spokes,
> turns the heavy concave Nysarian millstone
> cited in Usher (1954, p. 165).

11. Even Vitruvius, the most original classical writer discussing technical matters, who was fully aware of the importance of inventions for living standards, only spent the last of the ten books that comprise his *De Architectura* on mechanical devices, and close to half of that book is devoted to military machines. Moreover, he adds "there are innumerable mechanical devices about which it does not seem needful to enlarge because they are too handy in our daily use such as millstones, bellows, waggons, two-wheeled chariots, lathes, and so forth" (Book X, ch. 1, #5; 1962, p. 279).

12. The ships also contained a bucket-chain pump that, some scholars believe, employed a fly wheel and crank handle. Lynn White (1962, p. 106) has dubbed this interpretation an archaeological fantasy.

Figure 3. Byzantine mosaic, depicting a Roman watermill as described by Vitruvius.
Source: Judith Newcomer, from *Technology in the Ancient World,* by Henry Hodges, Penguin.

"The technology was there, and it has just not survived like the great marble buildings . . . and the constantly recopied literary works of high culture" (Price, 1975, p. 48). Yet a recent summary (Greene, 1986, p. 170) has concluded that "the [Roman] economy does not show signs of advance or evolution, simply an intensification of everything that already existed in Greek and Roman Republican times." Moreover, as Finley (1965, p. 29) points out, there is a reverse bias as well. The first appearance of an invention does not correspond necessarily with a major effect on productivity, for which widespread adoption is necessary. Classical society was inventive, original, and inquisitive. But it was not particularly technologically creative. It built waterwheels, but it did not really exploit water-power. It constructed fore-and-aft rigged ships, but only on a small scale. It had a good understanding of glassmaking, it was a comparatively literate society, and it realized that light rays could be manipulated, yet it never produced spectacles.[13]

When classical civilization succeeded in creating a novel technique it was often unable or unwilling to take it to its logical conclusion and to extract anything approximating the maximum economic benefits from it. Many inventions that could have led to major economic changes were underdeveloped, forgotten, or lost. In some cases, breakthroughs that failed to spread had to be reinvented independently.[14] The paradox is that whereas it might have been expected that these losses would occur in illiterate societies with low geographical mobility, classical civilization was relatively literate and mobile, and ideas of all kinds disseminated through the movement of people and books. Of course, factors other than a lack of interest in matters technological might be relevant. First, it may well be that many inventions did not work properly or could not be produced on a sufficient scale because of limitations in workmanship or materials. Second, many ancient civilizations were so thoroughly destroyed by savage marauders that the technologies they developed, embodied in implements or recorded in books, were simply lost. Third, as De Camp (1960, p. 180) points out, the number of engineers and inventors was small and they often tried to keep their inventions secret, taking their ideas with them to their grave. As we have seen, the crank, the lateen sail, reaping machines, and ball bearings are all ideas that may have existed at some time but failed to affect the economy. This is not to say that the ancient economy was primitive, poor, or incapable of growth. But its growth derived from those aspects for which the Greeks and Romans are famous: organization, trade, order, the use of money, and law. This kind of growth can take an economy a long way, and it did. When the political foundation on which it is built becomes shaky, however, the prosperity based on Smithian growth alone is rapidly lost.

When all is said and done, it remains something of an enigma why a commercial and sophisticated economy, heavily dependent on

13. Seneca observed that letters were enlarged and made more distinct when seen through a glass globe full of water (cited in De Camp, 1960, p. 274).

14. Central heating was apparently already in use in southwestern Anatolia in 1200 B.C. and was independently reinvented by the Romans. Seneca complained that "many of the discoveries of older generations are being lost" (cited in De Camp, 1960, pp. 180, 275).

transportation and animate power, in which handicrafts and food-processing industries catered to a large urban population, failed to arrive at some rather obvious solutions to technical problems that must have bothered them. Many of these problems were resolved in the first centuries of what we now call the Middle Ages. I shall return to this question in Part III.

CHAPTER THREE

The Middle Ages

It is customary to divide the Middle Ages into an early period, from 500 A.D. until about 1150, and a late period, between 1150 and about 1500. Early medieval Europe, sometimes still referred to as a "dark" age, managed to break through a number of technological barriers that had held the Romans back. The achievements of early medieval Europe are all the more amazing because many of the ingredients that are usually thought of as essential to technological progress were absent. Particularly between 500 and 800 A.D., the economic and cultural environment in Europe was primitive compared to the classical period. Literacy had become rare, and the upper classes devoted themselves to the subtle art of hacking each other to pieces with even greater dedication than the Romans had. Commerce and communications, both short- and long-distance, declined to almost nothing. The roads, bridges, aqueducts, ports, villas, and cities of the Roman Empire fell into disrepair. Law enforcement and the security of life and property became precarious, as predators from near and afar descended upon Europe with a level of violence and frequency that Roman citizens had not known. And yet toward the end of the Dark Ages, in the eighth and ninth centuries, European society began to show the first signs of what eventually became a torrent of technological creativity. Not the amusing toys of Alexandria's engineers or the war engines of Archimedes, but useful tools and ideas that reduced daily toil and increased the material comfort of the masses, even when population began to expand after 900 A.D., began to emerge. When we compare the technological progress achieved in the seven centuries between 300 B.C. and 400 A.D., with that of the seven centuries between 700 and 1400, prejudice against the Middle Ages dissipates rapidly.

Medieval Western technology drew from three sources: classical antiquity, Islamic and Asian societies, and its own original creativity. Medieval engineers did not seem to care much where ideas came from—in the majority of cases we do not know—as long as they worked. Sometimes what seems like the adoption of a known technique may, in fact, have been an independent invention. It is also possible that what seem like original ideas were in fact borrowed from other civilizations in which no record of the invention has survived. In any event, diffusion of new technology was usually slow, and the old techniques often stubbornly survived and coexisted with the new for decades and even centuries.

In terms of their direct contribution to aggregate output, changes in agricultural technology were particularly important, as the bulk of the population was engaged in farming. The transformation of agriculture that began in the early Middle Ages took many centuries to complete, but eventually it shaped European history. Yet changes here were especially slow. Agricultural technology differs from manufacturing, transportation, or information technology in that it tends to be highly site-specific. Different crops have different requirements, and the same crop will use different inputs and technology depending on elevation, rainfall, soil type, and so on. As a result, improvements have to be modified and adapted infinitely. A significant part of the development cost is thus imposed on the user of an innovation, and the additional experimentation slows down the process.

The essential elements of the agricultural evolution were the introduction of the heavy plow and the creation of the three-field system. It has taken the combined geniuses of Marc Bloch (1966) and Lynn White (1962) to make historians fully recognize the importance of the heavy plow, or *carruca*. The ancient plow used in the Mediterranean economies scratched the soil with a wooden- or iron cutting point, or "stock," which cut and pulverized the soil, preventing the evaporation of moisture and bringing subsoil minerals to the surface by capillary action. This *aratrum* was ill-suited to the heavy and moist clay soils of the plains north of the Alps. In its ultimate form, the heavy plow moved on wheels, and had a coulter that cut the soil vertically, a flat plowshare that cut it horizontally, and a mouldboard that turned the cut sods aside to create a deep furrow. The heavy plow made possible the cultivation of huge tracts of fertile land that in Roman times were either uncultivated or exploited by a primitive slash-and-burn technique. Although some Roman plows had wheels, the complete heavy plow did not make its appearance before the sixth century (White, 1962, p. 53).[1] The heavy plow created the peculiarly long and narrow strips that characterized European open fields. But its impact was especially momentous because

1. The exact timing of the introduction of the complete wheeled plow into the plains of northern and western Europe is a matter of some dispute. The evidence supporting Lynn White's claims has been disputed by other scholars, especially regarding the mouldboard. See, for example, Wailes (1972).

Figure 4. Two inventions that revolutionized medieval agriculture: the wheeled plow and the horse collar.
Source: Biblioteca Apostolica Vaticana, Rome.

it required a team of oxen to pull it.[2] Few peasants could afford to own such an expensive capital good, and in part in an attempt to solve the fixed cost problem, medieval society developed a semicooperative organization sometimes referred to as the manorial system.

The dependency of plowing on draft animals created the technical problem with which European agriculture grappled for many centuries: how to feed its animals. The solution found in the early Middle Ages combined three elements, though not all three were necessarily present simultaneously. First, under the new three-field system of crop rotation that spread slowly through Europe in the early Middle Ages, one third of the arable land was left fallow. The animals were let to graze on the fallow land, thus feeding themselves while at the same time fertilizing the soil with their droppings. Under the rotation system, each plot of land would rotate between fallow, winter crops, and spring crops. Second, the fields under crops were opened up to stubble grazing after the harvest, a custom known as "the right of common stock," or *vaine pâture.* Third, the village usually had a separate common field, not part of the rotation system, on which animals grazed. The right of common stock and the commons, together meant that farmers' individual plots could not be separated by fences, and hence the system is sometimes known as open-field agriculture. The open-field system was not a technical invention strictu

2. Pliny's wheeled plow was pulled by eight oxen. In the Middle Ages the normal number of oxen was four to six.

sensu as much as a brilliant organizational solution to a technical problem that combined private and public property rights in an ingenious fashion. The introduction of the three-field system permitted the expansion of the cultivation of additional crops beside the main staples, such as wheat and rye. The second field was often used to grow oats (an ideal horse fodder), barley (an industrial raw material as well as a human food), and legumes (an important dietary supplement). None of these crops was novel, but their expansion depended on the gradual replacement of the biennial with a triennial rotation in the areas north of the Alps.

A second area in which early medieval Europe was successful was energy utilization. Energy takes two forms, kinetic and thermal. Kinetic energy could be derived from animate power (including human muscles), transmitting the energy of the sun through living bodies, and nonanimate power, using solar power directly through water- or wind motion. Wind power had been used in sailing ships, but had not been harnessed in the West in other ways until the first windmills were built there in the twelfth century. In waterpower, radical improvements came early. During the Merovingian and Carolingian eras (seventh to tenth centuries) better and bigger waterwheels spread through Europe. Medieval Europe not only produced the more efficient overshot wheel, but also adapted and improved the gearing of both horizontal and vertical waterwheels, making it possible to use wheels on both rapidly flowing and slower flowing streams. Medieval engineers made much progress in the construction of dams, allowing controlled usage of water power through storage, and diverted streams to mill races. They applied cams, and later cranks, to convert the circular motion of waterwheels into the reciprocating motion needed for hammering, fulling, and crushing. The cam had been known in antiquity but had apparently not been combined with the waterwheel. The crank was in all likelihood a medieval invention. The result was that the waterwheel was transformed from an occasional device used for grinding flour into a ubiquitous source of energy operating on rivers of every type. By about 1100, waterpower was used to drive fulling mills, breweries (to prepare beer mash), trip hammers, bellows, bark crushers, hemp treatment mills, cutlery grinders, wire drawers, and sawmills. In 1086, Domesday Book listed 5,624 watermills in England south of the Severn river, or roughly 1 for every 50 households. Unlike their Roman ancestors, medieval men and women were surrounded by water-driven machines doing the more arduous work for them.[3] The waterwheel

3. In the eleventh century, there were even tidal mills in Europe (around Venice, in the south of England, and on the west coast of France), the first use in history of power from nonsolar origins (Minchinton, 1979). Although this source of power never

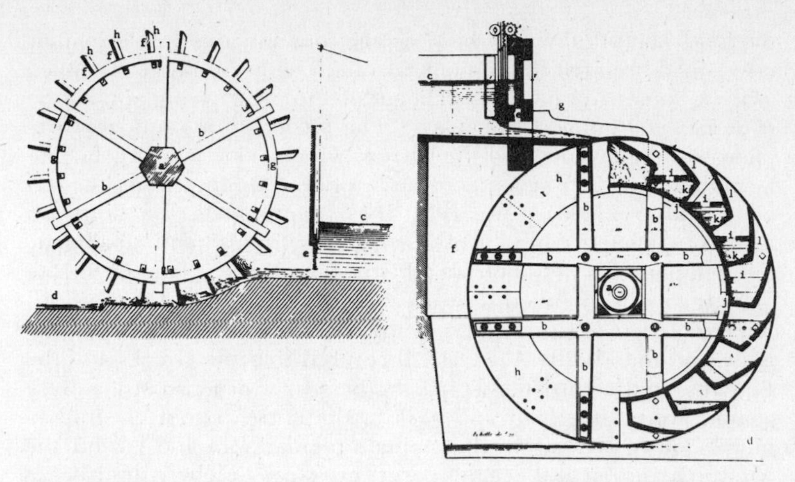

Figure 5. Main types of the waterwheel: undershot wheel (left) and overshot wheel (right).
Source: (a) Arthur Morin, *Experiences sur les roues hydrauliques à aubes planes, et sur les roues hydrauliques à augets* (Metz and Paris: Thield 1836), pl. 1, Fig. 5. (b) [Jacques] Armengaud, *Moteurs hydrauliques* (Paris: Baudry et Cie. and Armengaud Aine, n.d.), pl. 14.

may not have been invented in medieval Europe, but it was there that its use spread far beyond anything seen in earlier times. As Lynn White has remarked, medieval Europe was perhaps the first society to build an economy on nonhuman power rather than on the backs of slaves and coolies.

In animate energy, too, there was progress. For millennia, horses had been used in war and peace. In the centuries following the fall of Rome, three improvements vastly increased the efficiency of horses. The first was the nailed horseshoe. A hipposandal protecting the horse's hooves had been used in antiquity, but from every point of view the horseshoe represented an improvement. The horseshoe was especially useful in the moist soils prevalent north of the Alps and for the heavier horses used in the later Middle Ages. Horseshoes protected the hooves from the soil moisture that wore out hooves quickly and caused them to splinter. The exact dating of the emergence of horseshoes is still a matter of controversy, as the archaeological evidence is ambiguous, but there can be no question that by

became important, it is symptomatic of the medieval drive to harness inanimate energy wherever possible. See Derry and Williams (1960, pp. 252–53) and Gimpel (1976, p. 23).

the ninth century their use had become common and their economic effects had been felt throughout Europe. The horseshoe was applied to pack horses and mules as well, and led to the growing application of horses to commercial haulage. The second important improvement was the invention of the stirrup, whose value was largely military, though it also benefited civilian riders.[4] By increasing the stability and comfort of the ride, the stirrup made the horse an increasingly important part of European transportation. Finally, the third important innovation was the modern horse collar. At the beginning of the twentieth century a retired French cavalry officer, Richard Lefebvre des Noëttes, compared the use of horses in antiquity and the Middle Ages. He discovered that the Greeks and the Romans used a throat-and-girth harness in which two straps were wound around the belly and neck of the horse. The neck strap simultaneously pressed on the animal's jugular vein and cut off the windpipe as soon as it began to exert pressure. Lefebvre des Noëttes found by experiments that a horse strapped this way lost about 80 percent of its efficiency.[5] In early medieval times, by way of contrast, such an easily corrected waste of valuable energy was not tolerated. The solution to this problem emerged when the breast strap, which rested against the horse's chest, and the collar harness, which rested against the shoulders, were invented. Both eliminated the yoke, and thus avoided the main shortcoming of the Roman harness. The breast strap appeared somewhat earlier than the collar, but both were more or less in place in the ninth century. Consequently, horses gradually

4. In a famous but controversial argument, Lynn White (1962, p. 28) maintained that the stirrup was directly responsible for the emergence of feudalism. If true, this is a remarkable example of technological determinism. The stirrup established for centuries the unassailable superiority of the horseman over the foot soldier, leading to the need to equip and arm large numbers of knights. Because both horses and iron were scarce, the entire economic system had to be geared toward paying for these armies. This theory has been criticized, and it is likely that feudal society had other roots as well.

5. An attempt to explain this failure of classical technology is provided by Landels (1978, pp. 176–77), but the lameness of his argument is self-evident. He notes that the chariots pulled by horses in antiquity were small and light, so that no great effort on their part was required. Here he places his own horses before the cart: the chariots were small *because* there was no way to pull them efficiently by horses. Needham (1965, p. 314) points out that in antiquity the horse harness was often felt to be unsatisfactory, and documents a list of unsuccessful attempts to replace it. It is true, however, that the use of horses as draft animals also depended on the ability to feed the horses adequately and to breed horses large enough to provide enough power (Barclay, 1980, p. 109). Moreover, Spruytte (1977) has argued that the ancient horse harness was not as inefficient as has been claimed. In his view, the classical world used *two* different harnesses, a neck collar in which the traction was carried out by the horse's shoulders, and a dorsal yoke in which a breast strap pulled the carriage by the horse's breast. Spruytte maintains that Lefebvre des Noëttes confused the two and thus erroneously inferred that the harness of antiquity was ineffective. It is also im-

Figure 6. Earliest depiction of horse harness, dating from about 800 A.D.
Source: Trier, *Apocalypse,* City Library, MS. 31, fol. 58r.

acquired major economic importance in agriculture and in the hauling of wagons. The horse harness was supplemented by other advances in horse technology. In the early Middle Ages the tandem harnessing of horses (in a row rather than one beside the other) came into use. The whippletree, a wooden bar connecting the collar to the wagon or harrow, appeared in the eleventh century. Thus, errors in the most basic applications of animate power that the sophisticated civilizations of the Mediterranean had made for centuries

portant not to exaggerate the importance of inventions, even significant ones. Lefebvre des Noëttes claimed that the horse collar was the chief cause of the disappearance of slavery, a theory that has not been widely accepted. His other argument, that heavy land transport was rare and unimportant in the classical world, has also been questioned, as he neglected the importance of oxen as draft animals (Burford, 1960).

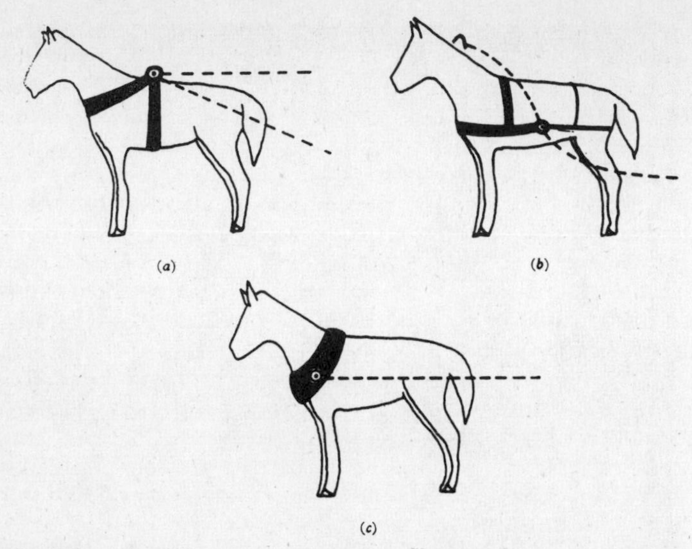

Figure 7. Three forms of horse harness: (a) depicts the inefficient throat-and-girdle harness used in antiquity; (b) shows the breast strap employed in the early middle ages; (c) shows the shoulder collar introduced in the later middle ages.
Sources: Joseph Needham, *Science and Civilisation in China,* Vol. 4, part 2, Cambridge University Press.

were corrected in early medieval times. These innovations, according to a recent work (Langdon, 1986, p. 19) opened the door to "the substantial, even massive, introduction of the horse to general draught work." The importance of these improvements is underlined by the computation that by the late eleventh century, 70 percent of all energy consumed by English society came from animals, the rest coming from water mills (ibid., p. 20).

What was the importance of the changes in technology that turned the horse into a major source of energy to the medieval economy? Horses were increasingly used for agricultural tasks, primarily plowing, but their effect on agricultural productivity is not entirely clear. Oxen remained of great importance to agricultural production and were not wholly supplanted by horses. Where horses were truly indispensable, in addition to military uses, was in land transportation. The increased speed and range of horse-pulled wagons, coupled with improvements in the design of the carts themselves, were instrumental in the revival of land transport and medium-distance trade. Here, then, is a prime example of how Schumpeterian growth feeds Smithian growth.

Similar processes occurred in water transport, although somewhat later. In ship design, around 800 A.D. the Vikings developed a truly seaworthy ship with which they crisscrossed Europe and reached Greenland and America. The improved Viking ships were equipped with keels and masts, and were small and light and thus extremely flexible and maneuverable. But because they could not be made larger or made to carry bulky cargo, their economic significance was limited. In the long run, the more important ship was the *cog,* which gradually developed from primitive Celtic boats and dominated the seas of Northern Europe by the twelfth century. Rather tubby looking, slow, and awkward to maneuver, the cog proved eminently seaworthy, capable of carrying large cargoes at low cost, and easy to defend against pirates.

Many other inventions date from the early Middle Ages, although timing and placing them with any accuracy is impossible. European society learned gradually to protect itself better from cold temperatures, especially when winters grew harsher after the thirteenth century. Changes in construction techniques, such as the use of shingles, the use of plaster for better insulation, the use of fossil fuels, and above all the development of chimneys, testify to the ability of medieval society to struggle against the harsh winters of northern Europe (Dresbeck, 1976). The chimney also facilitated home cooking, thus making warm foods more easily accessible. Soap, which was known in the late Roman Empire, became widespread in Europe around 800 A.D. and was improved upon in the tenth and eleventh centuries in both the Christian and Islamic regions of the Mediterranean. Cakes of hard soap first appeared in the twelfth century. Inventions that affected daily life, such as butter, strong distilled liquors, skis, wheelbarrows, the use of hops in brewing, improved window glass, and crank-driven grindstones originated in western Europe in the early Middle Ages.

During the early Middle Ages, the cultural and technological center of gravity of Europe remained to a large extent in the Mediterranean region. Spain, North Africa, and the Middle East were ruled by Islam. Between the eighth and the twelfth centuries, the sophistication and culture of the Islamic world made it the suitable heir of classical civilization. The medieval Islamic world was a highly mobile society, and traveling was a normal activity, enjoyed by rich and poor alike. Travelers were eager to learn from other societies, past and present. The culture and technology of Islam constituted a synthesis between Hellenistic and Roman elements, adorned with ideas from central Asia, India, Africa, and even China. Early Islamic society collected, compiled, and catalogued knowledge avidly. It was a society literate beyond Europe's wildest dreams. Not only the rulers, but many important mosques and even some individuals maintained large

Figure 8. Twelfth century illustration of the wheelbarrow. Note the contrast between wheelbarrow and stretcher, which required double the labor.
Source: Umberto Eco and G. B. Zorzoli, *The Picture History of Inventions,* Bompiani.

libraries. Between 700 and 1200 the Moslems knew more about the different parts of the known world than any other civilization. Their ability to preserve, adapt, and develop techniques borrowed from others is a lasting testimony to their creativity. Yet when it came to original creativity and invention, the achievements of Islam appear strangely truncated. The Moslems were enthusiastic collectors, but they offered little in the way of interpretation or theory; unlike the Christian West they were not driven, apparently, "by a strange urge to peer beneath the surface of things and see how they worked" (Watson, 1983, pp. 94, 146).

Still, in the centuries between 750 and 1100, Islam had some impressive technological achievements to its credit, in addition to being a more tolerant and cultured society (Singer, 1958, pp. 755–61). The Moslems, together with the Byzantines, are credited with the development of the lateen (i.e., Mediterranean) sail, though earlier versions of it were known to classical civilization. By the ninth century the problems of shifting the sail spar over the mast during tacking, which had prevented the Romans from building larger ships using fore-and-aft riggings, had been solved, and lateen sails were adapted to larger merchant ships. With a lateen sail a ship could sail 60–65 degrees off the wind, which was "very much better than the performance of square-rigged ships" (Parry, 1974, p. 13). By the eleventh century the improved sails were used throughout southern Europe and the Middle East by both Moslems and Christians. In power technology, the Moslems were the first to use a tidal mill (exploiting the motion of ebb and flood), in Basra around 1000 A.D. (Minchinton, 1979), and windmills are mentioned in Islamic sources from the ninth century on (Al Hassan and Hill, 1986, pp. 54–55). Waterpower was used in sugar mills and sawmills. Yet overall Islamic society appears to have relied on inanimate energy much less than Europe did (Hill, 1984b, pp. 169–72).

The Moslems were also responsible for the introduction of paper into the Middle East and Europe. The need for a substitute for the expensive parchments (made of dried calf or kid skins) used by medieval scribes was evident. Paper was invented by the Chinese before 100 A.D. and was widely used in the Orient. According to traditional accounts, when the Arabs conquered the city of Samarkand in 753 A.D., they learned the secret from captured Chinese workmen. In about 793 A.D., the first paper factory was set up in Baghdad, and by 1000 the entire Islamic world was enjoying bound books, wrapping paper, and paper napkins. The penetration of paper into northern and western Europe was slow. Once it was introduced, in the thirteenth century, the West characteristically began to use waterpower to process the pulp.

In textile production, the Islamic world made substantial advances in fabric quality. Here etymological evidence is suggestive: many names for European fabrics betray their Islamic origins. Damasks, a fine silklike linen, came from Damascus; mousselins, or muslins, a fine cotton fabric, from Mosul (in Iraq); fustians, a fabric made of a linen warp with a cotton weft, were named after Fustat (a Cairo suburb). Many of these fabrics may have originated from Persia, the cradle of many textile technologies, and moved slowly westward during the tenth century. Cotton production was introduced by the Moslems into the Middle East and from there into Spain and Sicily. The cotton variety they introduced into the Mediterranean was capable of maturing in a harsher and colder climate. By the end of the Middle

Ages, cotton had become the main raw material for clothes and other fabrics in southern Europe and the Middle East. In leather production, the Islamic economies were famous for producing high-quality goods. Morocco and Cordoba, whose importance to the leather-processing industry has been immortalized in Western languages, were centers of this industry.

The Moslems' most original contribution was in chemical technology. Al Jabir and Al Razi (known in the West as Geber and Rhazes, respectively) wrote books that for centuries were the recognized standard works in the field. Much work in chemistry was aimed at the search for gold and had a mystical element to it, but by the eleventh century Avicenna or Ibn Sina doubted that gold could be made at all. The Moslems invented alkalis, and greatly improved the quality of glass and ceramic products. They produced *naphta,* a flammable petroleum derivative akin to kerosene, and their perfume and acid industries were far superior to anything known before.[6]

In mechanical engineering, from water mills to clocks, the Moslems were for centuries far ahead of the West. The Banu Musa brothers, writing in Baghdad in around 850 A.D. summarized and extended Hellenistic mechanical engineering (Hill, 1977). Al Jazari's *Book of Knowledge of Ingenious Mechanical Devices,* written in the early thirteenth century in Diyarbakir (Asia Minor), has been dubbed "the most remarkable engineering document to have survived from . . . pre-Renaissance times" (Hill, 1984b, p. 128). In metallurgy, too, Islamic technology thrived for a while. Using Indian steel, craftsmen in Toledo and Damascus produced swords of a quality that was proverbial in the West. The Moslems also learned to grow and refine sugar and to make confectionery. Egypt in particular became famous for its sugar products, though they remained luxury goods until the seventeenth century. Many Asian crops were absorbed by and then transmitted through the Islamic world to improve Western diets: cereals such as sorghum and rice, and the hard durum wheat of which pasta is made; fruits such as oranges, lemons, bananas, and watermelons; and vegetables such as asparagus, artichokes, spinach, and eggplant (Watson, 1983). Indeed, agricultural progress in the Islamic world between about 700 and 1100 A.D. was such that the term "agricultural revolution" has been used. In many cases, the adoption of new crops required irrigation, and the Moslems were masters in the utilization and modification of ancient hydraulic technology, most of which was unknown in the western Mediterranean before the advent of Islam. From Persia to the Pyrenees, there is

6. There is some dispute as to whether Islamic scientists discovered alcohol distillation (Al-Hassan and Hill, 1986, p. 141), but see Lynn White (1978, pp. 7, 115).

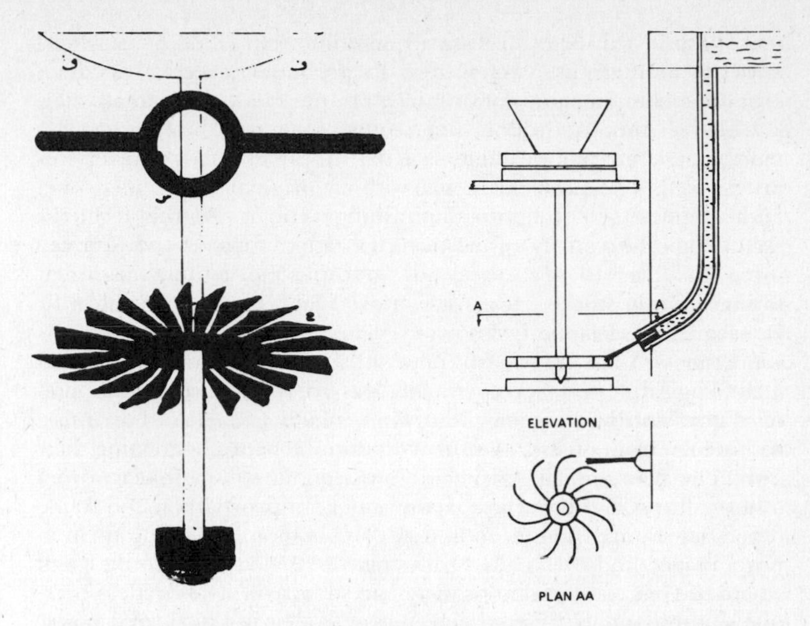

Figure 9. Horizontal waterwheel as described by the Islamic writer al-Jazari. The left shows a thirteenth-century illustration; the right shows a modern reconstruction.
Source: Bodleian Library, Oxford.

evidence of a proliferation of *norias,* Archimedes screws, *shadufs* (a pole with a pail on one end and a counterweight on the other), underground canals, and so on. For the most part, the Moslems used techniques known in the classical world: their contribution was "less in the invention of new devices than in the application on a much wider scale of devices which in pre-Islamic times had been used only over limited areas" (Watson, 1983, p. 108). From that point of view, their achievement was comparable to that of the West, which did the same for waterpower and heavy agricultural implements.

Assessing the technological achievements of the Islamic world is difficult. Not all Moslems were Arabs, and the so-called Islamic world contained at different times large numbers of Christians and Jews. Large variances also appear to have existed within the Islamic world, with Persia and Spain, at the geographical peripheries of that world, appearing on the whole to have been more creative than Mesopotamia and Egypt, although the information is lacking to make such judgments with certainty. Until recently, the verdict of technological historians on medieval Islam was harsh. Wiet (1969, p. 371) contends that it is difficult to discover anything that could be called an enrichment of acquired knowledge. Yet technological creativity as an economic phenomenon does not require the creation of totally new knowledge: innovation does not require invention. Borrowing, extending, and adapting will increase the supply of goods and services just as well. The problem is that without original ideas, borrowing and adapting will run into diminishing returns. At this level the technological creativity of Islamic civilization for some reason ran into difficulties. Islamic society was not capable of adding much new to the existing stock of ideas it retrieved and applied so brilliantly. Consequently, Islamic technology eventually ran out of steam. At some time around the twelfth century, Islam lost its momentum, and technological supremacy eventually passed to their Christian enemies north of the Pyrenees, who were equally capable at borrowing technology from others, and turned out to be better at creating their own. The West, initially far behind, was equally willing to learn from other cultures, but never lost its capacity to improve upon and refine others' ideas, applying them in new combinations, adapting them to novel usages, and eventually surpassing the original ideas to the point where the original inventing society had to borrow its own ideas back, often unrecognizably altered and improved. Having pioneered power and chemical technologies, by the thirteenth century Islamic society began to show signs of backwardness.[7]

The reasons for the slowdown are unclear. The destruction of the eastern part of the Islamic world by Mongol invasions was a dreadful blow; yet the loss of Spain and the decline in the western Mediterranean were themselves the result of internal weaknesses and divisions. It is tempting to relate the difference in creativity between the Christian world and the Islamic world to religious and political differences. I shall make an attempt in this direction in Part III.

By about 1200, the economies of western Europe had absorbed most of what Islam and the Orient had to offer. From then on, they pulled ahead mostly on their own steam. Despite a temporary setback in the fourteenth century, a string of brilliant inventions between 1200 and 1500 prepared the way for Europe's eventual technological leadership. The unique character of European technological change was determined by both the ingenuity displayed in making production more efficient, and in the speed with which some of these innovations were diffused throughout western Europe.

In power technology, the most important invention of the later Middle Ages was the windmill. The windmill combined the ideas of

7. Al-Jazari wrote in 1206 that windmills would never work, because the wind is too unreliable to serve as a source of power (Lynn White, 1978, p. 223).

Figure 10. Medieval drawing of an early windmill.
Source: Bodleian Library, Oxford.

the water mill and the sail. It, too, may have been imported to Europe by Moslems (from central Asia) but in spite of its apparent advantages in arid climates, it was not used widely in the Islamic world.[8] The first windmills that can be documented with certainty were in Yorkshire in 1185. Whatever their exact place of origin, within a few years windmills were used throughout Europe, and in 1195 the pope imposed a tithe on them. Because of the variability of the wind direction in Europe, engineers had to mount the windmills on a pivot so that they could be turned to face the wind at an optimal angle. Later, brakes and inclined sail beams were introduced to increase stability and efficiency. The windmill owed much to the water mill, from which it adopted its horizontal axle, gearing, and transmission machinery (Reynolds, 1983, p. 48). Yet as Usher points out (1954, p. 176), the windmill was a more complex mechanism than the water mill and its development was therefore slower. The early windmills

8. The design of the European mill differed radically from that of Asian mills, as it used gears and a horizontal axle. It is thus possible that it was an independent invention. The windmill was reportedly introduced by "German soldiers" (Crusaders) into Syria in 1192. Until the twelfth century, the windmill was confined to Persia and Afghanistan (Forbes, 1956c, p. 617).

were post mills, in which the whole engine house turns towards an optimal position into the wind using a single fulcrum.

Wind power was also exploited more efficiently in shipping as a result of substantial changes in the design of ships. In northern Europe, the gradual adoption of the lateen sail was complemented by the addition of a foremast and a mizzenmast to produce an entirely new type of ship known as a "carrack." Full three-masted rigging combined the advantages of the lateen and the square sails, namely, maneuverability in sailing against the wind and speed in sailing with it. The fully rigged ship has been termed "the Great Invention," not because of its revolutionary nature, but simply because "it could do more than any of its predecessors and could do so with considerably less risk" (Unger, 1980, p. 219). The mainmast and foremast carried square sails that drove the ship forward. The mizzenmast had a lateen sail rigged fore and aft, which helped in beating against the wind and maneuvering ships in narrow waters. It was first developed around 1400 in the Basque region around the Bay of Biscay and spread rapidly to northern Europe and the Mediterranean. During this time horizontal treadle looms produced higher quality and stronger sails. A second innovation, possibly introduced from China, was the sternpost rudder, which replaced the steering oars used in antiquity. The rudder appeared in the late twelfth century, and greatly reduced drifting and the physical effort required to keep a ship on course.

Ship construction also improved. By about 1300 the so-called carvel construction technique was adopted by northern European shipbuilders. This technique placed boards edge to edge along a skeleton of beams, with caulking between the planks to preserve watertightness. Carvel-built ships such as carracks did not entirely replace ships built by the age-old "clinker-planking" technology, in which the construction was based on overlapping planking. Both methods have advantages: clinker planking produced much sturdier and heavier ships, because lateral strains were absorbed by the entire body of the ship, whereas lateral strain was absorbed in carvel-built ships by the heavy beams that held the skeleton together. Carvel construction saved on wood and could make much lighter and larger ships than clinker planking; by 1400, ships of 1,000 tons were already in existence. Larger ships meant cheaper and faster transportation and improved seaworthiness (Rosenberg and Birdzell, 1986, pp. 80–96). Clinker construction survived, however, for smaller ships, and in some instances for medium-sized ocean-going ships as well. By the middle of the fifteenth century the Portuguese *caravel* had emerged. It is this type of ship that was used by Da Gama, Columbus, and Magellan. It was of a carvel construction, lateen rigged, had two or three masts, and a sternpost rudder. It measured a burden of perhaps 100–200 tons; and required a crew of about 20 (see Parry, 1974, pp. 17–22). It was a flexible ship, designed primarily for coastal shipping but sufficiently seaworthy to enable the Portuguese to lead the effort to discover the non-European world.

The Europeans also developed better navigational tools than existed before. There is no evidence that Greek and Roman ships used any navigational aides at all, other than watching the stars and staying within sight of the coastline wherever possible. Cases such as the wreck of St. Paul's ship en route to Rome, which lost its way because "neither sun nor stars appeared for many a day" (Acts 27:20) must have been common. The Viking sailors, too, used landmarks for direction and often fell victim to *hafvilla,* the loss of all direction at sea. All this changed in the closing centuries of the Middle Ages. Marine charts and navigating tables of unprecedented accuracy covering the entire Mediterranean and Black seas came into use in the thirteenth century. The compass, probably an independent European invention that was also known in China, was first mentioned by the Englishman Alexander Neckam in his *De Utensilibus* (about 1180). The device was at first little more than a magnetized needle afloat in a bowl of water. It took until about 1300 before this crude instrument was turned into a self-contained unit we would call a compass today, complete with the 16-point wind rose. One specialist (Kruetz, 1973, pp. 372–73) believes that it was only after 1410 that anyone really steered by compass. In the Mediterranean, the compass was especially important because its deep waters did not allow navigation by sounding (widely practiced in the North and Baltic seas). The effects on seafaring were profound. Until the thirteenth century, the Mediterranean was closed for navigation in the winter months; the new navigation technology made year-round shipping possible from around 1280.

The revival of geography culminated in the translation in 1409 of Ptolemy's *Geography* into Latin, which taught Europeans how to divide the world into lines of latitude and longitude. To measure latitude, the Portuguese navigators learned to adapt astronomical instruments developed by Hellenistic and later Islamic astronomers. The most famous of these instruments was the astrolabe, invented by the Hellenistic astronomer Hipparchus for astronomical purposes. The astrolabe measured the altitude of the polar star, thus indicating the observer's latitude. This much had been known since Pytheas of Massalia (third century B.C.) but the application of this knowledge to navigation was entirely novel. Astrolabes were often too complex for use at sea, and their first documented use on board occurred in 1456. Somewhat earlier the cross-staff, which served similar purposes, emerged. In the Southern Hemisphere, where Europeans ventured after 1450, they had no recourse to the polar star and used the astrolabe to measure the sun's altitude (Taylor, 1957). For navigational purposes, the quadrant, a simplified version of the astrolabe invented by Moslems, was found most useful. Although there was thus progress in Europe, it is not clear that by 1450 their navigational techniques were as yet superior to those of China. The Europeans were not alone in being able to build ships that could cross the oceans: the fifteenth century witnessed a short-lived attempt by the Chinese to explore distant regions. The difference was that the Europeans were able to keep the momentum of innovation going.

A third area of progress was metallurgical engineering. Mining and metallurgy were especially highly developed in southern Germany, Austria, and Bohemia. The landmark achievement here was the construction of larger furnaces that could achieve higher temperatures thanks to water-driven bellows known as blast furnaces. Although genuine blast furnaces do not appear until the end of the fifteenth century, the improved furnaces appeared somewhat earlier. Higher temperatures permitted the casting of iron, revolutionizing the production of goods requiring the brittle and hard qualities of cast iron.[9] The blast furnace differed from the primitive bloomeries of antiquity and the Middle Ages in that its higher temperatures allowed the iron to absorb carbon (about three percent). A higher carbon content lowered the melting point, so that the mixture could actually be poured, or cast. The output of blast furnaces became known as pig iron, which could either be cast into molds, or into the more malleable wrought iron by subsequent decarburization. Blast furnaces produced iron at a far lower cost than the antiquated bloomeries. Most of the other crafts of metalworking—such as welding, riveting, and forging—remained by and large the same. In nonferrous metallurgy, a process to separate silver from copper ores by means of lead, developed around 1450, was the most important discovery of this age.

Metallurgy was important to technological developments in other industries. An example of such a spillover effect can be found in that most famous of all medieval inventions, the printing press. The

9. Here, too, the Europeans were preceded by the Chinese, who used cast iron "almost as soon as they knew about iron at all" (Needham, 1969, p. 101). Needham rejects the idea that the Europeans invented cast iron independently, but concedes that he has no idea how the transmission was brought about. It is possible that he is mistaken and that cast iron was invented independently in Europe. The application of waterpower to bellows generated much higher furnace temperatures, which made cast iron almost a natural next stage.

printing press is justly famous, because it is the first medieval invention for which the name of the inventor and the time of invention are reasonably well established, and for which we know that the inventor single-handedly solved the entire problem to the point that within a short period the new gadget was in use all over Europe. Printing itself was not unknown in Europe by the early fifteenth century. Even before 1400 playing cards were stamped in Europe, and coins had been stamped for two millennia by this time. Yet Johann Gutenberg's invention of moveable type (1453) was an achievement of profound brilliance, made possible inter alia by his knowledge of metallurgy (his father had been goldsmith to the Archbishop of Mainz). Casting the moveable type was a difficult problem: all letter units had to be of equal length and thickness but of varying width. The mechanical solution Gutenberg found—a mold consisting of two overlapping L-shaped parts—was ingenious. The type was made of an alloy of tin, zinc, and lead, while the molds were made of iron and copper (Cardwell, 1972, pp. 20–25). Like the windmill, the printing press spread with dazzling speed. By 1480, there were over 380 working presses in Europe, and in the 50 years following the invention more books were produced than in the preceding thousand years.

Of equal technical brilliance was the weight-driven mechanical clock. Its inventor is unknown, but its first appearance can be dated to the end of the thirteenth century. Earlier in that century, Europeans and Arabs had been able to build sophisticated water clocks. But water clocks were unreliable because of freezing and evaporation. European mechanics searched doggedly for a device that would allow them to use the force of gravity.[10] Around the year 1300 the verge-and-foliot escapement mechanism appeared, which succeeded in converting the continuous but variable force exerted by a falling object into the regular oscillating motion required for the accurate operation of a clock. Here was a macroinvention if ever there was one. In his recent work on the history of clocks, Landes (1983) refers to it as the Great Invention. Progress was breathtaking: by the middle of the fourteenth century clockmakers such as the Dondi family and Richard of Wallingford were making complex devices that indicated not just the time, but also every astronomical motion then known. Clocks spread rapidly throughout Europe. Landes (1983, p. 57) points to a "clear sense of excitement and pride" in the new mechanisms. Every town felt that it had to possess this marvel. In White's (1962,

10. In 1271 Robert the Englishman stated that "clockmakers are trying to make a wheel or disc which will move exactly as the equinoctial circle does, but they cannot quite manage it" (cited by Lynn White, 1962, p. 122).

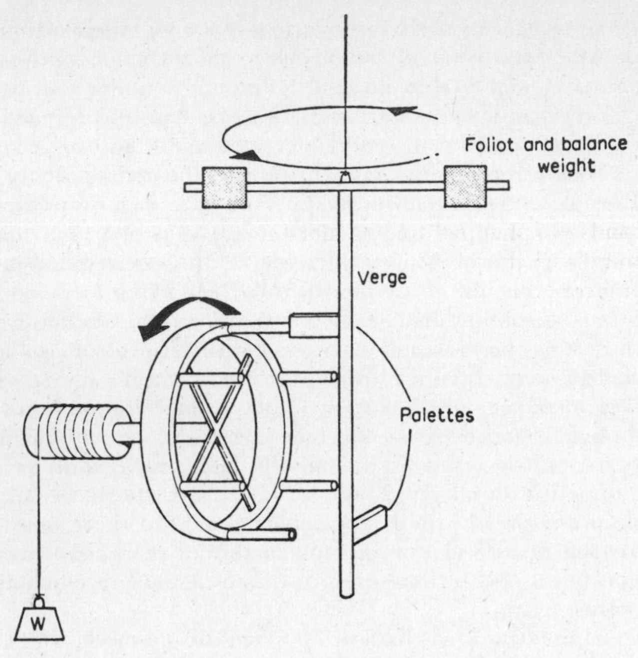

Figure 11. Principle of the weight-driven clock. The weights on top of the foliot ensure that the power of gravity is uniform. The palettes help convert the uniform pull of the weight into the oscillating motion needed for the clock.
Source: D. S. L. Cardwell, *Turning Points in Western Technology,* Science History Publications.

p. 124) words, "No European community felt able to hold up its head unless in its midst planets wheeled, angels trumpeted, cocks crew, and apostles, kings, and prophets marched and countermarched at the booming of the hours." In the middle of the fifteenth century spring-driven clocks and watches appeared, as new devices to regulate the uneven force of an uncoiling spring were discovered. The best solution was found to be the *fusee,* which used a conical axle to equalize the uneven force. The first *fusee*-driven watches were made around 1430, and watches became a popular consumer good among the better-off. The advances in clockmaking made the miniaturization of clocks feasible, and led to the democratization of time measurement.

Many scholars have stressed the importance of mechanical clocks as both a symptom and a further cause of technological progressiveness. "The clock, not the steam machine," writes Mumford ([1934] 1963, p. 14) with some exaggeration "is the key machine of the modern industrial age." It is mechanical, automatic, and demands a high level of precision in design and maintenance and thus served as an example for all other machinery. It created order and organization and a shared set of objective information. By the middle of the fourteenth century, the custom of dividing the hour into 60 minutes of 60 seconds each had become standard. Four o'clock was four o'clock for all individuals, an hour was an hour. This communicability of facts and concepts, the "I-see-what-you-see" stage of information diffusion, was an important element in the diffusion of innovations. Moreover, it permitted a more accurate measurement of productivity. After all, implicit in our notion of efficiency is the need to measure time: productivity is a flow concept. Clocks brought home differences in efficiency: more productive workers and better implements and tools could be seen to produce more output per hour. Productivity comparisons became easier, and with them the choice between the faster and the slower. Clockmakers brought new standards of accuracy and complexity to the construction of mechanical contrivances, and many played important roles in subsequent inventions in other industries.

Another important mechanical idea was the flywheel, which had previously been used only for grindstones. Its most important application was the spinning wheel, which first appeared in the woolen industry in the twelfth century (Munro, 1988). The spinning wheel is also the first known application of a belt-driven transmission. At first the spinning wheel was turned with one hand and the yarn was twisted with the other. It increased output relative to the age-old distaff-and-spindle method because the spindle was mounted in a fixed place and turned much faster. The spinning wheel greatly increased productivity in the Flemish cloth industry.[11] Nonetheless, the old drop-spindle technique survived for many centuries. The yarn produced by early spinning wheels was inferior and could not serve as warp in high-quality fabrics. Moreover, the spinning wheel confined the spinner to his or her home, while the spindle could be taken anywhere, allowing work to progress during other activities. In the fifteenth century the so-called Saxony wheel was invented, in which a U-shaped flyer was mounted on top of the spindle, so that it could spin and wind fibers simultaneously. By the sixteenth century the foot-operated treadle was added.

11. Adam Smith believed that the spinning wheel doubled the productivity of labor. See Smith ([1776] 1976, p. 273). Munro (1988, p. 698) estimates the increase in productivity at a factor of three. Kuhn (1988, p. 202), in his discussion of the Chinese spindle-wheel also concluded that a factor of three is reasonable.

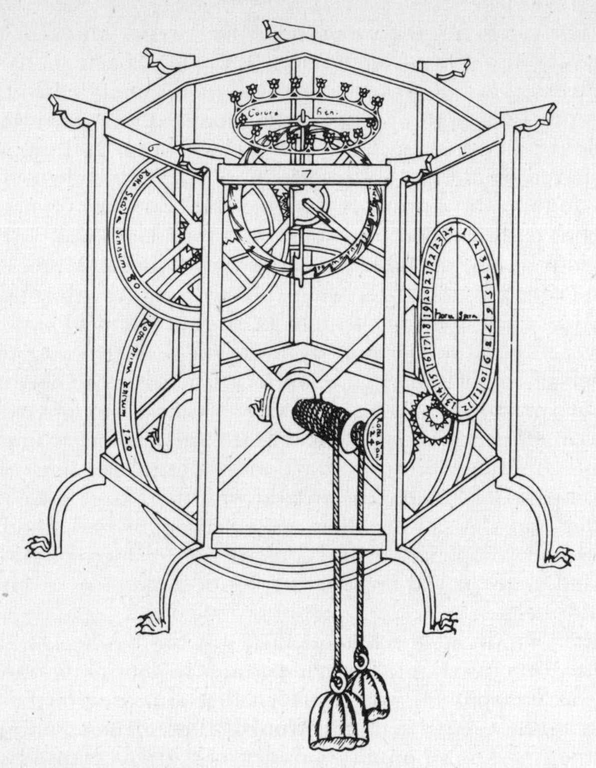

Figure 12. Astronomical clock designed by Giovanni Dondi in 1364.
Source: Bodleian Library, Oxford.

In weaving, the vertical looms of classical times had used weights to stretch the warp, or had spanned the warp between two beams. This technology was overthrown by the appearance of the horizontal loom. The first mention of the horizontal loom is found in the Talmudic commentaries of Rashi, a Jewish sage who lived in Troyes in the eleventh century, who indicates that such a loom was used by professional weavers. It employed a foot-operated treadle that raised and lowered the wires guiding the warp threads (heddles) (White, 1978, p. 274n.). By producing much longer cloths and stretching the warp more tautly, the new loom increased labor productivity by an estimated 325 percent (Munro, 1988, p. 704). In the thirteenth century, Europeans also mastered the processes of silk throwing, with Lucca in Tuscany becoming the center of the industry. The Lucca

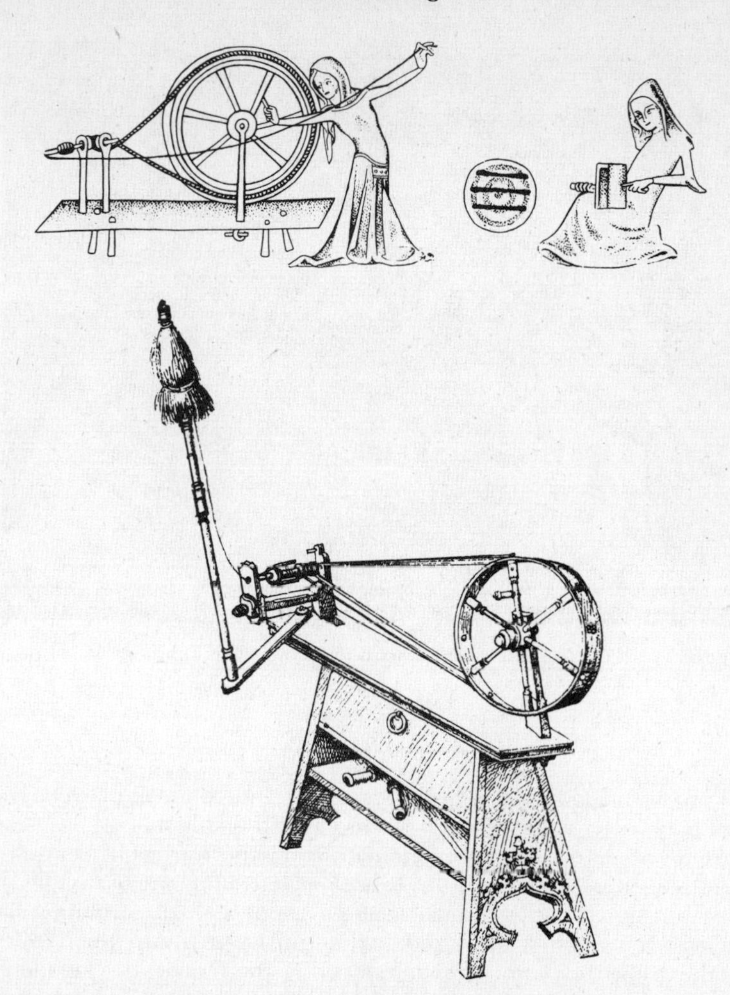

Figure 13. Medieval spinning wheels. The top one dates from about 1338 and is far less sophisticated than the one on the bottom, which dates from the late fifteenth century.
Source: London, British Museum, Add. MS. 42130, fol. 193. After 'The Luttrell Psalter.' Facs. edition by E. G. Millar, Pl. CXXXVIII London British Museum, 1932. By courtesy of the Trustees. D. E. Woodall. (bottom) 'Das mittelalterliche Hausbuch im Besitz des Fürsten von Waldburg-Wolfegg-Waldsee.' Facs. edition by H. T. Bossert and W. F. Storck, Pl. XXXV (fol. 34a). Leipzig, Seemann, 1912.

Figure 14. Fourteenth-century horizontal loom, with spinning wheel next to it.
Source: Science Museum Library, London.

silk works made heavy use of waterpower and complex machinery, and may be regarded as the first mechanized textile mills.

Some other useful inventions date from the thirteenth century. Spectacles were invented around 1285 in Italy and spread rapidly.[12] The functional button, which brought about a small revolution in apparel design, first appeared in Central Germany in the 1230s. Nothing like the button ever developed in the Orient: the Japanese were delighted with them when they first saw them worn by Portuguese traders, and retained the Portuguese word for button (White, 1968, pp. 129–30). In hydraulic engineering, the Middle Ages witnessed the invention of the canal lock. Canals and lift locks are known to have existed in the Low Countries in the twelfth century, and were widespread by the end of the fourteenth century (Forbes, 1956d, p. 688). In the fifteenth century the leadership in hydraulic engineering moved to northern Italy, where a variety of locks was built,

12. It is telling that Islamic scientists (e.g., Alhazen, who lived around 1000 A.D.) studied the reflection of light in curved mirrors and glass spheres, yet the application of optics to better the human lot came from a Western society.

culminating in the mitre-gate lock invented by Leonardo Da Vinci. Dike construction improved, leading to greater durability. In the fifteenth century, some urban areas began to worry about water supply, and books discussing water supply and hydraulics began to appear. The German military engineer Konrad Kyeser, who published a book on military engineering entitled *De Bellefortis* (1405), dedicated an entire chapter to hydraulic engineering. Some of the knowledge was used for purposes we think of today as frivolous, such as the famous pleasure gardens of the Count of Artois, in which automatic sprinklers were used to suddenly wet the clothes of stately ladies from below (Price, 1975, p. 65). But the willingness to apply knowledge was there, and when the tools and skills allowed it, medieval engineering easily surpassed the achievements of classical civilization.

The later Middle Ages also witnessed the increasing use of chemicals in Europe's economy. The Moslems had been better chemists; it was a long time before Europe produced a chemist of the stature of Rhazes or Avicenna. But whatever the Europeans knew and learned about chemicals, they used in production. Alcohol, dyes, alum, saltpeter, mercury, and acids were all used wherever possible and necessary. Gunpowder may not have been a European invention, but the Europeans soon designed and built guns that left Islam and the Orient at their mercy. Most chemicals were used for peaceful ends, such as staining windows, dyeing, tanning, oil painting, medicine, and metallurgy. It was progress without science, chemicals without chemistry, but it worked. Progress was attained by thousands of forgotten tinkerers and craftsmen, often replicating each other, many of them wasting their creative energy in the fruitless pursuit of alchemy and other dead ends. Yet progress there was: slow perhaps, but inexorable in the long run. It may well be true, as Alfred North Whitehead has stated, that as far as science is concerned, Europe still knew less in 1500 than Archimedes knew in 212 B.C. As far as technology is concerned, this assessment is definitely false. By 1500, technology in Europe had advanced far beyond anything known in antiquity. Although Europeans may not have been wiser or more enlightened in 1500 than in 600, they had become incomparably better at producing the goods and services that determine material living standards.

By 1500, Europe had more or less achieved technological parity with the most advanced parts of the Islamic and Oriental worlds. Indeed, in the assessment of some historians, by that time Europeans already controlled more energy, machinery, and organizational skill than any civilization, ancient or contemporary (Lach, 1977, p. 400). It was soon to turn from borrower to lender. Much of the achievement in technology preceded the beginning of European science. Systematic learning had little to do with technological progress. Medieval technology differed from classical and modern technology in another important respect. Cardwell (1972, pp. 9–10) has pointed out that unlike classical technology, medieval technology was not grandiose or extravagant. Apart from a few imposing church buildings and castles, it was concentrated largely in the private sector. It was carried by peasants, wheelwrights, masons, silversmiths, miners, and monks. It was, above all, practical, aimed at modest goals that eventually transformed daily existence. It produced more and better food, transportation, clothes, gadgets, and shelter. It was the stuff of Schumpeterian growth.

CHAPTER FOUR

The Renaissance and Beyond: Technology 1500–1750

As noted, by 1500 Europe was no longer the technological backwater it had been in 900, nor was it the upstart imitator of 1200. It is clear that Europe owed China a great deal, as Needham has argued tirelessly.[1] Yet in the two centuries before 1500, Europe's technological creativity had become increasingly original. In the later Middle Ages Chinese technology had become, in Landes's phrase, a "magnificent dead end." After 1500 China ceases to be of much interest to the historian of technology. Its use of iron and waterpower did not lead to a Chinese Manchester any more than its knowledge of printing led to a massive outpouring of printed books in China; Su Sung's famous water clock did not cause a large clock to be erected in the center of every town in China. In Chapter 9, I shall examine the Chinese experience in some detail.[2]

In the centuries after 1500, the gap between Europe and the rest of the world gradually widened, even though the age witnessed relatively few macroinventions. Technological progress in the conventional sense continued unabated. The increase in productivity, however, became more gradual and consisted largely of sequences of microinventions and modifications to existing techniques. One ex-

1. Much of Needham's persistent defense of the Chinese origins of Western technology is a needed antidote to the Eurocentric histories of technology that are now falling out of style. But Needham may have exaggerated the Chinese influence. It may well be true, as his widely quoted phrase notes, that "the world owes far more to the relatively silent craftsmen of ancient and medieval China than to the Alexandrian mechanics" (Needham, 1969, p. 58). But by 1500 Europe already owed far more to its own silent craftsmen and engineers.

2. Europe's debts to the Orient were not confined to China. India, Tibet, and even Malaya contributed to European power technology, metallurgy, and textiles (Lynn White, 1978, pp. 43–58).

planation for the absence of discontinuous breakthroughs between 1500 and 1750 is that although there was no scarcity of bold and novel technical ideas, the constraints of workmanship and materials to turn them into reality became binding. If inventions were dated according to the first time they occurred to anyone, rather than the first time they were actually constructed, this period may indeed be regarded just as creative as the Industrial Revolution. But the paddle-wheel boats, calculating machines, parachutes, fountain pens, steam-operated wheels, power looms, and ball bearings envisaged in this age—interesting as they are to the historian of ideas—had no economic impact because they could not be made practical. The paradigmatic inventor of this period was the Dutch-born engineer Cornelis Drebbel (1573–1633), who made minor contributions in a host of areas, including chemical dyes, clockmaking, and furnacemaking, but whose main claim to fame rests on the demonstration of the idea of the submarine in 1624, two-and-a-half centuries before submarines became practicable.

From a purely economic point of view, the most important technological change in terms of its potential contribution to material welfare can hardly be termed an invention at all. The "new husbandry," as it is now called, was a set of modifications in agricultural practice that made its first appearance in the Low Countries by the closing of the Middle Ages. These changes spread, ever so slowly, to England and eastward, but by 1750 their adoption was far from complete and in some areas, including most of France, had hardly begun. Yet the principles of the new husbandry were revolutionary, and their adoption led eventually to increases in agricultural output. The three elements of the new husbandry were all closely related: new crops, stall feeding of cattle, and the elimination of fallowing. The result was that farmers were able to maintain more and better-fed cattle, thus increasing the supply of animal products. Better-fed animals produced more fertilizer, which helped to increase cereal yields. The new fodder crops, such as alfalfa, clover, artificial grasses, turnips, and mangel-wurzels, also turned out to be useful as alternating crops to cereals in new rotations. Some of these crops were nitrogen fixers and all of them broke disease and pest cycles. With the increased supplies of fertilizer and the need to hoe some of the new crops, such as turnips, fallowing the land became less necessary and the practice began gradually to disappear in some regions, increasing the effective supply of arable land. The new husbandry was a tale of complementarities, of mutually reinforcing and symbiotic changes, but it was slow to unfold. Some of the new crops were not suitable to heavy clay soils, others needed better drainage than was available. Capital scarcities, the scattering of plots, and hostility to novel practices by those who were frightened or threatened by them, slowed diffusion. Some scholars believe that the adoption of the new husbandry depended on the enclosure of open fields, but this is now disputed. In any event, the often-used term "agricultural revolution" to denote the introduction of the new husbandry is misleading. There was nothing abrupt about it.

The effect of the new husbandry on living standards is hard to quantify. In many areas its full-scale adoption took place only in the nineteenth century. Even in areas where it was adopted, it is difficult to know exactly how much of the increased food production was attributable to this new technology. Yet most experts agree that in the long term it had profound consequences for the economic well-being of most Europeans. Nevertheless, although it permitted larger quantities of food to be produced, technological progress in farming did little to reduce the toil of the men and women working in the fields. Even the new implements introduced at this time were, by and large, capital- and land saving rather than labor saving. The seed drill is a case in point. In traditional European agriculture, sowing was carried out by *broadcasting,* spreading the seeds by hand. This technique not only wasted seed and led to an uneven utilization of the soil, it also made weeding difficult because of the uneven germination of the plants. In the sixteenth century the practice of "setting" seeds (by using sticks to make holes in which seeds were dropped) was known. The modern seed drill, which deposited seeds into equally spaced holes, greatly increasing the yield seed ratio, is traditionally associated with Jethro Tull, who built and demonstrated the first prototypes around 1700, though the implement was little used before the nineteenth century. Tull also suggested the introduction of horse hoeing in 1714. New iron plows, introduced first in the Low Countries and then in England, also probably saved more capital than labor. The new plows reduced the friction with the soil by curving the shape of the mouldboard that cut the furrows, and were thus easier to manipulate and handle. It was difficult, however, to shape wood into exactly the desired form, and so after 1650 mouldboards were increasingly made of iron. This led to the disappearance of wheels and the reduction of the number of draft animals required for plowing. In 1730 the Rotherham, or Dutch, plow was patented in England.

In the area of energy use, medieval techniques were improved but not revolutionized. The windmill continued on its tortuous road to ever greater efficiency when Dutch and Italian engineers in the sixteenth century introduced the tower mill, which left the structure permanent and made the axis and roof pivot toward an optimal angle. By the seventeenth century the windmill supplied the Dutch economy, at that time the economic *wunderkind* of Europe, with a cheap, clean, and inexhaustible source of energy that is the envy of today's ecologists. The Dutch were often able to increase the efficiency of manufacturing using wind power through technical ingenuity, as was the case with the *Hollander,* a device invented around 1670 that was used in papermaking. It consisted of horizontal rollers with spikes and mallets that ripped up the rags used for pulping (Hunter, 1930, pp. 170–71). Thanks to the *Hollander,* Dutch paper was of a higher quality than paper made elsewhere in Europe. Windpower was also adapted to drive the sawmills in the Zaan area, where for many decades the Dutch shipyards produced the best ships of Europe. Waterpower generation and transmission, too, became more sophisticated. Yet, in spite of the attractive features of wind-

Figure 15. Corn-grinding windmill. This type of mill is known as a *post-mill,* as the entire structure swivels.
Source: The Various and Ingenious Machines of Agostino Ramelli, translated by Martha Teach Gnudi, Dover Publications, Inc.

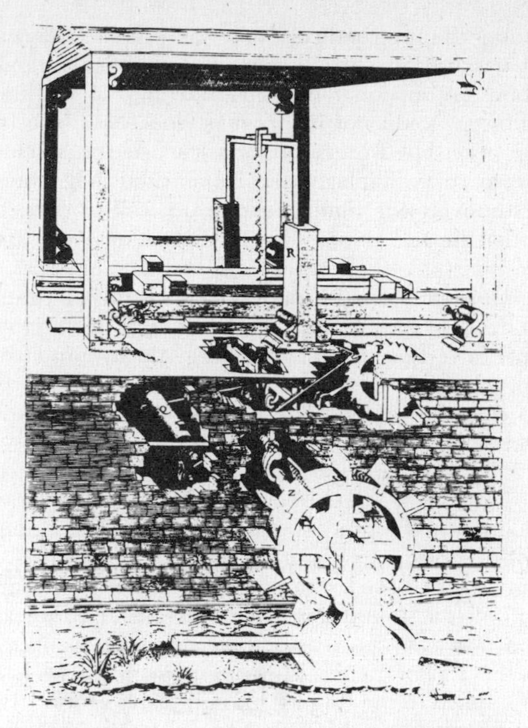

Figure 16. Water-driven sawmill as depicted by Ramelli in 1588.
Source: The Various and Ingenious Machines of Agostino Ramelli, translated by Martha Teach Gnudi, Dover Publications, Inc.

and waterpower, there was a perceptible need for a machine that would not depend on the vicissitudes of European weather.

On or immediately below the surface of Europe lay vast supplies of stored-up solar energy in the form of peat and coal. The use of peat and coal in Europe was not new in 1500, but its geographic expansion and the increase in the number of its uses have prompted some to consider it as important as the greatest inventions of the Industrial Revolution.[3] In Britain and in a few places on the Conti-

3. Consider John U. Nef's statement (1964, p. 170): "by the mid-seventeenth century a new industrial structure was being built in England on coal, and this structure provided the basis for the industrialized Great Britain of the nineteenth century."

nent, including the Principality of Liège in what is today eastern Belgium, coal was used in iron forges, glassmaking, saltmaking, soap boiling, alum production, and lime burning.[4] The fuel-intensive brewing industry could not use regular coal, because it ruined the taste of the beer, but British brewers learned to use charred coal, later known as coke. Similarly, the Dutch used their abundant peat supplies extensively for home heating, as well as for a myriad of industries that helped create the Dutch Golden Age. Brickmakers, madder producers, kiln operators, salt refiners, bakeries, bleachers, tilemakers, and many others made extensive use of peat. Only two major uses of fossil fuels remained elusive: the smelting of iron and the efficient conversion of thermal energy (heat) into kinetic energy (work).

The two centuries after 1500 witnessed major improvements in the use of blast furnaces. Their size and efficiency increased substantially: in 1500 a best-practice blast furnace could produce 1200 kg/day, by 1700 an average figure was more than 2,000 kg/day, and fuel consumption fell in the process. An important improvement was the adoption of a continuous smelting process, in which ore and fuel were fed into the furnace continuously, producing a continuous flow of pig iron. Such a "found day" could last up to 40 weeks by 1700. This period also saw the use of reverberatory furnaces, first described by the Italian Vanoccio Biringuccio in his *De la Pirotechnia* (1540) and applied to the English glass industry around 1610. These furnaces eliminated the chimney and used underground pipes to draw in fresh air. A dome-shaped roof lined with refractory clay reflected the heat back into the oven, generating very high temperatures. Another improvement was the *trompe,* dating from the mid-sixteenth century, which used flowing water to blow air into the forge like a reversed air-lift pump. Yet there is some evidence (Smith and Forbes, 1957, p. 30) that these technical improvements were insufficient to prevent rising fuel and labor costs from raising the price of iron. In the refining and shaping sectors of the iron industry, the most important innovation was the introduction of rotary action, usually waterpowered. Rolling mills, which produced flat sheets of wrought iron, and slitting mills, which cut them into narrow strips for the manufacture of nails, wire, pins, cutlery, and other final products were operating in the Liège region around 1600, where they were a significant factor in the growth of the industry (Gutmann, 1988, p. 62). In England rolling was applied to lead frames

4. Alum is a double sulfate of ammonium and a trivalent metal. It is used widely in a range of chemical industries, including dyeing, tanning, paper making, and pharmaceuticals.

Figure 17. Iron-smelting furnace and forge, mid-sixteenth century.
Source: Georgius Agricola, *De re metallica,* 1556.

used for windows in 1568, and in the iron industry in the middle of the seventeenth century, although the final products were of a low quality (Tylecote, 1976, p. 90).

Economic activity expanded not only sideways but also downward, into the earth itself. From about 1450, mining, especially in central Europe, entered an age of progress unlike anything ever seen before (Molenda, 1988). Here, too, we have no famous inventors, just an endless succession of anonymous improvements on the margin. We do have, however, a hero of sorts, namely Georg Bauer, who, under his latinized name of Georgius Agricola, wrote *De Re Metallica,* published posthumously in 1556. *De Re Metallica* is one of the finest and most detailed books on mining engineering ever written.[5] From it we can infer the improvements introduced into mining after 1450. Agricola describes the machines used for drainage and ventilation, the cranes used for hauling the ore, the construction of shafts, even the sampling of ore quality. The technical problems in mining appear to be universal: flooding, explosions, and vertical haulage lead the list. Germans led Europe and the world in mining technology, developing the transmission of waterpower to high-elevation mines from waterwheels in the valleys by means of overland rod systems; applying gunpowder for blasting rocks; pioneering the use of rails for underground transport; using horse-operated treadmills to run windlasses; and above all developing a variety of pumping devices (that were subsequently applied to fire fighting and other uses).[6] Of comparable influence was the Bohemian mining engineer, Lazarus Ercker, whose magnum opus on mineral ores and mining techniques, published in 1574, was used for generations as a manual on assaying and sampling. Agricola and Ercker were both empiricists, not scientists. There was no theory in their work, just descriptions of things that worked: mining engineering remained almost entirely an empirical body of knowledge. Neither Agricola nor Ercker paid much attention to iron, by far the most important industrial material of the time. Nor were their insights uniformly valuable. Ercker ([1580] 1951, p. 223) explains, for example, that he has reluctantly been forced into the conclusion that iron turns into copper after being treated with vitriol (sulfuric acid). Cyril Stanley Smith, his editor, remarks that this error is a good example of how difficult it is to understand chemical processes without good quantitative measurements. Still, the contribution of scientists, if not science, to mining technology was substantial. The greatest minds of the seventeenth century, from Galileo to Newton, were concerned with the problems of air circulation, safety, pumping, mineralogy and assaying, and the raising of coal and ore from the mines (Merton, 1938, pp. 147–59).

The large number of technical "how-to" books published after 1450 provided a vehicle through which technology was diffused through Europe. Renaissance engineers wrote about a variety of machines and contraptions, many of them serving architectural and military

5. Agricola's work was all but forgotten until his work was rediscovered by a young American mining engineer named Herbert Hoover who, with his wife, translated the work from Latin and published it in 1912, before moving on to other matters.

6. It is with only a little exaggeration that one historian refers to this period as "the age of the pump" (Burstall, 1965, p. 144). Some of the writers on technical matters of the time devoted much attention to describing a variety of pumps. In addition to Agricola, there was Agostino Ramelli, whose massive book *Dell' Artificiose Machine* ([1588] 1976), contained descriptions of over 100 pumps.

Figure 18. Furnace for the melting of copper and lead as depicted by Lazarus Ercker in 1580.
Source: Lazarus Ercker, *Treatise on ores and assaying,* 1580.

purposes. Thus, a technical literature emerged, written by engineers for engineers, and technical knowledge became increasingly communicable and thus cumulative. One of the earliest and greatest of the technical writers was the Sienese engineer Marianus Jacobus Taccola, whose *De Machinis Libri* summarized the state of the art in machine technology in the middle of the fifteenth century. His influence was such that he has been called "the Sienese Archimedes." Jacques Besson's *Theatrum Instrumentarum et Machinarum,* published in Latin and French in 1569, went through three translations and seven editions in the following 35 years. The German Hieronymus Brunschwygk published a book on distilleries, *Liber de Arte Distillandi* (1500), which went through numerous editions and translations. This literature illustrates the growing respect shown by Europeans for machines and the people who made them. Yet outside a few areas, it is unclear that these writings had much effect on the industrial practices of the time (Gille, 1966). Cipolla (1972) has pointed out that Vittorio Zonca's *Nuovo Teatro di Machine et Edificii,* first published in 1607, contained a detailed description of the supposedly secret silk throwing machine in use in northern Italy, and that this book was available in Britain from 1620. Yet silk throwing did not come to Britain until a century later, after John Lombe, one of the greatest industrial spies of history, spent two years in Italy studying the intricacies of the technology.

The meticulous description in books on engineering and mining may leave a misleading impression that the machines described by Agricola, Ramelli, and others were standard equipment in Renaissance Europe. In fact, the gap between the best-practice technique and the average-practice technique was large. For one thing, many of the complex machines described were simply too expensive; even if they would eventually pay for themselves, it was often difficult for a machine builder or engineer to cover the costs of construction or to borrow the necessary funds. In other cases, lack of local skilled labor and mechanics made it difficult to adapt a machine that worked well on one site to operate on another under different circumstances. Innovation remained a live force, but its effects on productivity came only slowly. It may well be that most of the increases in labor productivity in engineering industries and mining were the result of better tools, economies of scale, and a more efficient organization of labor.

Among the successes of Renaissance technology were its achievements in hydraulic engineering, an area in which the classical world had excelled, but which had been neglected for many centuries. Italian engineers, including Leonardo Da Vinci, wrote extensively about dams, pumps, conduits, and tunnels. The leading practical hydraulic engineers of the time were Dutch. After the disastrous floods of 1421, Dutch engineers gradually reclaimed their land, protecting it from the large rivers and the sea, employing power-driven scoop wheels and screw pumps. They used the experience they accumulated to help solve similar problems in the marshes of Poitou and to drain the English Fens in the seventeenth century. From Venice to Lübeck, hydraulic engineers struggled with oceans, rivers, and swamps. Advances in hydraulic engineering helped Europeans make progress in one of the areas in which they were still much behind the Romans, the supply of fresh water. Mechanical-powered water supply systems were installed in Toledo (1526), Augsburg (1548), and London (1582). One of the most famous engineering feats of the baroque era was the Marly pumping plant, built by the Walloon engineer Arnold de Ville between 1678 and 1685, to provide water to

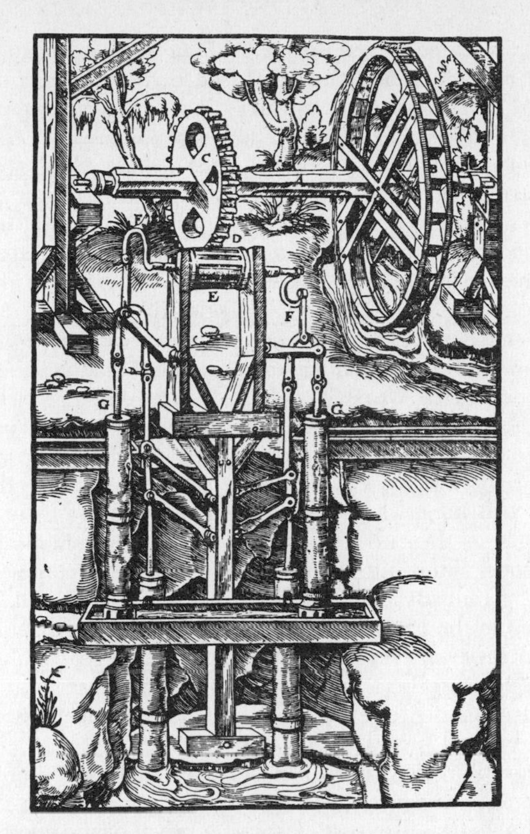

Figure 19. Sixteenth-century water-powered mining pump as depicted by Agricola.
Source: Georgius Argicola, *De re metallica,* 1556.

the royal palaces at Versailles, Marly, and Trianon. Fourteen huge waterwheels drove 221 pumps, and while the project had flaws (the pumps were inefficient and noisy), it nevertheless stands out as a triumph of seventeenth-century hydraulic engineering. In terms of sheer size, the largest hydraulic project was the Languedoc Canal, completed in 1681, which connected the Atlantic Ocean with the Mediterranean, using 26 locks for the rising portion from Toulouse to the summit and another 74 downhill to Sète.

In textiles, the spinning wheel was provided with a treadle and a crank that increased its ease of operation. The foot-operated drive freed both hands of the operator to tend the yarn. The stocking frame, a hand-operated knitting machine consisting of hooks built on a wooden frame, invented by William Lee, a Nottinghamshire clergyman in 1589. The frame diffused throughout Europe in the first half of the seventeenth century, and counts as one of the few true macroinventions of this age. The Dutch loom, which could weave up to twenty-four ribbons simultaneously, dates from 1604, its invention attributed to the Dutchman Willem Dircxz van Sonnevelt. New fabrics were also introduced around this time and it is in this period that Europeans first started to make cotton products themselves, rather than importing them from the Orient. With the predominance of cotton in Western Europe still far in the future, the early modern period saw a great expansion of the production of worsteds, a woolen product made of coarse wool that had been combed rather than carded. Worsted did not need fulling, which made it relatively cheap, and it was lighter than regular wool, making it more attractive to consumers in warmer climates. Although no great technological breakthroughs were made, the techniques spread from the Continent to Britain, where it became a major part of the industrial sector known as the "new draperies." From a mechanical point of view, the most interesting innovation was the development of the silk-throwing mill. By the seventeenth century, large silk mills had been erected in the Piedmont and Tuscany regions of Italy that could be called factories in every respect. The Bologna-type mills set up by the Lombe half-brothers in Derby in 1717 consisted of a huge waterwheel driving no fewer than 25,000 small wheels that simultaneously threw and reeled silk on a vast scale.

The geographic discoveries were in many ways the dominant feature of this age. In some ways, the discoveries may have slowed the rate of technological progress, absorbing much of the energies of the more adventurous and resourceful Europeans. Yet technological and geographical discovery were often complementary, and the interplay between Smithian growth and Schumpeterian growth took many forms. One of these was the design of ships. Although few Renaissance innovations in shipbuilding and seafaring techniques were as dramatic as those of the later Middle Ages, progress was made in less spectacular but economically crucial areas, leading to significant reductions in the cost of transport. Dutch shipbuilding yards were at the center of progress. In 1570, a Dutch sailor came up with the idea of a separate topmast, fitted in a cap on top of the mainmast. The advantage was twofold: the tall and expensive spar trees used to make mainmasts could be dispensed with, and the topmast could be removed in bad weather, reducing the chance of damage to the mainmast (Unger, 1978, p. 28). The Dutch shipbuilding industry led the movement toward specialization, building at least 39 different types of canal- and riverboats alone. The culmination of this trend was the Dutch *fluytschip,* or "flute," according to tradition first built in 1595. The *fluytschip* was the crowning achievement of a century of continuous rationalization and improvement. It was cheap to build and operate, carried a small crew and large cargoes, and until imitation caught up with them, enabled the Dutch to undercut the French and British carrying trade by 30–50 percent (Derry and Williams, 1960, pp. 209–10). With the introduction of heavy naval guns, ships in northern Europe became specialized again, as they had been in Roman times and in the Mediterranean in the Middle Ages. Cargo ships and naval vessels were differentiated, and large unarmed merchant ships sailed in convoys, protected by heavily armed men-of-war (Unger, 1981). In navigation, too, improvement was marginal, with better quadrants and maps, but the greatest difficulty seamen faced—determining accurate longitude while at sea—was not satisfactorily solved until the eighteenth century.[7]

Transportation over the land remained slow and awkward, but gradually became less so during the Renaissance. Better carriages, using leather straps as primitive springs, appear during this period, reputedly first in the Hungarian town of Kocsi, or Kocs, hence the English term "coach." By the late seventeenth century stagecoaches had steel springs, glass windows, and brakes. Renaissance Europe also experienced a revival of an international postal system. As Europe became more integrated and unified, technology could spread faster as people and ideas moved about more easily.

Discoveries in the New World and elsewhere had a clear and visible impact on Europe. Crops from other continents were introduced into Europe, or were cultivated abroad by European entrepreneurs for the sake of European consumers. Maize (corn), tobacco, and potatoes were brought to Europe from the New World. Tobacco was grown successfully in many places in Europe, though the quality rarely matched that of the best varieties grown in America. Maize was introduced in southern France and Italy as a lower-class food. The potato eventually had the greatest impact on European diets as a nutritious and cheap food, first in Ireland, then in the Low Countries, and after 1800 throughout most of Europe. Similarly, Europeans were exposed to new industrial goods that led to the growth

7. The problem was essentially one of building an accurate clock that would withstand motion on sailing ships, so that one could compare the time on board as measured from solar charts with the time at a fixed point, such as Greenwich; a comparison between the two allows the inference of longitude. The best scientific minds of the seventeenth century became involved in the problem, but the technical difficulties proved stubborn. In 1714 a specially established Board of Longitude promised the huge sum of £20,000 to "such person as shall discern the longitude at sea." A clock accurate enough to solve the problem was built by John Harrison in 1762.

of import substitutes. One of these was the chinaware industry, begun in Saxony in 1712. Another was the British cotton industry, which emerged in response to a desire to compete with the high-quality Indian cotton goods known as calicoes.

The age of discoveries was thus the age of exposure effects, in which technological change primarily took the form of observing alien technologies and crops and transplanting them elsewhere. The aggressive Europeans adopted crops from America in exchange for the livestock, wheat, and grapes they transplanted into the New World. Furthermore, they also transplanted non-European flora from America into Africa and Asia and back in a massive act of what could be called ecological arbitrage.[8] Thus, they introduced bananas, sugar, and rice into the New World, and cassava (also known as manioc) into Africa, where it eventually became the staple crop in many areas (Crosby, 1972, p. 187). Sweet potatoes and peanuts were a great success in China after Portuguese traders brought them there from America in the sixteenth century.

The discoveries, together with improved technology, also increased the supply of fish, an important part of the European diet. Because domesticated animals were expensive to raise and the supply of game was small in most parts of Europe, fish was indispensable as a source of protein. The problem was one of preservation. In medieval times, fresh fish could reach only those living close to sources of supply. Herring was the main catch in Europe, first in the Baltic and later in the North Sea. In the late fourteenth century, Dutch fishermen discovered the technique of gutting and salting fresh herring, which allowed preservation for long periods. By about 1415, they had introduced drift nets, which were towed alongside ships and increased catches substantially. The ships that carried these nets, introduced at about the same time, resembled little floating factories. They carried coopers and salters aboard who processed the fish immediately. These fishing ships, known as *busses,* helped the Dutch establish a domination in North Sea fishing that lasted for centuries. The other major crop provided by the sea was gutted and dried cod, known as stockfish and sold throughout Renaissance Europe. The discovery of huge supplies of codfish off the banks of Newfoundland in 1497 by John Cabot, and the use of a new type of line with thousands of hooks, gave the Europeans a new and unexpected free lunch of dried cod, not appetizing perhaps by modern standards, but rich in protein.

Above all, the age of discoveries was one of instruments. Instru-

8. The ecological consequences of the age of discoveries are explored masterfully by Crosby (1972, 1986).

ments in Western technology came before machines. The affinity of Europeans for gadgets derived to a large extent from the clockmaking industry. Clockmakers revealed the wonders that precision-built spring-driven gears and cogs could achieve. By the middle of the fifteenth century the German town of Nuremberg had become the world's center for gadgets. Its fame is immortalized by E. T. A. Hoffmann and Jacques Offenbach in the tale of the Doll of Nuremberg. Not all instruments were toys, however. Astronomical instruments and compasses were crucial to the worldwide navigation in which Europeans became involved.[9] Military technology required precision for the calibration and sighting of guns. Commerce required precision scales, real estate required odometers. A special branch of the instrument making industry was optics. The earliest spectacle lenses were convex and could aid only the far-sighted. A little before 1500, concave lenses were developed that corrected shortsightedness as well. The telescope was invented by Dutch opticians in the early 1600s. Though of limited direct economic use, the telescope nicely illustrates the pragmatic bent of the European mentality at the time. Within a few years, Prince Maurice of Nassau used the telescope to gaze at the Spanish armies and his sea captains used it to look for cliffs and hostile galleons at sea, while in Padua a mathematician by the name of Galileo used it to gaze at the moons of Jupiter. Technical ideas and gadgets that worked, and worked well regardless of the environment, spread more rapidly than ever before.

The precision instrument industry produced important spillover effects in the manufacturing sector. The main breakthroughs had to await the Industrial Revolution, but the lathe, one of the oldest carpenter's tools in use, underwent improvements as clock- and instrument makers needed precision parts and accurately cut screws, and opticians needed precision-ground lenses. In the sixteenth century the fly wheel and the crank, those irrepressible medieval ideas eternally in search of applications, were applied to the lathe. The greatest lathemaker of the age was Jacques Besson, a French engineer employed by the court of King Charles IX, who built, around 1569, an ingenious and sophisticated screw-cutting machine. The Besson

9. From Nuremberg and Augsburg the art of instrumentmaking spread to Louvain in the southern Netherlands and from there to London. The London instrumentmaker Humfray Cole was apprenticed to the Liège craftsman Thomas Gemini. Among Cole's customers were Francis Drake and Martin Frobisher. Gemini himself had studied in the south of Germany. Another German instrumentmaker, Nicholas Kratzer, lived in England for many years. It is interesting to note that although the Germans themselves were little involved in discoveries, the instruments they made were. The European world by 1500 had become sufficiently integrated that knowledge and technique could spread across boundaries without difficulty.

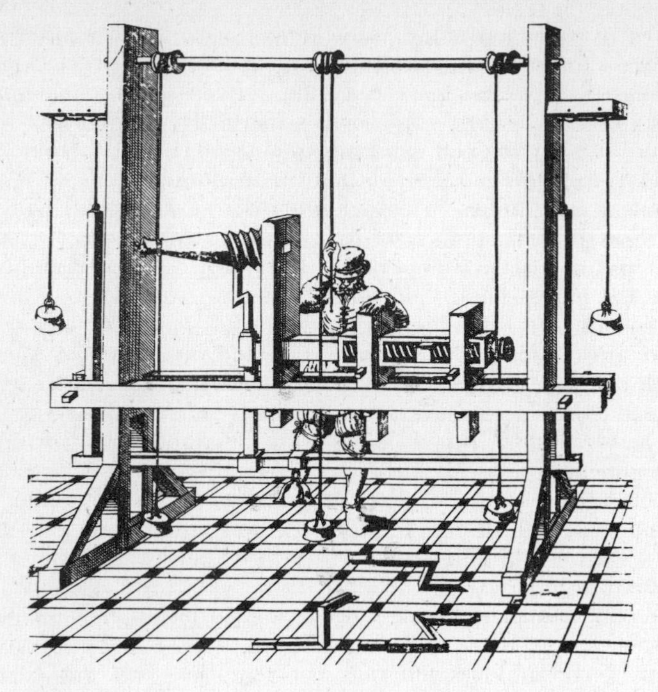

Figure 20. Jacques Besson's screw-cutting lathe, dating from 1579.
Source: From J. Besson. "Theatre des instrumens," Fig. 9. Lyons, 1579.

machine was semiautomatic in that the operator needed only to pull and release a cord. Its construction was complex and the lathe was probably not widely used (Woodbury, 1972, p. 57).

Instrument making in the sixteenth and seventeenth centuries was an art, not a standardized technique. Most improvements were the result of serendipity and trial-and-error searches. Learning and training took place mostly through apprenticing and informal contact. Mechanics had to build their own parts, and often the gap between the visionary who saw what *might* be done and the craftsmen whose material and tools limited what *could* be done was too wide to be bridged. The most famous of these visionaries was of course Leonardo Da Vinci, whose mechanical brilliance was on a par with his other talents. Leonardo left us with 5,000 pages of unpublished notebooks, many of which dealt with machinery. Yet the Last Supper notwithstanding, Da Vinci's creativity produced few free lunches, and few of his technical insights were realized in his lifetime.[10] Nor were the equally prophetic technical dreams of Leonardo's precursor, Francesco di Giorgio (1439–1502). As we shall see, the Industrial Revolution became possible when mechanics and machine tools could translate ideas and blueprints into accurate and reliable prototypes. Until then, instruments and tools were handmade, expensive to make and repair, and limited in their uses.

The period 1500-1750 is better known for its scientific achievements than its technological breakthroughs. The interaction between the two is the subject of an extensive literature, which cannot be done justice to here. What is striking, however, is that during the Renaissance, the classical dichotomy between thinkers and makers had all but disappeared in Europe, whereas the modern distinction between scientist and engineer had not yet appeared. Many scientists made their own instruments and contributed to the solution of practical problems associated with their manufacture. Galileo built his own telescopes and supplemented his salary as a professor at the University of Padua by making and repairing instruments. In England, Robert Hooke, the brilliant and eccentric physicist and biologist, pioneered the use of balance springs for watches and invented the Hooke's joint, an elegant device used for power transmission. Together with Robert Boyle, one of the most versatile scientists of his age, he produced a superior air pump, originally invented by another scientist of fame, Otto von Guericke. The physicist and mathematician Christiaan Huygens invented the pendulum clock and was one of the first to suggest an engine using an internal combustion chamber, for which he wanted to use gunpowder. Gottfried Leibniz worked on a wind-driven pump to remove water from mines in the Harz mountains. Even Isaac Newton, despite his professed (and unusual) lack of interest in technology, made a contribution to the perfection of the marine sextant, and was deeply interested in the problem of determining longitude at sea (Merton, 1938, pp. 172–73).[11] Thus, scientists may have been more important to technological change than science itself. All the same, their role was not decisive. The number of truly important technological breakthroughs

10. Not all of Leonardo's brilliant ideas were of academic interest only. His invention of mitred gates, fitted with small wickets for letting the water through, was applied to the Milan canal, and then to the Exeter canal in 1564.

11. Historians of technology refer to seventeenth-century science as "experimental philosophy," an oxymoron that nicely captures the essence of advances in knowledge in those years. As late as the closing years of the eighteenth century, the physicist Count Rumford asserted, with some exaggeration, that "invention seems to be particularly the province of men of science" (cited in Hall, 1967, p. 115).

that the world owes to men renowned for their scientific contributions is not large (Kuhn, 1977, p. 142).

The Renaissance and the baroque period also witnessed the beginning of the application of mathematics to engineering in a variety of areas. During the Middle Ages, Europe had made few major contributions to mathematics. The Arabs adopted and perfected the decimal numeral system that still (unjustly) bears their name. They learned algebra from India and preserved and extended classical geometry. In the later Middle Ages the Europeans first saw, then learned, then imitated, then applied, then improved, then eventually took over the field, so that modern mathematics is by and large a European product. Mathematics was discovered to be useful to all economic activity, not just to engineering. The use of Arabic numerals was first introduced to Europeans by Leonardo Fibonacci of Pisa, whose *Liber Abaci* was published in 1202. The system caught on slowly, but it had enormous advantages for accounting, measuring, and calculating, and it was doubtless instrumental in the development of double-entry bookkeeping, which appeared in the middle of the fourteenth century.[12] Italian boys aspiring to become merchants had to attend the *scuola d'abbaco,* or schools of arithmetic (Swetz, 1987, p. 20). It soon became clear that mathematics was useful in more than just accounting. In the fifteenth century, Italian mathematicians showed how navigation could be aided by mathematics and Venice created a university chair of mathematics devoted to navigation. Niccoló Tartaglia, a mathematician living in the first half of the sixteenth century, dispensed mathematical advice to military engineers, surveyors, ore assayers, and merchants. Simon Stevin, a Flemish engineer, suggested in 1585 the use of what was to become known as the decimal point. About a decade later the Scottish mathematician John Napier discovered logarithms, making business calculations such as compound interest easy and accurate. William Oughtred, the Rector of Albury in England, invented the slide rule in 1621. At about the same time a Dutch mathematician, Willebrord Snell, developed the technique of trigonometric triangulation, which proved invaluable in determining distances and revolutionized mapmaking. Twenty years later, Blaise Pascal built a machine that could add and subtract, though his machine, like Leibniz's (which could also multiply and divide), remained too expensive to be practical. John Graunt, an English merchant, published the first life tables in 1662, and should be considered the founder of demography. Soon afterwards

12. The papers of the merchant of Prato, who after his death in 1410 left his estate (including his papers) to his hometown, include an astonishing amount of accounting material, including 500 account books that bear witness to the usefulness of the decimal system (Burstall, 1965, p. 112).

the English political arithmeticians William Petty and Gregory King pioneered the idea of observing a society through aggregate statistics.

The main applications of mathematics were in mechanical engineering. Mathematicians and engineers discovered that they needed each other. Mathematics was needed in measurement, civil engineering, ballistics, optics, navigation, and hydraulic systems.[13] Clockmakers wanted to know the optimal shape of the teeth on gear wheels that would minimize friction: the mathematicians Ole Roemer and Christiaan Huygens were able to show that the epicycloid was the curve satisfying this condition. In other areas, such as shipbuilding and machinery, the application of mathematics was more difficult, because much of the mathematics still needed to be developed. Galileo's development of mechanical physics and the later invention of calculus were necessary for those further advances. From the viewpoint of the history of technology, Galileo is particularly important because his theory of mechanics and concept of force lies at the basis of all machines. Until Galileo, the idea that general laws governed all machines was not recognized; each machine was described as if it were unique. Galileo realized that all machines transmitted and applied force as special cases of the lever and fulcrum principle.

As Cardwell points out, Galileo's theory of mechanics is interesting to the economist because the concept governing it is one of efficiency: "The function of a machine is to deploy and use the powers that nature makes available in the best possible way for man's pur poses . . . the criterion is the amount of work done—however that is evaluated—and not a subjective assessment of the effort put into accomplishing it" (Cardwell, 1972, pp. 38–39). In the writings of Galileo, the leading scientist of his time, economic efficiency is linked with science. In *Motion and Mechanics,* he wrote that the advantage of machines was to harness cheap sources of energy because "the fall of a river costs little or nothing." In this he differed radically from his inspiration, Archimedes, and this difference between the two scientific giants who established the science of mechanics epitomizes the difference between classical and early modern society.

The period between 1500 and 1750 was thus one of technological development, but not one of revolutions. Considering the handicaps and obstacles that new technology faced in this period, it is surprising that the process did not grind to a halt altogether. As noted, the great discoveries may have been a substitute for technological frontiers: the challenging and possibly enriching opportunities lay over-

13. "Mechanics is the paradise of the mathematical sciences, because it is in mechanics that the latter find their realization . . . when none of the mathematical sciences can be applied, there is no certitude," wrote Leonardo (cited in Bertrand Gille, 1969, pp. 135–36).

seas. Moreover, the effects of the Reformation on the rate of technological progress were probably on the whole negative. That Protestantism itself was conducive to technological changes is doubtful. What matters to innovation is not only what one believes per se but to what extent society tolerates deviation and nonconformism (Goldstone, 1987), an issue to which I shall return in Part III. The Reformation, and its natural sequel the Counter-Reformation, made Europe a more bigoted place than it had been since the Crusades: Giordano Bruno was burned by the Catholic Inquisition, Miguel Servetus by its Calvinist counterpart in Geneva. Throughout Europe in the sixteenth- and early seventeenth centuries, the authorities' patience for people who thought for themselves and were critical of dogma was wearing thin. This pressure to conform slowed down technological change, though it is difficult to assess to what extent. In southern Europe, which came increasingly under the domination of the reactionary power of the Counter-Reformation, the climate for technological creativity changed for the worse.[14] Moreover, religious differences helped trigger wars that destroyed some of the most active centers of technological change in Europe, especially in the southern Netherlands (1568–90) and most of Germany (1618–48). Europe's ability to maintain its momentum, despite worsening circumstances, serves as a testimony to the resilience of the forces of technological progress. If Antwerp and Augsburg were destroyed, there was always Amsterdam and London.

Between 1500 and 1750 important changes in the form of industrial organization occurred in Europe, and these changes are likely to have affected the rate of technological progress. The driving force in these changes was the de-urbanization of industry. Cities were unhealthy places, plagued by sieges, epidemics, fires, overcrowding, poor water supply and sewage, and consequently high mortality rates. During the Religious Wars, many cities were besieged and sacked. The poor quality of life and short life expectancy in cities raised urban wages relative to the countryside, since towns had to continue to attract migrants. Another cause behind manufacturing's move to

14. Around 1588, Giambattista della Porta constructed in Naples a successful incubator for hatching eggs. His innovative proclivities drew the wrath of the Holy Inquisition, who threatened to persecute him as a sorcerer. Della Porta wisely gave up his experiments. They were later resumed by the Dutchman Cornelis Drebbel and the Frenchman René Réaumur in societies where nonconformist ideas were better tolerated. France became less tolerant during the reign of Louis XIV, and many technologically skilled Huguenots emigrated. Among them was Denis Papin, who became Huygens' assistant in Holland and who made a fundamental contribution to the invention of the steam engine, as we shall see.

the countryside was the tight corporate structure of craft guilds, which restricted entry and imposed strict rules on the quality and price of output. It may well be the case that by the sixteenth-century town guilds had begun to stifle technological progress to protect their monopolistic position and vested interests. There are many documented cases of the authorities trying to suppress innovations in established industries, doubtlessly instigated by lobbies of vested interests. We shall return to this phenomenon of resistance to technology in Part III.

The stifling environment of Renaissance and baroque cities should not be exaggerated. Many cities, such as Leyden, had no guilds at all, and putting-out organization was by no means confined to the countryside. Design, finishing, marketing, and the production of custom-made upmarket goods remained largely in urban areas. But in town after town complaints surged that manufacturing was hurt by low-cost rural competition. As manufacturing in urban areas became more expensive, industry discovered the countryside. For goods whose production required relatively low skills, so that cheap and unsupervised labor could be profitably employed, rural workers in slack seasons were gradually recognized as an efficient source of labor. Much of this rural industry was organized by urban entrepreneurs who broke the production process into simple discrete stages and gradually developed a division of labor despite the dispersion of production sites. The rural-cottage industries were capitalistic, integrated into world markets, and devoid of the tight controls and regulations of urban industries.

The effect of this transformation on technological change has not been much explored. In the Middle Ages and the early Renaissance, cities, by multiplying human contact and facilitating the exchange of information, had been important in generating and diffusing technology, but by the sixteenth century this positive effect came to be dominated by the ability of organized vested interests to throttle new ideas. By removing much of industry from the cities, the guilds were faced with a source of competition that weakened their conservative influence. Some technological improvements clearly catered to the new modes of production: the stocking frame was widely adopted in rural cottages, new and superior spinning wheels were designed, and the flying shuttle, invented by the Englishman John Kay in 1733, increased the productivity of the domestic handloom weaver, who by that time was also most likely a rural resident. Above all, rural industry in many areas was the first attempt toward something akin to mass production. Although mass production without standardization and supervision had its limits, the merchant-entrepreneurs who

Figure 21. Pin factory dating from the time of Adam Smith. Smith used the pin factory as an example of the advantages of the division of labor. *Source:* Réaumur, *L'art de l'épinglier,* 1762.

ran the putting-out system increasingly realized the potential of cheap goods produced on a large scale, and learned to appreciate the profits inherent in cost-reducing technological advances.

The rise of nation-states between 1450 and 1750 also had important effects on technology. Although government officials rarely participated directly, many governments adopted policies that encouraged new technology. The objectives of these policies were, of course, frequently political and military, such as the design of fortifications, the casting of cannon, and the construction of men-of-war. Yet, in this period a mercantilist outlook led governments to follow an active industrial policy. States increasingly employed and subsidized engineers, and awarded monopolies, patents, and pensions to inventors deemed to have made important contributions to the welfare of the realm. When a nation felt left behind, it sometimes made a deliberate effort to catch up. The most famous example of government-inspired technological diffusion is Czar Peter the Great's stint as a carpenter in a Dutch shipyard. More than 200 years earlier, however, the czars were already sending for architects, miners, printers, and metal workers from the West. Similarly, King Gustav Adolph of Sweden invited the Dutch industrialist Louis De Geer, a Walloon by birth, to set up blast furnaces in Sweden, which—thanks to Sweden's high-quality ores and large forests—soon became a leading iron producer. These initiatives were not confined to the peripheral areas of Europe. The Lyons silk industry was founded when Louis XI enticed some Italian craftsmen to settle there, and the Sforza family attracted some of the best engineers of the time to work in Milan in the first half of the sixteenth century.

Some European governments discovered that protecting the property rights of the inventor encouraged technological change. The idea of granting an inventor a temporary monopoly position through a patent to reward inventive activity emerged from customs in mining activity. Mining contractors were awarded monopoly rights over discoveries of new mineral resources. These arrangements were subsequently adopted in other activities, such as grain milling, and eventually applied to new inventions. This custom appears in northern Italy in the first quarter of the fifteenth century. In 1460, the Republic of Venice granted two inventors a privilege stating that no one could reproduce their inventions without their permission. In 1474 a formal patent system was enacted in Venice, the preamble of which noted that if "provisions were made for the works and devices discovered by men of great genius, so that others who may see them could not build them and take the inventor's honor away, more men would apply their genius . . . and build devices of great utility to our commonwealth" (Kaufer, 1989, p. 5). Although few patents were actually awarded in Venice, its example was followed widely and by the middle of the sixteenth century the idea had penetrated much of Europe.[15] The most effective and famous patent law was the Statute of Monopolies, in England, passed in 1624. States also sponsored scientific societies, such as the Royal Society in Britain (chartered 1662) and the Académie Royale des Sciences in Paris (1666). These societies soon gravitated toward pure science, and the more important societies from the point of view of "facilitating the manual arts" (as Robert Hooke, one of the founders of the Royal Society, put it) were private.

Despite the absence of macroinventions, then, the late Renaissance and baroque periods were ages in which Western society became permeated with technology. As Bertrand Gille (1969, p. 146) points out, everything about the Renaissance was technological, including its art and its political philosophy. Medieval natural philosophy had pictured the universe primarily through biological metaphors. These organic images gradually yielded ground to a more mechanistic approach. Philosophers of the period increasingly adopted the view that technology was inherently virtuous and that knowledge of nature should be converted into control over nature for the purpose of increasing material production. Although such views were already implicit in medieval times, they are expressed with ever increasing

15. According to Bertrand Gille (1969, p. 146) more patents were granted in Germany in the sixteenth century than in the eighteenth century.

clarity and vigor in the sixteenth- and seventeenth centuries.[16] Europeans were becoming conscious of the infinite possibilities that technology promised for human welfare, and realizing that by accepting change as a way of life, they could have access to a never-ending stream of free lunches.

16. The German physician Paracelsus argued in the 1530s, for instance, that God desires that "we do not simply accept an object as an object but investigate why it has been created. Then we can . . . cook raw food so that it tasteth good in the mouth and build for ourselves winter apartments and roofs against the rain." Francis Bacon, writing three quarters of a century later, insisted that knowledge should be made useful, and science be put in the service of technology in the quest for an "empire of man over nature." Galileo hoped that his practical philosophy would help "in the invention of an infinitude of artifices which would allow us to enjoy without trouble the fruits of the earth and all its commodities" (cited in Klemm, 1964, pp. 144, 174, 180).

CHAPTER FIVE

The Years of Miracles: The Industrial Revolution 1750–1830

By 1750, Europe had consolidated its technological superiority over the rest of the world. From the Bosporus to Tokyo Bay, the Oriental empires were falling behind by isolating themselves from the West and experiencing a slowdown in their own technological progress. Some of them, like India, were already coming under Western domination. Yet, it seems plausible that if European technology had stopped dead in its tracks—as Islam's had by about 1200, China's had by 1450, and Japan's had by 1600—a global equilibrium would have settled in that would have left the status quo intact, with few exogenous forces to upset it. Instead, the last two centuries have been a period of ever accelerating change, a disequilibrium of epic proportions unlike anything that came before it. In two centuries daily life changed more than it had in the 7,000 years before. The destabilizing agent in this dizzying tale was technology, and Western technology alone. Of course, technological progress did not start in 1750, and the difference between the period after 1750 and the period before it was one of degree; but degree was everything. The effects of the gains in productivity allowed Europe to expand its population manifold in blatant defiance of Malthusian constraints; to provide Europeans with a quality of life incomparably higher than that of traditional societies; to extend, for a while, political control over most of mankind; and to reshape technology elsewhere in the European image.

In recent years, the concept of the Industrial Revolution has come under serious scrutiny. Authors such as Jones (1988) have argued that there was little economic growth in Britain in the second half of the eighteenth century. Since it has been customary to identify the Industrial Revolution with growth of per capita income (see E. A. Wrigley, 1987, for a recent restatement), the implication seems to be that the concept of an Industrial Revolution is dispensable. Yet recent calls to ban the concept from our research papers and lecture notes seem misplaced, to say the least. The Industrial Revolution was not primarily a macroeconomic event that led to a sudden acceleration of the rate of growth, although growth eventually became its inevitable corollary. The identification of the Industrial Revolution with economic growth suffers from a number of serious defects. First and foremost, by focusing on a per capita variable it glosses over changes in a ratio, per capita income, in which both the numerator (income) and the denominator (population size) were increasing more or less pari passu. As it happened, the years of the Industrial Revolution were years of rapid population growth, and per capita changes were swamped by demographic changes. Second, economic growth need not be a result of industrial change at all; it could be (and often was) rooted initially in agricultural or commercial developments. Third, per capital income is notoriously hard to measure accurately during a period in which the economy undergoes rapid changes in the way its markets operate. Commercialization implies that goods previously produced by households are now purchased on the market. Unless meticulous adjustments are made, these changes tend to bias the measures of economic growth. Furthermore, a second bias is introduced when technological change introduces new goods or improves the quality of existing ones.

Thus, even if aggregate statistics do not reveal a sudden leap, there is room for the concept of an Industrial Revolution (Mokyr, 1991). It is appropriate to think about the Industrial Revolution primarily in terms of accelerating and unprecedented technological change. In the words of T. S. Ashton's (1948, p. 42) famous schoolboy, it was first and foremost a "wave of gadgets" that swept over Britain after 1760, a string of novel ideas and insights that made it possible to produce more and better goods and do so more efficiently. To return to the terminology introduced earlier, a clustering of macroinventions occurred, leading to intensified work in improvement and adjustment, and thus creating a complementary flow of microinventions. The result was a sharp increase in patenting activity. Patent statistics do not permit us to distinguish between radical and minor inventions. The propensity to patent varied widely from industry to industry, from location to location, and even from individual to individual (MacLeod, 1988, pp. 75–114). Yet dismissing the volume of patents altogether as an indicator of inventive activity is premature. The sharp increase in the rate of patenting after 1760 requires an explanation (Sullivan, 1989). Something profound changed in the role of technology in the British economy around this time, although it is yet far from clear whether the rise in patenting was a response to perceived needs and opportunities or a consequence of deeper change, affecting the technological creativity of Britain as a whole.

The Industrial Revolution is usually dated between about 1760 and 1830. Britain is usually thought of as its locus, but a large part of the new technology was the result of work done in other European countries and later in the United States. The fruits of the Industrial Revolution were slow in coming. Per capita consumption and living standards increased little initially, but production technologies changed dramatically in many industries and sectors, preparing the way for sustained Schumpeterian growth in the second half of the nineteenth century, when technological progress spread to previously unaffected industries. It is not easy to generalize about the kind of technological change that occurred. Some scholars have proposed that the main feature of technological change in this period was the substitution of inorganic for organic materials (E. A. Wrigley, 1987). Others try to define the Industrial Revolution as an increase in energy inputs, especially inanimate energy, and focus on steam power as the most significant advance (Cipolla, 1965a). Still others focus on the use of machines instead of hand tools (Paulinyi, 1986). Yet these generalizations fail to do justice to the rich diversity of progress in these years. The growth of cotton at the expense of wool and linen, the improvements in the efficiency of waterpower, the development of gaslighting, the advances in the machine tool industry, and the invention of food canning, to mention just a few examples, really share few common characteristics, save their ability to increase both the quantity and quality of the supplies of goods and services. As McCloskey (1981, p. 118) put it, the Industrial Revolution was not the Age of Cotton, or the Age of Steam; it was the age of improvement. Yet improvement was not ubiquitous. Large sectors of the economy, employing the majority of the labor force and accounting for at least half of gross national product were, for all practical purposes, unaffected by innovation before the middle of the nineteenth century. In services, construction, food processing, and apparel making techniques changed little or not at all before 1850. The reason some industries changed and others did not has little to do with either the demand side of the economy or the supply of raw materials and coal. Technological opportunities and constraints by and large determined where and when improvements were to occur.

During the Industrial Revolution, technological progress was usually the result of the joint and cumulative efforts of many individuals. A typical innovator in those years was a dexterous and mechanically inclined person who became aware of a technical problem to be solved and guessed approximately how to go about solving it. The successful inventors were those who put the pieces together better than their colleagues, or those who managed to resolve one final stubborn difficulty blocking the realization of a new technique.

It is useful to divide the technological changes during the Industrial Revolution into four main groups: power technology, metallurgy, textiles, and a miscellaneous category of other industries and services.

POWER TECHNOLOGY

The protestations of some economic historians notwithstanding, the steam engine is still widely regarded as the quintessential invention of the Industrial Revolution. Its background was not purely British; it is more accurate to think of it as the result of an international joint effort. The basic idea for the construction of an atmospheric engine was based on the realization that an atmosphere exists. What seems today a commonplace insight was the fruit of the work of Evangelista Torricelli, a student of Galileo's, and Otto von Guericke, the mayor of Magdeburg, famous for his experiment in which two teams of horses could not separate two hemispheres enclosing a vacuum.[1] The existence of an atmosphere and its pressure may well have been known to the Chinese, but what followed in the second half of the seventeenth century was a typically European story. It occurred to many who had grasped the newly discovered phenomenon that if a vacuum could be created repeatedly, the force of atmospheric pressure could yield a novel source of power. The Marquis of Worcester, among others, suggested in 1663 a machine utilizing condensation for this purpose. The first known model was built in 1691 by Denis Papin, an assistant to Christiaan Huygens, who showed in a prototype how a piston could be moved up and down a cylinder using steam. Application to a useful purpose followed suit. Thomas Savery built the first working steam engine in 1698, though this device was really a suction pump that condensed steam in a closed vessel and sucked up the water by means of the vacuum. A different version was perfected in the first decade of the eighteenth century by Thomas Newcomen, who, unlike Savery, used atmospheric pressure in a ma-

1. The progress from the discovery of the atmosphere to the construction of atmospheric engines is sometimes regarded as an example of how science influenced technology at this early stage. The relationship was, however, fully reciprocal. The discovery of the atmosphere was prompted by the puzzling observation that vacuum pumps could not raise water more than 32 ft. Galileo struggled with this problem, and Torricelli was probably influenced by his master's interest in it.

chine that was alternately heated and cooled, so as to create repeated vacua by condensation. The first economically successful engine, known as the Dudley Castle Machine, was installed in a coal mine near Wolverhampton in 1712. Newcomen's engine was far more complex and sophisticated than Papin's prototype, yet it was within the ability of the craftsmen of the time and it was safe. It was powerful enough to pump water out of mines, and despite its awkward dimensions, its voracious appetite for fuel, and the difficulty early eighteenth-century mechanics had in achieving hermetic sealing, the Newcomen machine was widely adopted. Within a few years of its inception, it spread to France, Germany, and Belgium, and by 1730 it was operating in Spain, Hungary, and Sweden and later in the American colonies. The machine solved drainage problems in the Cornish tin mines, as well as in the deep coal mines in the north of England. But above all, it was the first economically useful transformation of thermal energy (heat) into kinetic energy (work).[2]

Yet the steam engine will forever remain associated with the name of James Watt. The basic improvement that Watt introduced was to separate the condenser from the piston cylinder, so that the latter could be kept hot constantly. This separation greatly reduced the fuel requirements of the machine and permitted it to be used almost anywhere. Watt's ingenuity provided many further improvements on the steam engine, including steam-jacketing, to keep the cylinder hot; a transmission mechanism known as the "sun-and-planets" gears, which converted the reciprocating motion of the atmospheric engine to the rotative motion needed in textile mills and other industrial applications; and a parallel motion gear that allowed steam to be introduced alternately into both ends of the cylinder, thus creating a double-acting engine that utilized the push as well as the pull of the end of its beam. The double-acting expansion machine used the steam above the piston to drive it down, but cut off the steam after the piston had moved part of the way in order to save fuel. As a result, fuel efficiency was raised from less than 1 percent in the Newcomen engine to around 4.5 percent in Watt's design. The principle of double-acting engines had been known in Europe since the fifteenth cen-

2. Strictly speaking, firearms are also based on this principle. The cannon is the first known example of a controlled conversion of thermal energy to kinetic energy; it is a one-cylinder internal combustion machine. Yet it took another half millenium to perfect the internal combustion engine. In 1673 Huygens suggested exploding minute amounts of gunpowder in a metal cylinder to create an engine based on high pressure. Huygens's insight was theoretically correct, but the internal combustion engine he suggested could not be made practical for another two centuries, and ultimately did not use explosives. The debt of the steam engine, or any other form of power technology, to explosives is doubtful.

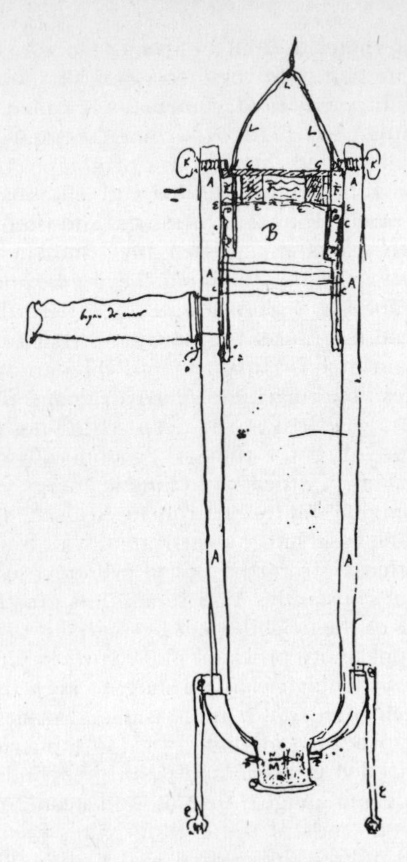

Figure 22. Rough sketch by Christian Huygens of his gunpowder-fueled internal combustion engine.
Source: Chr. Huygens, *Oeuvres completes,* 1763.

tury, and used in pumps and bellows, but the steam engine was by far its most successful application (Reti, 1970). The utilization of the expansive power of steam was regarded by Usher (1954, p. 354) as the difference between an atmospheric and a genuine steam engine. Equally important was the "governor" for regulating the speed of engines, which foreshadowed twentieth-century feedback servomechanisms, which form the basis for cybernetics. The device consisted of two balls that were pushed outward when the speed of operation increased and that lifted an arm attached to the valve in the steampipe. Similar automatic regulators had been applied earlier to windmills, but Watt's invention symbolizes the desire for full control and automation that increasingly permeated the techniques of the time.

Watt's work, which combined inventive genius with a desire to cut costs, minimize wear and tear, and extract "the last drop of 'duty' from the last puff of steam in his engine" (Cardwell, 1972, p. 93), was paradigmatic of the kind of mind that helped make the Industrial Revolution. Watt himself, in his oddly written third person autobiography, wrote that "his mind ran upon making engines *cheap* as well as *good*." The search for economic value in addition to functionality and beauty represents the culmination of a millennium of de-

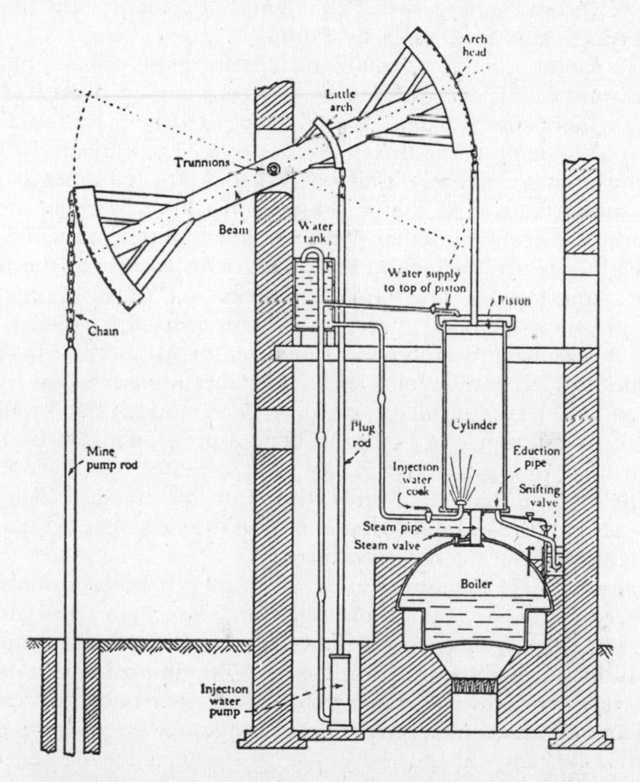

Figure 23. Diagram of Newcomen's atmospheric engine.
Source: From H. W. Dickinson. "A Short History of the Steam Engine," Fig. 7. Cambridge University Press, 1939.

velopment of European technological rationality. Yet rationality meant nothing without technical ability, and Watt's mechanical talents bordered on the virtuoso. In short, in the history of power technology, Watt is comparable to, say, Pasteur in biology, Newton in physics, or Beethoven in music. Some individuals did matter. We should bear in mind, however, that Watt stood on the shoulders of Papin and Newcomen; of John Wilkinson, whose new boring machines supplied Boulton & Watt with cylinders of great accuracy; and of his partner Matthew Boulton, with whom Watt formed a classic inventor–entrepreneur team (Scherer, 1984, pp. 8–31). The steam engine became a familiar sight in eighteenth-century Britain. It is now reckoned that close to 2,500 engines were built in the eighteenth century, of which about 30 percent were made by Watt, by far the largest producer (Kanefsky and Robey, 1980). The most important user of steam power was mining, with 828 engines in collieries and another 209 in copper and lead mines by 1800.

Watt's patent expired in 1800, and a new genius applied himself to the construction of a revolutionary steam engine. Watt had felt that a high-pressure machine—already suggested by Savery—was too dangerous to be practical. But in 1802 another Englishman, Richard Trevithick, built a machine that created pressure ten times as high as the atmosphere. These high-pressure machines were smaller in size and more economical than Watt's engines. In the mines of Cornwall, high-pressure engines were applied with success to the beam engine pumps used for drainage, known as "Cornish" engines. Moreover, these engines could be placed on boats and horseless carriages. A steamboat prototype was built by the Marquis de Jouffroy in France in 1783 and by John Fitch and James Rumsay in the United States in 1787. It was made practical by the American Robert Fulton in 1807. Within ten years Fulton's boats were dominated by high-pressure engines, pioneered in the United States by Oliver Evans (Hindle, 1981, p. 55). Meanwhile, the Watt low-pressure stationary engine also underwent improvements, and the two types existed side by side throughout the nineteenth century.

The next step in the development of steam power was compounding, that is, the use of the same steam in more than one cylinder, one of which is high pressure. In these engines, after the steam does its duty in the high-pressure cylinder, it is admitted into a larger cylinder, where it drives down a piston using the principle of expansion.[3] The first successful application of compound engines was made

3. The original inventor of the compound steam engine was Jonathan Hornblower, who patented a compound engine in 1781. His invention was determined to be an infringement on Watt's patent, however, and compounding was made to wait another two decades.

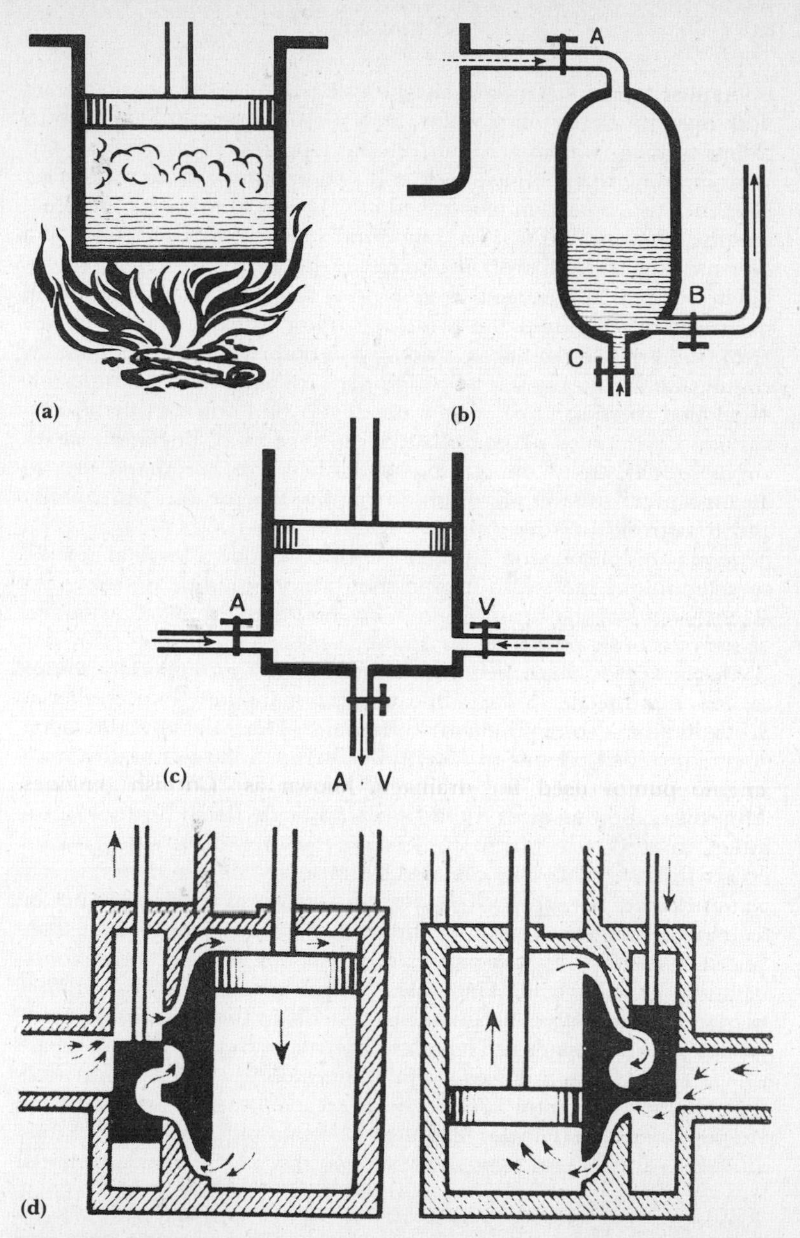

Figure 24. Principles of early steam engines: (a) Papin's prototype, (b) Savery's pistonless suction pump, (c) Newcomen's engine, and (d) Watt's double-acting engine. The machine exploits the contrary motion of the piston and the slide-valve above it, with the steam entering alternately from the left and the right.

Source: Umberto Eco and G. B. Zorzoli, *The Picture History of Inventions.* © Gruppo Editoriale Fabbr: Bompiani Sonzongo Etas S.p.A.

by Arthur Woolf in England in 1803.[4] Compounding became practical only in 1845, when John McNaught (after whom the verb "Mcnaughting" was coined) perfected the process. The advantage of compounding was fuel saving: Woolf's compound engine raised fuel efficiency to 7.5 percent (compared to 4.5 percent in Watt's engines), and the sophisticated Corliss compound steam engine of 1878 had a thermal efficiency of over 17 percent.

The success of the steam engine preceded the establishment of a science that formalized the principles upon which it was based. In 1824, the Frenchman Sadi Carnot, upon observing a working steam engine and asking himself why high-pressure engines were superior, developed the kernel of what was later to be known as thermodynamics. Carnot also maintained that "to take away England's steam engines today would amount to robbing her of her iron and coal, to drying up her sources of wealth, to ruining her means of prosperity and destroying her great power" (cited in Cardwell, 1972, p. 130). Economic historians would probably disagree. The assessment of the contribution of the steam engine should not be based on the gross achievements of the steam engines, but on the marginal contribution of steam over its next best alternative.

That next best alternative was waterpower, still an important source of power in Britain in 1830, and the dominant source of energy in Switzerland and New England at that time. The gains that the steam engine provided relative to waterpower before 1850 were fairly small (von Tunzelmann, 1978, pp. 285–92), but this should not take away from the achievement of the people who made the Industrial Revolution. Indeed, it confirms it, because the slow diffusion of steam power in many places is explained by improvements in the efficiency of waterpower. In other words, the wide range of progress in power technology was responsible for the relatively slow growth of one specific technique. This was particularly true for the European continent and New England. The improvements in waterpower after 1750 were associated with constant improvements in the understanding of the theory of hydraulics. The most important advance was the breast wheel, which was introduced by John Smeaton in the 1750s and soon spread all over Britain. Breast wheels receive water at an intermediate point between the summit and the bottom of the wheel, and thus in a sense are a compromise between overshot and undershot wheels.

4. Woolf based his compound engine on a scientific "law" that, he claimed, showed a linear relation between excess pressure and the increase in the volume of steam when allowed to expand to atmospheric pressure.This erroneous theory, quite at variance with Boyle's law, was the basis for a sound mechanical design. Cardwell remarks, quite correctly, that "not for the first time in history a useful advance was made on the basis of a defective theory" (Cardwell, 1971, p. 155).

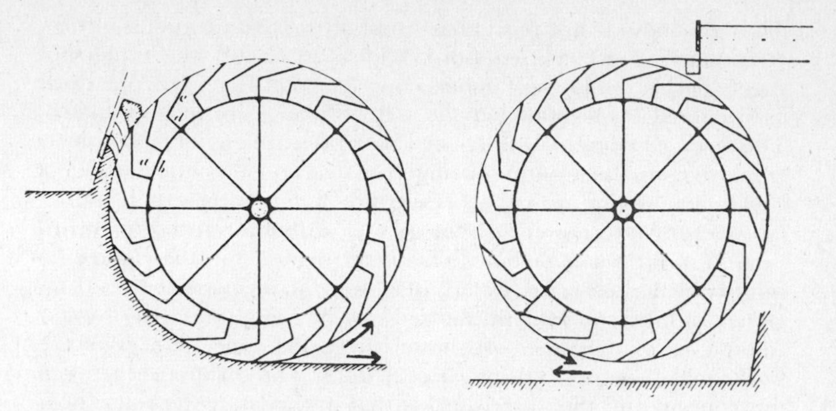

Figure 25. Comparison of high breast waterwheel (left) and traditional waterwheel (on the right). Note that in the breast-wheel the water moves in the same direction as the wheel, in contrast with the traditional wheel.
Source: F. Redtenbacher, *Theorie und Bau der Wasserrader* (Mannheim: Friedrich Basserman, 1846), pl. 1, Figs. 5–6.

The breast wheel was as efficient as the overshot wheel, but had the advantage that the wheel turned in the same direction as the flow of water in the tail race, which allowed it to work under flooding conditions. Smeaton's work was improved by the introduction of the sliding hatch, introduced by John Rennie in the 1780s, which allowed the breast wheels to adapt to varying water levels. Later it was discovered that efficiency could be further improved by setting the blades such that they entered the water at an angle of 45 degrees to the surface (Daumas and Gille, 1979a, p. 28). Like the engineers working on steam engines, their colleagues working on the "traditional" technology were determined to extract as much energy as possible from every moving drop of water. In the nineteenth century, waterwheels were increasingly made of iron parts, which reduced wear and tear. Smeaton, Rennie, and their colleagues were practical engineers, not scientists, and Smeaton was known for his distrust of scientific theory. This distrust was not altogether misplaced: it was not until the 1750s that scientists realized that different principles applied to the gravity-driven wheels of overshot mills and the impulse-driven wheels of undershot mills. For a long time, practical men without formal training in hydraulics kept making the important improvements.

Across the Channel, French engineers were equally successful. Jean Charles Borda was the first to attack the theoretical problems of waterpower in 1766, though his work was too abstract to be of immediate use and was not recognized until after 1810 (Reynolds, 1979). Jean Victor Poncelet used Borda's ideas to modify the undershot waterwheel to build the famous Poncelet waterwheel (1823), which used curved blades and fed the water through an inclined hatch. The so-called column of water or water-pressure engine, extensively employed on the Continent, ingeniously combined the idea of a Newcomen engine with the pressure of water (rather than the atmosphere). Waterpower technology was further advanced by the invention of the water turbine. The idea originated in the eighteenth century with Leonhard Euler, the Swiss mathematician, who, together with his son Johann, showed that by using the force of water coming *out* of the vanes of a waterwheel, the entire energy of the flow could be converted into useful work. The difference between this concept and the waterwheel is that in the latter the water does not move relative to the buckets or vanes, while in the former the power is derived from the water flowing rapidly through curved passages driving the engine. The practical implementation of the idea took many years of tedious work by many engineers (mostly in France), culminating in the success of the Frenchman Benoît Fourneyron in 1837. The turbine was introduced into the New England textile industry and "delayed for decades the dominance of steam power in that industry" (Rae, 1967b, p. 338). By this time, advances in waterpower technology had been so impressive that one historian notes that in theory, and to a large extent in practice, engineers had complete command over waterpower (Cardwell, 1971, p. 184). Nonetheless, the utilization of waterpower remained constrained, not so much by the water mills themselves as by the lack of a scientific understanding of watershed hydrology and the requisite data on rainfall (Gordon, 1985).

The Industrial Revolution was an age of power technology, and the prospects must have seemed unlimited. The economic impact of the steam engine during the Industrial Revolution may not have been initially as large as Carnot thought, but in the second half of the nineteenth century steam power penetrated every aspect of economic life in the Western world and beyond. In conjunction with other inventions, power technology created the gap between Europe and the rest of the world, a temporary disequilibrium that allowed the Europeans to establish global political and military domination.

METALLURGY

Prior to the Industrial Revolution, metallurgy had been an empirical and experimental art in which gifted amateurs and semiprofessionals tried to solve complex chemical and physical problems that baffled those engaged in the day-to-day manufacture of metals. Their success is a monument to their determination to produce materials that satisfied their needs better and more cheaply. Among these materials, iron ruled supreme. There were no substitutes for it in terms of durability, versatility, and malleability. We have seen that the later Middle Ages witnessed the development of cast iron, which has a relatively high carbon content and a low melting point. It was brittle and hard, and could therefore not be shaped using the blacksmith's traditional tools, but it could be cast to make pots, ovens, and cannon. Most products, including machine parts, nails, locks, and tools, needed to be shaped in forges, and for these wrought iron was needed. The process of turning pig iron, the output of blast furnaces, into wrought iron remained a major bottleneck in the metal industry. During the eighteenth century an extensive search was conducted into this problem in Britain. The Wood brothers pioneered the so-called potting process, using crucibles, or "pots," to heat the pig iron (Hyde, 1977, pp. 83–88). The problem's ultimate solution by Henry Cort in 1784 was a skillful combination of a number of elements, such as the coal-burning reverberatory furnace that had been used in glassmaking for a long time, and the rolling of heated metal using grooved rollers. Cort's puddling and rolling process was typical of many great inventions of the Industrial Revolution in that it was the culmination of a dispersed and drawn out search for the solution of a difficult but economically important problem. After several improvements in the late 1780s, Cort's process took the British world of metallurgy by storm. The small independent forge, until then the source of all wrought iron, vanished, to be replaced by the larger puddling furnaces. The supply of high quality and cheap wrought iron grew dramatically, making iron almost literally the building block of the Industrial Revolution.

The eighteenth century also witnessed another invention comparable in fame: the use of coke in blast furnaces. Coke is purified bituminous coal, and its use in smelting was pioneered by Abraham Darby. Coke had been employed earlier in industry, but its use in blast furnaces dates from about 1709. New research on the iron industry has refuted the widespread myth that coke smelting was triggered by a scarcity of wood, and that the diffusion of the invention was impeded by the secrecy of the ironmaster's family (Flinn, 1978; Hyde, 1977, pp. 25–29). The simple problem was that for a long time coked iron contained silicone, which made it more costly to convert coked pig iron into wrought iron (Tylecote, 1976, pp. 108–9). The growth of coking after 1750 has been explained by further technological progress, especially Darby's son's success in remelting the pig iron in a so-called foundry furnace, to remove the silicone.

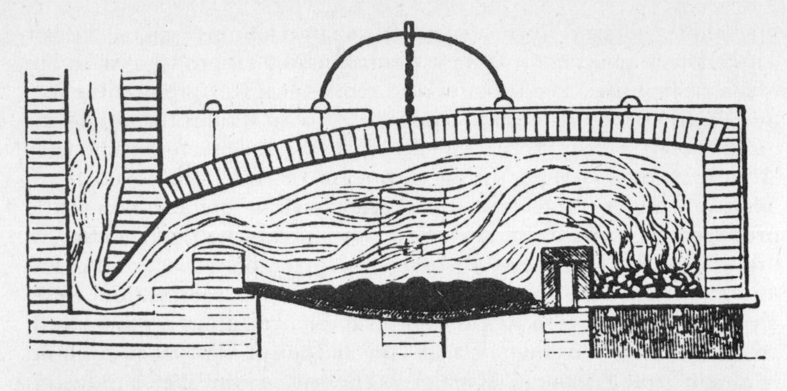

Figure 26. Section of Cort's puddling furnace, which produced wrought iron.
Source: Archibald Clow and Nan L. Clow, *The Chemical Revolution,* Batchworth Press.

Other improvements included the making of cokes in closed ovens shaped like beehives, and the replacement of old-fashioned bellows by new waterpowered blowing cylinders, invented around 1760 by the versatile John Smeaton. Between 1760 and 1790, coke replaced charcoal in British iron smelting and was gradually introduced on the Continent. Blast furnaces became bigger and more efficient, producing better-quality iron at lower prices. Another major breakthrough was achieved by the Scotsman James Neilson, who hit upon the idea of using the blast furnace's own gases to preheat the air inside (1829). The hot blast procedure was cheap to install and simultaneously reduced fuel requirements by a factor of three. It created temperatures hot enough to use fuels other than coked bituminous coal, such as anthracite. Between 1828 and 1840, Scottish producers experienced a cost-reduction of almost two thirds, but as the technique spread through Britain, pig iron prices fell, and profits with them (Hyde, 1977, pp. 151–52). It was this process that made it possible to exploit the black-band iron ore deposits in Scotland and allowed Scotland to compete with the Black Country (in the English Midlands) and Wales in pig iron production.

The product that resisted innovation most stubbornly was steel. Chemically, steel is an intermediate product, halfway between the almost carbonless wrought iron and high-carbon pig iron. Steel can be made from iron by adding carbon to low-carbon wrought iron (cementation or carburization); by removing carbon from high-carbon cast iron (decarburization); or by mixing high and low carbon scraps of iron together (cofusion). A fourth process produced steel directly from ore by packing it in crucibles with pieces of special wood. This last product, known as *wootz* steel, originated in India (in Hyderabad). The desirable physical properties of steel (resilience, tenacity, and flexibility) made it ideal for razors, weapons, shears, springs, and machine parts, but it was prohibitively expensive. In the West, steel was produced by cementation, a process first known in Asia Minor around 1000 B.C. The production of this "blister steel" entailed "baking" the wrought iron by heating it in direct contact with charcoal and hammering it for long periods to spread the carbon through the metal. The blast furnace offered new opportunities for making steel by refining the high-carbon cast iron, or by immersing pieces of low-carbon wrought iron in molten cast iron. Furthermore, ores high in manganese allowed a better control of the carbon removal, so that some residual of carbon could be left in the iron. By the seventeenth century, Europeans had learned that steel could be improved by remelting and hammering small pieces of it at very high temperatures, thus spreading the carbon somewhat more evenly. The production of high-quality steel was perfected in about 1740 by Benjamin Huntsman, who used coke and reverberatory ovens to generate sufficiently high temperatures to enable him to heat blister steel to its melting point. In this way he produced a crucible, or cast, steel that was soon in high demand by instrument- and clockmakers. The importance of crucible steel was that it could be cast, and eventually the production of larger ingots became possible through the coordinated operation of many crucibles. The German steel manufacturer Alfred Krupp pioneered these techniques in the casting of steel cannon. His six-pounder was one of the great sensations of the Crystal Palace Exhibition of 1851, as was a gigantic steel casting weighing 4,300 lb. Yet steel remained too expensive to be of widespread use during the critical years of the Industrial Revolution. Wrought iron rather than steel was the main material until 1860.

TEXTILES

The central technical problem in textiles was that of spinning. Since time immemorial, the crucial operating part in the spinning process had been the human finger, the thumbs and index fingers of millions of women who gave the raw material in the rovings the "twist" that made it into yarn. The spinning wheel increased the efficiency of the spinner's work, but did not replace the human finger as the tool that transformed the material. The search for a replacement for human fingers in cotton spinning was taken up by Lewis Paul, an Englishman who pioneered the idea of using rollers to replace the fingers in drafting out the fibers. His patent was taken out in 1738, but it is Richard Arkwright to whom the mechanization of spinning is usually credited, more than 30 years later. Arkwright's machine, the "throstle" or "water frame," differed from Paul's in one respect: he used *two* pairs of rollers moving at different speeds and separated by a distance about equal to the length of the longest fiber to be spun. The result was that Arkwright's machine worked, whereas Paul's did not. The water frame was incapable of spinning the finer yarns, as these would have snapped when they were wound on the bobbins.

The water frame was complemented by another invention, the spinning jenny, which was patented a year after the throstle, but actually invented in 1764. Its inventor, James Hargreaves, reputedly hit upon the idea after watching a spinning wheel fall on its side and continue to spin for a few more seconds. He realized that it was possible to "draft against the twist," that is, to impart the twist not by the movement of the fingers but by the correct turning of the wheel itself. The jenny twisted the yarn by rotating spindles that pulled the rovings from their bobbins, with metal draw bars playing the role of human fingers guiding the spun yarn onto the spindles by means of a faller wire. Instead of the single spindle turned by the spinning wheel, Hargreaves' machine used many spindles and thus allowed a large number of threads to be spun at the same time. The quality of the yarn was rather uneven, however, and it was suitable only for weft. Moreover, the jenny was an extremely uncomfortable machine to work with, forcing adult spinners to bend over nearly

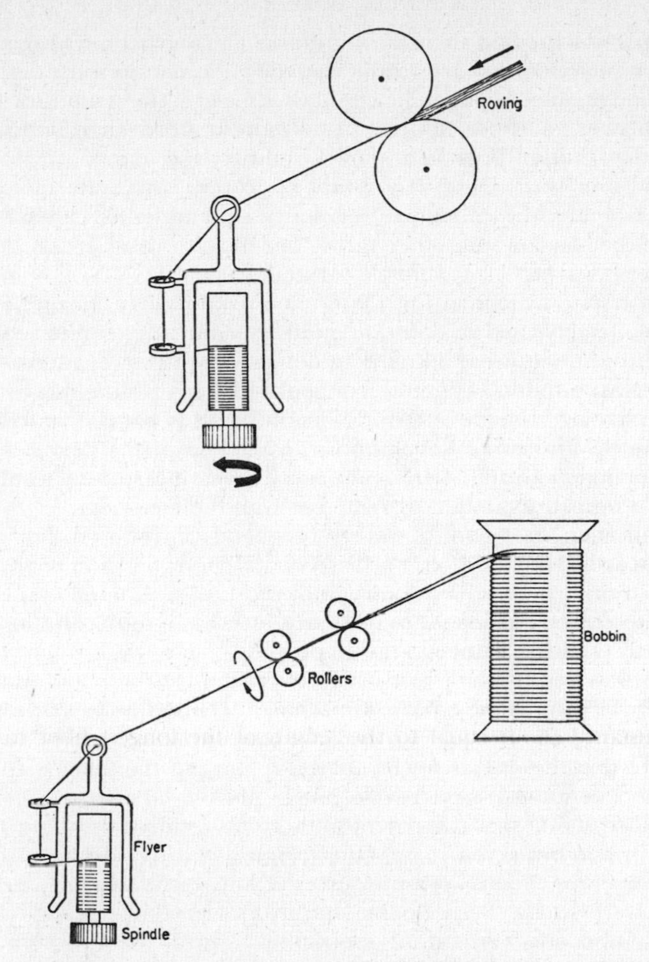

Figure 27. The mechanical principle behind the Wyatt-Paul spinning machine, and Arkwright's throstle.
Source: D. S. L. Cardwell, *Turning Points in Western Technology,* Science History Publications.

double (Hills, 1979). Combining the throstle's rollers with the multiple spindles of the jenny led to the *mule,* the ultimate spinning machine, invented by Samuel Crompton in 1779. One of the most famous inventions of all times, the mule consisted of a carriage that was driven back and forth. In so doing, the spindles mounted on it turned quickly and together with the rollers imparted the twist on the yarn, which could then be wound on bobbins. At no stage was the yarn subjected to much strain, and thus the chances for breaking were much reduced. The mule could thus make cotton yarn that was both cheaper and finer, stronger, and more uniform than hitherto. As a result, cotton became a growth industry the like of which no one had ever seen. Until Crompton, the cotton yarn spun in England was not strong enough to serve as warp and hence cotton was used in combination with other yarns. The mule made all-cotton cloth possible. It was especially suitable for finer yarn; coarse yarns continued to be spun by jennies for a long time, as they were cheaper and could be readily used in domestic industry. Spinning jennies, water frames, and mules were all tried in domestic industry, but soon the factory was found to be a more congenial location for the new spinning technology (Landes, 1986). Nonetheless, it is not warranted to associate the Industrial Revolution with the rise of the factory system; domestic industry, too, experienced some measure of technological progress, some of it in symbiosis with the factories.

Modifications in spinning were subsequently introduced, but the main breakthroughs had been achieved by 1780. Application of steam power to the new machine followed in the 1780s, although animal- and waterpower dominated in the early years. The self-acting mule, patented by Richard Roberts (an improved version was brought out in 1830) was a triumph of British engineering, in the words of Mann (1958, p. 290) "an almost perfect machine." The self-actor made the movements of the carriage which pulled and wound the yarn automatic, so that the operator who moved it and put the faller wire in place became unnecessary. Yet despite its obvious advantages, it was adopted so slowly that the patent authorities decided to extend the patent by another seven years. One reason seems to have been that self-actors were expensive, and sources of long-term credit for fixed capital were scarce. Second, the self actors were better suited for coarse (low-count) yarns until about 1860. Third, the structure of the labor force established on the common mules, in which a male "minder" carried out certain managerial and supervisory functions, may have had a momentum of its own as the minders had an obvious interest to resist the introduction of a device that would weaken their authority in the workplace because it turned the spinner from a skilled operator into little more than a machine tender. The invention of the self-actor was directly aimed at reducing the bargaining power of these workers, something in which it was unsuccessful (Lazonick, 1979). Thus a limit was imposed on the cost saving made possible by the self-actor, slowing down its diffusion. Some idea of the magnitude of the improvements attained can be gained from Chapman's (1972) calculations of the number of hours needed to

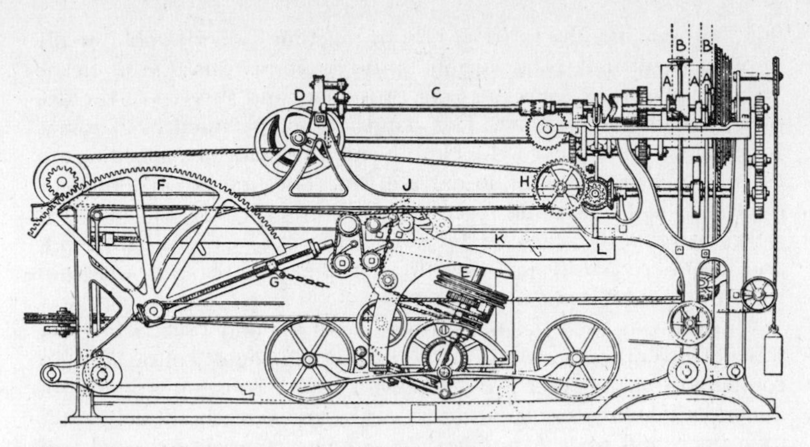

Figure 28. Side view of Richard Robert's self-acting mule.
Source: W. S. Murphy. *The Textile Industries* Vol. 3, Fig. 130. London, Gresham, 1910– . E. Norman.

spin 100 lbs. of cotton. The "old" technology was the Indian hand-spinner, who took about 50,000 hours. Arkwright's rollers and the mule brought that number down to around 300 hours in the 1790s, and the self-actor reduced the figure to 135.

Although the improvements in spinning were the most spectacular, improvements in other stages of cotton production were also impressive. The cotton gin, invented by Eli Whitney in 1793, ensured the supply of cheap raw cotton to Britain's mills. A carding machine was patented by Lewis Paul in 1742 and later improved by Arkwright, who pioneered the use of large rollers to prepare the rovings for the water frame. The finished yarn was bleached using chlorine, a process invented in 1784 by the French chemist Claude L. Berthollet and improved in 1799 by Charles Tennant by combining chlorine with slaked lime to make bleaching powder. Chlorine bleaching meant a fundamental change in process, as it works through oxygenation rather than washing of the color products, and historians of the chemical industry assessed that "in the last quarter of the eighteenth century, no greater advance was made in the finishing sections of the textile trade than the art of bleaching" (Clow and Clow, 1952, p. 186). In 1783 Thomas Bell invented metal printing cylinders that printed patterns on the finished cloth. Edward Baines, one of the earliest historians of the British cotton industry, judged that Bell's cylinder printing bore the same relation to block printing (the manual technique used until then in the calico printing industry) as throstle spinning did to the spinning wheel (Baines, 1835, p.

265). In weaving, the introduction of machines was slower. Initially improvements took place largely in the equipment used in domestic industry, which explains the long prosperity and survival of the domestic handloom weavers. The dandy loom, patented by Thomas Johnson in 1805, moved the cloth beam automatically, speeding up the weaving process with no extra effort. The first power loom was built in 1785 by Edmund Cartwright, an enterprising and idealistic clergyman, but the machine did not work properly until about 1815, and the finer yarns (which were more subject to breakage) were not woven by power looms until the 1830s.

Thus, during the brief period between 1760 and 1800, a feverish wave of inventions focused on the manufacturing of cotton. Cotton combined qualities that are attractive to both consumers and producers: it takes dyes well, launders easily, and ventilates much better than linen and wool. Compared to its main competitors, wool and linen, cotton fibers lent themselves easily to mechanization. Moreover, the supply of the raw material was elastic. No wonder, then, that cotton grew at a rate never before witnessed in textiles, and is regarded as the quintessential growth industry of the early stages of the Industrial Revolution.

Since the sixteenth century the English woolen industry had consisted of two branches, the woolens and the worsteds. The preliminary combing of the wool in worsted production was the slowest to become mechanized. Although a number of patents were taken out on combing machines, successful combing machines were not available before 1827 and not perfected until Josué Heilmann, a Frenchman, patented his combing machine in 1845. On the other hand, Arkwright's double-roller device was well-suited to the spinning of worsted yarns. In wool the order of mechanization was reversed: preparation was mechanized before spinning. Water-driven carding machines were operating in Yorkshire in the 1770s, whereas the adaptation of the mule to the spinning of wool was not achieved satisfactorily until 1816. In weaving, power looms were applied to worsteds after 1820, but the diffusion of these machines was slower than in cotton. In wool, the yarns were too fragile for the power looms, and mechanization did not occur until the 1840s. Handloom weavers in the woolen industry survived longer than in cotton or worsteds. In the finishing processes of the woolen industry, innovation encountered some resistance, but here too mechanization proved inevitable.

Like silk, worsteds were a delicate and relatively up-market fabric woven into patterns using the so-called Jacquard loom, one of the most sophisticated technological breakthroughs of the time. The Jacquard loom was perfected in 1801 by Joseph Marie Jacquard, a Lyons

Figure 29. Cylinder printing press for the printing of calicoes, invented by Thomas Bell in 1785.
Source: Archibald Clow and Nan L. Clow, *The Chemical Revolution,* Batchworth Press, London.

silk weaver, following a century of efforts by French inventors to devise a loom that could automatically weave patterns into fabric (Usher, 1954, pp. 288–95).[5] The patterns were coded on cards by means of holes representing the information in a binary code. The cards were probed by rods connected to wires, transmitting the information embedded in them. The Jacquard loom saved labor, as the draw boy was replaced by an automatic device, and it made the

5. What became known as the Jacquard loom was in fact invented by two Frenchmen in the late 1720s. One of them, Basile Bouchon, was the son of an organmaker, which is significant because the organ is the first direct application of the binary coding of information. Bouchon's invention was improved in 1728 by Jean Baptiste Falcon. Bouchon coded his patterns on perforated paper, whereas the Falcon loom used separated perforated cardboard cards. In 1775 the great engineer and inventor Jacques de Vaucanson turned his attention to the loom and suggested using a cylinder with holes punched in it, and attaching the warp threads directly to a hook manipulated by needles probing the perforated cylinder. De Vaucanson's ingenious device was never put into operation. Jacquard's own contribution to the machine was to combine the best aspects of the Falcon and the de Vaucanson devices and to dispense with the drawboy, who traditionally helped in directing the pattern woven.

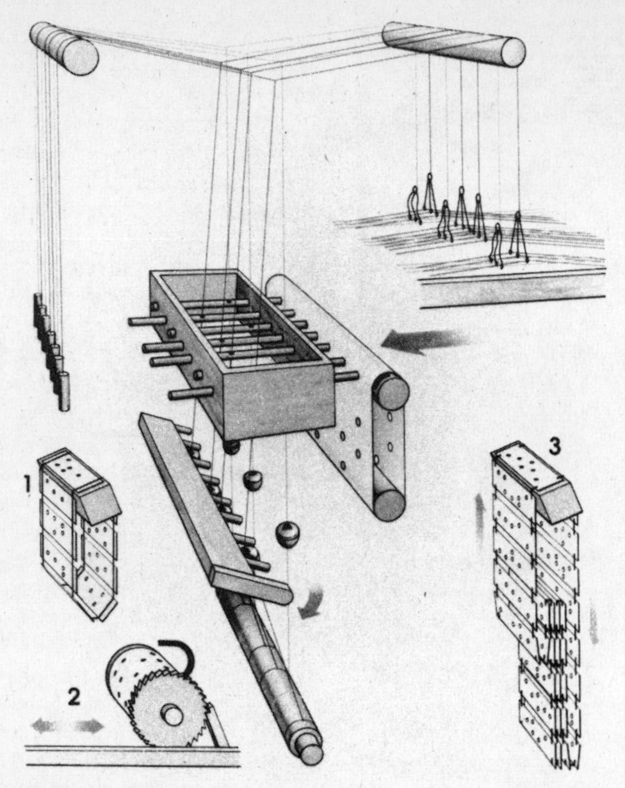

Figure 30. Stages in the development of the Jacquard Loom: (1) is the Falcon paper roll, (2) de Vaucanson's cylinder and (3) Jacquard's endless chain of cards. The control mechanism is shown in the center.
Source: Macmillan London Ltd.

design of brocades much easier, permitting the weaving of more varied and richer patterns. Moreover, it eliminated the frequent and costly errors made by traditional draw looms. Despite resistance among French weavers, the Jacquard loom spread rapidly in the Lyons region. A decade after its invention, 11,000 such looms were operating in France. After 1820, its diffusion in Britain began, accelerating in the late 1830s (Rothstein, 1977).[6] Apart from the Jacquard loom, the

6. The Jacquard loom inspired Charles Babbage in the design of his famous Analytical Engine and Richard Roberts in the design of a multiple-spindle machine, which used a Jacquard-type control mechanism for drilling rivet holes in the wrought-iron plates used on the Britannia tubular bridge (Rosenberg and Vincenti, 1978, p. 39). It

silk industry changed little. Silk throwing, once the most mechanically progressive of all textile industries, experienced no progress, and by 1825 was regarded as backward (English, 1958, p. 311).

In the ancient and venerable linen industry, mechanization proved to be difficult. In flax, the filaments contain a rubbery substance that needs to be dissolved before spinning can take place. The search was intensified during the Napoleonic Wars, when the Continental textile industries were cut off from their sources of raw cotton supply. Napoleon, who had a keen interest in industry, offered a large financial reward to the inventor who would do for linen what Arkwright and Crompton had done for cotton. In 1810 the Frenchman Philippe de Girard perfected the idea of "wet spinning," soaking the raw flax in a hot alkaline solution before it went to the spindles. This idea was introduced in Leeds in 1825, and the mechanization of linen spinning spread rapidly. The preparatory stages of flax processing, such as the scutching and heckling of the material (roughly corresponding to carding in wool or cotton, or combing in worsteds), which had been a highly labor-intensive cottage industry, were mechanized in the 1830s, though the finest yarns remained those heckled by hand. Moreover, the adaptation of the power loom to linen weaving was difficult because the lack of elasticity in linen caused the yarn to snap under strain. As late as 1850, there were only slightly over 1,000 power looms weaving linen in the United Kingdom, as opposed to 42,000 in wool and worsted and a quarter of a million in cotton. The technological difficulties in mechanizing the manufacture of linen cloth led to the sharp decline of this industry, with devastating effects on regions that had historically specialized in it, such as Ireland and Western Belgium.

MISCELLANEOUS TECHNOLOGICAL PROGRESS

Among the factors responsible for making the Industrial Revolution possible in the late eighteenth century rather than a century or two earlier must surely be the existence of a small but vital high-precision machine-tool making industry. In 1774, John Wilkinson patented a machine, originally designed to bore cast-iron cannon, in which the drill and material were independently manipulated. This technique greatly increased accuracy, and within two years he was hired by Boulton and Watt to finish cylinders and condensers. It is only a mild exaggeration to say that Wilkinson and his colleagues, by actually being able to manufacture the required parts as specified by

also inspired the American Herman Hollerith to use punchcards to store information from the 1890 U.S. census.

the inventor, made the difference between Watt and Trevithick on the one hand and Leonardo Da Vinci on the other. Machine tools such as planing machines, milling machines, lathes, screw-cutting machines, and so on permitted the creation of precise geometric metal forms, essential to machinemaking and uniformity. It was, in the words of one technological historian (Paulinyi, 1986, p. 277), "the most important step on the way to the production of machines by machines . . . it [became] possible to use iron and steel as the material whenever it appeared functional to do so."

Unlike in textiles, the masters of the engineering and machine-tool industries were a closely knit group, whose members taught each other the secrets of the trade. Father-and-son dynasties, such as George and Robert Stephenson or Marc and Isambard Brunel, were complemented by master-and-apprentice dynasties. The most famous example of the latter was the one started by Joseph Bramah, who had eighteen patents to his name, including an improved water closet, a wood-planing machine, sophisticated locks, and a spring-winding machine. In 1797, Bramah's foreman, Henry Maudslay, left the firm to start out on his own, devising a screw-cutting lathe that produced screws with unprecedented accuracy and at an affordable price. He built numerous machine tools used for planing, sawing, boring, mortising, and so on (Woodbury, 1972). The famous Portsmouth block-making machines, devised by Maudslay together with Marc Brunel around 1801 to produce wooden gears and pulleys for the British Navy, were automatic and in their close coordination and fine division of labor, resembled a modern mass-production process in which a labor force of ten workers produced a larger output than the traditional technique that had employed more than 10 times as many (Cooper, 1984). Maudslay in his turn trained three other toolmakers—Richard Roberts, James Nasmyth, and Joseph Whitworth—all of whom made major contributions to the machine-tool industry. Nasmyth invented the steam hammer and milling and planing machines. Whitworth had no fewer than 23 exhibits at the famous Crystal Palace Exhibition of 1851 where the glories of British engineering were extolled. To his credit are, among others, a measuring machine that could measure up to one millionth of an inch and the standardization of screw threads. Richard Roberts was one of the most brilliant engineers of his time, making fundamental contributions to half a dozen inventions, including a multiple-spindle drilling machine controlled by a binary logic mechanism similar to that used in the Jacquard loom. Roberts's improvements to Cartwright's power loom helped transform it from a curiosum invented by a well-meaning eccentric into the backbone of the British cotton industry (and the doom of Britain's handloom weavers). His most famous inven-

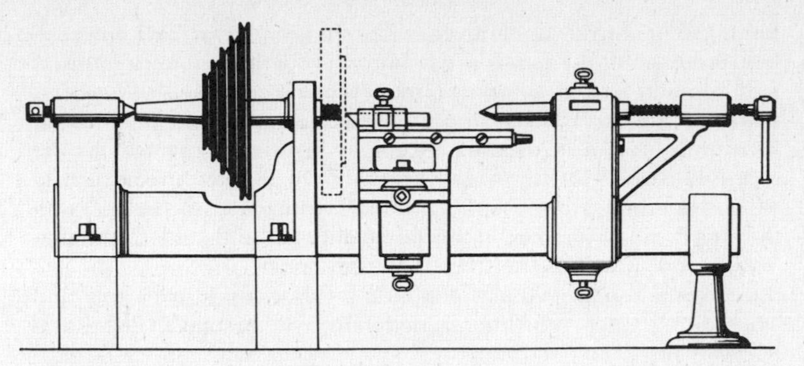

Figure 31. Henry Maudslay's slide and finishing lathe, dating from 1800. *Source:* Maurice Daumas, *A History of Technology and Invention,* Vol. III, Crown Publishers Inc., New York.

tion was the self-actor (1830), which, as we have seen, wholly automated the mule-spinning process.

Other machine-tool makers, not all associated directly with the Bramah "dynasty," deserve mention. In 1763 Jesse Ramsden built a dividing machine for the accurate graduation of circles, crucial to the construction of navigational and surveying instruments. Matthew Murray and Bryan Donkin also contributed to the design of improved machine tools. William Fairbairn, a producer of improved waterwheels, steamboats, and boilers, was a worthy heir to John Smeaton. In the United States, Eli Whitney, John Hall, Simeon North, Thomas Blanchard, and others pioneered new machine tools, paving the way eventually for the American System of interchangeable parts.

In the ceramics industries, the location and timing of technological change were quite different than in textiles and metallurgy. Here the Continent played a major role, and many important technical developments occurred before 1750. In the seventeenth century Dutch craftsmen, struggling to imitate Chinese blue and white porcelain, hit upon the processes of making the tin-and-enamel wares for which Delft is famous. Porcelain was produced successfully in 1708 by J. F. Böttger, who was employed by August the Strong, Prince-Elector of Saxony. Dresden chinaware was made in Meissen, using a technique that was kept secret by the Saxons. It took the rest of the Continent several decades before they were able to copy the process. England came up with a close substitute in the salt-glazed earthenwares produced by John Astbury of Staffordshire in about 1720. By 1759, when the renowned Josiah Wedgwood started his factory in Burslem, he could draw upon a venerable tradition of making an excellent product. In the preceding century, the British pottery industry had adopted coal as a fuel for most purposes, as well as new raw materials, such as ball clay and flint. Wedgwood is justly famous for his rationalization of production employing steam engines, his use of a fine division of labor, and his ability to market a relatively inexpensive product to snobbish consumers, competing successfully with the more expensive hard-paste porcelains made on the Continent. Wedgwood was, however, also an important inventor, who pioneered the use of gold and platinum as glazes on lustre wares, and whose work on pyrometers earned him a fellowship in the Royal Society (1783).

In the glass industry, too, Britain followed the Continent. In or around 1688, French glassblowers adopted the cast-plate process, which produced flat glass (by far the most important output of the industry) with a much more even and flat surface than the cheaper method of blown "crown" glass. The great royal manufacture at St. Gobain led in the production of window glass and mirrors using this system. Its adoption in Britain dates from only the 1770s, and even then it remained dependent on French know-how. On the other hand, the British pioneered glassmaking using a coal-fired reverberatory furnace. Through covered crucibles, the smoke was prevented from discoloring the glass, a technique the French did not master until the Revolution (Scoville, 1950, p. 43). The most important breakthrough in the glass industry was made in 1798 by Pierre Louis Guinand, a Swiss, who invented the stirring process in which he stirred the molten glass in the crucible using a hollow cylinder of burnt fireclay, dispersing the air bubbles in the glass more evenly. The technique produced optical glass of unprecedented quality. Guinand kept his process secret, but his son sold the technique to a French manufacturer in 1827, who in turn sold it to the Chance Brothers Glass Company in Birmingham, which soon became one of the premier glassmakers in Europe.

In the technology of papermaking, machinery producing continuous sheets had been patented in 1798 by the Frenchman Nicholas Louis Robert. Robert's main idea was to produce the paper on an endless belt of woven wire, replacing the process that produced separate sheets made in molds used until then. The principles developed in Robert's machine were further developed by the British engineer Bryan Donkin and eventually came to be known as the Fourdrinier machine, after a London stationer who was the first to adopt the new machinery successfully (Coleman, 1958, pp. 179–90; Clapham, 1957, p. 416). Although the paper industry is not often thought of as a typical industry of the Industrial Revolution, the

Foudrinier machine was in fact a revolutionary device. It reduced the time involved in making a given piece of paper from three weeks to three minutes, and five workmen in a mill could furnish enough work to keep 3,000 workers busy. Chlorine bleaching, crucial to the development of the cotton industry, was also applied successfully to paper pulp from the 1790s on, thus permitting the use of dyed and printed rags in the production of paper.

Britain, then, had no monopoly on invention, but when it was behind, it shamelessly borrowed, imitated, and stole other nations' technological knowledge. Another good example of Britons applying new technology developed elsewhere was the chemical industry. Berthollet, the inventor of chlorine bleaching, was French, as was Nicholas Leblanc, who developed the soda making process named after him. Leblanc reacted salt and sulphuric acid to produce sodium sulphate, which after heating with lime or charcoal yielded raw soda together with hydrochloric acid, a noxious by-product. The Leblanc process became the basis of the modern chemical industry and is regarded as one of the most important inventions of the time.[7] In the adoption of soda, Britain was relatively slow, and only in the 1820s did it start to adopt Leblanc's process on a large scale. The explanation usually given for this delay is the high tax on salt, which made artificial soda more expensive than vegetable alkali. Once the salt tax was repealed, British soda production grew rapidly and by the 1850s exceeded French output by a factor of three (Haber, 1958, pp. 10–14). In other areas of the chemical industry, however, Britain led the Continent from the middle of the eighteenth century. Sulphuric acid, known as vitriol, used in bleaching and metallurgy, was obtained by adding saltpeter to sulphur in large glass vessels, a process invented in France in 1666. The French were unable to make much use of this invention, and in about 1740 the process was adopted by Joshua Ward, an English pharmacist who adopted the glass-chamber process, and the price of sulphuric acid fell from almost £9 to 10 s. per pound. In 1746, John Roebuck perfected his lead-chamber process for the production of sulphuric acid, and Britain remained

7. Leblanc's story is an illustration of the risks to which an inventor is subject. The French Academy had offered 100,000 francs to the person who could produce soda from sea salt. Leblanc solved the problem in 1787 but did not receive the prize. He did, however, secure the help of his employer, the Duke of Orleans, with whom he started a joint venture in St. Denis in 1789. As it happened, that year was a bad time to start a commercial enterprise in partnership with a prominent member of the ancien régime. The revolutionary government annulled Leblanc's patent under the pretext that soda manufacturing was a matter of national security. The price of soda fell precipitously, and Leblanc committed suicide in 1806 after having failed in his desperate attempts to get some compensation for his invention. Only in 1855 did his heirs receive an honorarium for his invention.

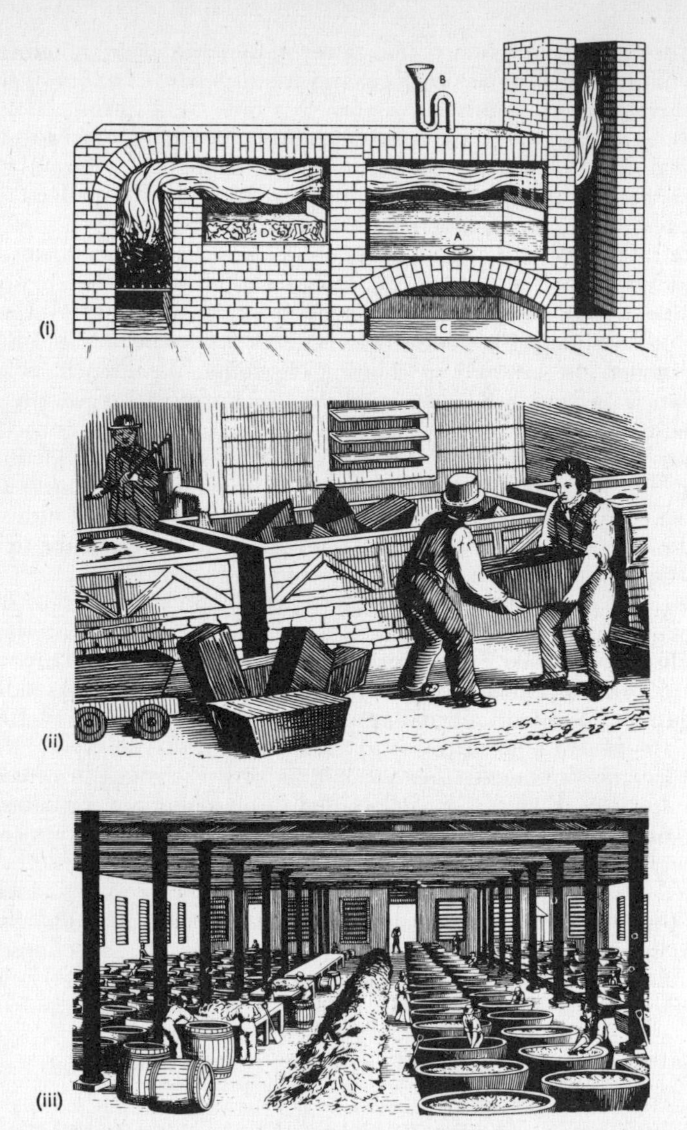

Figure 32. Soda manufacture using Leblanc's process: (i) shows the furnace in which the reaction takes place; (ii) shows the big vats in which the soda was dissolved out; (iii) shows the preparation of soda for the glass industry.

Source: (i) After C. Tomlinson. *The Useful Arts and Manufactures of Great Britain,* Part II, Section: "The Manufacture of Soda," p. 33. London, 1848. E. Norman. (ii, iii) After C. Tomlinson. Ibid., p. 26. D. E. Woodall.

the center of this industry. Only after the middle of the nineteenth century did the Germans establish firm leadership in chemistry, thanks to their superior chemists.

An interesting example of international collaboration in technological progress was the development of gaslighting. Although this invention was probably modest in its contribution to national income, it was important in its effect on the quality of life: gaslit streets were safer, gaslit homes promoted literacy, gaslit theatres made entertainment more sophisticated, and gaslit factories made night work cheaper and more efficient. Prior to 1780, lighting technology had changed little since antiquity. Lamps burning rapeseed and similar oils provided a smoky light. Candles were mostly made of tallow and produced a smoky and malodorous flame. Wax candles were superior, but too expensive to be used widely.

Gaslighting was a joint German-Anglo-French project. The use of gas as a source of light was first pointed out by the Belgian physicist Jean Pierre Minkelers and the German pharmacist J. G. Pickel in the late 1780s. In 1799, the Frenchman Philippe Lebon combined a wood-derived gas with the Argand burner, developed by the Frenchman Aimé Argand in the early 1780s.[8] In Lebon's so-called thermolamp, the gas and the air were introduced separately, and the heavier by-products were collected in a special receptacle. In 1798, the Scotsman William Murdock, who worked for Boulton & Watt (which had bought Argand's patent in Britain), lit his forge with gas derived from coal, which soon proved superior to gas made from wood. By 1807, Manchester's cotton mills and London's Pall Mall were illuminated by gas. The Englishman Samuel Clegg showed how gas could be distributed from a central production site to individual consumers using hydraulic gas mains. His son-in-law, John Malam, developed in 1819 a metering system that could monitor the quantities used by each customer. Following the return of peace in 1815, the idea spread rapidly through the Continent: the work of a German chemist appropriately named Wilhelm Lampadius led to the illumination of Berlin streets in 1826. Further innovations, such as filtering the gas through quicklime to remove the noisome odor caused by sulphurite hydrogen, helped to reduce the price and improve the quality of gaslighting.

In mining, the years of the Industrial Revolution were marked by

8. Argand's lamp contained three important new elements: a hollow, tube-shaped wick, providing an inside as well as an outside air supply; a cylinder-shaped glass enclosure acting as a chimney; and a mechanism to raise or lower the wick (Schivelbusch, 1988, pp. 10–11). Like his compatriot Leblanc, Argand failed to capitalize on his burner because of his inability to prevent widespread imitation. He died in London, bitter and impoverished, in 1803.

gradual progress but few spectacular advances. Coal became of ever-growing importance to the economy, in part because of its use in engines and metallurgy, and in part because a growing and more prosperous population demanded it for home heating. Yet aside from the use of steam engines in pumping water from the mines, the only radical invention in coal mining was the safety lamp, invented by Humphrey Davy in 1815. Better ventilation, the introduction of rails in underground hauling, and the redesign of shafts for increased safety were the main foci of progress in mining. By 1830 steam-driven fans came into use, further reducing the dangers of explosion. The technical problems involved in using power machinery inside a mine could not be solved, however. The increase in production here consisted primarily of a shift along the supply curve, rather than an outward movement of the curve, as was the case in textiles and metal. In other words, output in the mining sector increased primarily because more resources were allocated to it, not because new techniques allowed existing resources to produce cheaper or better.

Finally, one of the greatest macroinventions of all times occurred during the heyday of the Industrial Revolution, yet it is rarely mentioned in connection with it. The invention of ballooning by the Montgolfier brothers in France in 1783 is surely an epochal invention in terms of novelty and originality. Since time immemorial people had dreamed of human flight, but, some ill-fated attempts aside, it had never been achieved. Attempts to fly typically sought to imitate birds, equipping the flyer with wings and a tail, but without the faintest understanding of the principles of aeronautics. Joseph de Montgolfier was the first to take an entirely different tack, that proved amazingly successful. Montgolfier was aware of Henry Cavendish's discovery in 1766 of hydrogen, a gas lighter than air. He believed that fire gave off a similar gas, and that when the gas was captured in a closed vessel, that vessel would be lighter than air and thus rise. The reasoning was thus partly fallacious. What made the Montgolfier balloon rise was not a gas lighter than air but air itself, which, heated, expanded and thus reduced its specific weight. Success was immediate. Half a year after the emergence of the idea, the famous demonstration in Annonay took place. On November 21, 1783, the first two human beings lifted off, traveling through the air and living to tell about it. Except for a few military applications, ballooning was, of course, of limited direct economic effect. But there can be little doubt that the consequences of the invention of ballooning were far-reaching. Few inventions were more powerful in accustoming people to the idea of technological progress and alerting them to the ability of human ingenuity and creativity to control the forces of nature and do things never done before.

The technological developments in the British manufacturing sector increased output by hitherto unimagined factors. The price of cotton cloth declined by 85 percent between 1780 and 1850. As elementary economic analysis shows, in a competitive economy technological progress in existing goods is usually transmitted to consumers through lower prices. The role of demand in the process was largely passive, as consumers responded to lower prices by buying more or, in the terms of economics, slid down their demand curves.[9] Technological progress created both totally new goods and higher-quality old goods. The shift in supply thus either satisfied an already existing demand or created new needs previously not explicitly felt. Either way, it makes little sense to try to explain the timing or the location of the Industrial Revolution by exogenous changes in consumer demand for goods and services.

Why did these breakthroughs not occur earlier? Although the innovations that made the Industrial Revolution typically did not depend on new scientific knowledge, the fact remains that the technical problems that the engineers of the Industrial Revolution solved in metallurgy, power technology, and textiles were difficult. Given the tools, materials, and resources at the disposal of the most talented men in Europe, it is not surprising that it took a long time to tackle many of the challenges. Those who claim with Hobsbawm (1968, p. 38) that there was nothing inherent in the new technology of the Industrial Revolution that could not have developed 150 years earlier confuse scientific knowledge with technological ability. Easy as they may seem to us today, the problems were often simply hard. Even when all that was required was to combine previously known pieces of technical know-how into a new gadget that would actually work, the effort required by the inventor was often considerable. The "invention" and "development" stages of technological progress were not yet distinct. In recent years there has been a tendency to downgrade the role of key persons in economic history, and to tell the tale of the Industrial Revolution in terms of inexorable social forces. This approach was largely a reaction to the earlier simple-minded heroic tales of invention, in which a handful of brilliant individuals were credited with all technological progress. But it is possible to go too far in the other direction. The changes in the British economy during the Industrial Revolution were no doubt the result of profound economic, social, and demographic forces. But inge-

9. For more details, see Mokyr (1985), especially chapters 1, 2, and 4.

nious, practical, mechanically minded people came up with the ideas that changed the world. Ideas by themselves were not good enough, as a long line of frustrated inventors found out. Dexterity and perseverance were equally important. Samuel Crompton, the yeoman-inventor, wrote that he spent "four-and-a-half years at least wherein every moment of time and power of mind as well as expense which my other employment would permit were devoted to this one end" (cited by Baines, 1835, p. 199). Watt first conceived of his separate condenser in 1765; the first commercially successful machine began to operate only in 1776. The invention of the self-actor by Roberts involved a concentrated research effort (stimulated by the spinners' strike of 1824) that cost £12,000 and six years of hard work to complete. In addition to everything else, invention required people of unusual ability, such as Watt, Smeaton, Trevithick, and Roberts, as well as people of unusual energy, daring, and luck, such as Arkwright and Boulton. The supply of talent is surely not completely exogenous; it responds to incentives and attitudes. The question that must be confronted is why in some societies talent is unleashed upon technical problems that eventually change the entire productive economy, whereas in others this kind of talent is either repressed or directed elsewhere.

The story of the Industrial Revolution is, of course, far more than the tale of a handful of major inventors. The names mentioned above are but a small fraction of those who contributed to the new technology. Right below the "superstars" were hundreds and thousands of engineers, technicians, entrepreneurs, foremen, and gifted amateurs who made less spectacular contributions to technological progress. By adapting, modifying, improving, debugging, and extending earlier inventions, they were indispensable to the success of the uncoordinated, unconsciously joint project called the Industrial Revolution. In addition, we should not forget that there were even more numerous and more anonymous tinkerers and inventors who made things that did not work, or who were scooped by earlier or luckier competitors.

CHAPTER SIX

The Later Nineteenth Century: 1830–1914

It is widely believed that before the middle of the nineteenth century, technological progress moved more or less independently of scientific progress, and that since then the interaction between science and technology has gradually become tighter. As we have seen, this view is only partially correct. Science, and especially scientists, were not totally irrelevant to technological change before 1850. Between 1600 and 1850, technology learned some things from science, and more from scientists. In few cases, however, can we conclude that a particular invention depended crucially on a breakthrough of the scientific understanding of the chemical or physical, let alone biological, processes involved. After 1850, science became more important as a handmaiden of technology. A growing number of technologies, from waterpower to chemicals, depended on or were inspired by scientific advances. Yet the number of technological breakthroughs that were purely empirical has not declined, even if their *relative* importance has fallen. I shall return to these issues in Chapter 7.

In the years after 1850, technological change began to differ from that of earlier periods in another respect. Economies of scale were, of course, not new. Adam Smith had stressed the gains from the division of labor, and the most casual observer knew that many types of machinery could not be made as cheaply in small sizes and doses. Moreover, marketing, borrowing, the monitoring of workers, and the absorption of new information were less costly for large than for small firms. The factory system emerged when cottage industry, in which the firm size had been constrained by the size of the household, found it increasingly difficult to adopt new techniques as rapidly as the factories.[1] Mass production was slow in evolving, and was still quite rare by 1870. In the last third of the century, however, these effects became more pronounced. To an ever-growing degree, learning by doing, large fixed costs in plant and equipment, positive spillover effect (externalities) among different producers, network technologies, and purely technical factors, such as the inherent scale economies in railroads, in the metallurgical and chemical industries, and in mass production employing interchangeable parts and continuous flow processes, all operated together to reduce average costs at the level of the industry as well as the firm.

The relevance of this observation is that under conditions of increasing returns to scale the history of technology becomes a different tale. Simply put, increasing returns and economies of scale mean that larger firms are more efficient and can produce cheaper. It has been long known that increasing returns are incompatible with equilibrium economics. This holds, a fortiori, for the economics of technological change. In other words, the standard tools of economic analysis become inadequate in explaining the observed patterns of research and development, diffusion, and adoption. Brian Arthur (1989) has shown that under increasing returns, rational choice between competing techniques no longer guarantees that the most efficient technique will be chosen. Technological change becomes a disorderly process, better described by the concepts of modern evolutionary biology than by the static tools of economics. As Paul David points out, the process of technological development in such an untidy world necessarily calls forth the study of history (David, 1975, p. 16).

A special case of increasing returns coupled to external economies is the growing importance of technological systems and networks. Before 1850, technology consisted of more or less isolated chunks of

1. As I have noted earlier, domestic industries were quite capable of adopting new techniques and increasing productivity. Berg (1985) has argued tirelessly for the importance of these advances as a source of growth during the Industrial Revolution. But the issue is not really whether the factory replaced cottage industries because the factory could generate and adopt productivity growth and the cottages could not, but rather whether larger-scale industries were more appropriate to the new technologies. As Landes (1986) notes, "technology has its preferences." True, in some cases ambiguous evidence turns up. Berg (1985, p. 243) cites Hills (1979), who maintains that the first water frames were built on a small scale, perfectly suitable for domestic industry, but that Arkwright restricted the licenses to large units because of his fear of patent infringement. Yet Arkwright's patent expired in 1785; if the water frame had really been a domestically-sized machine, it would have returned to the cottages. It did not, because by the late 1780s the cost advantage of larger water- or steam-powered mules or water frames was beyond challenge and even jennies were increasingly located in factories.

knowledge in which sudden changes in production techniques could occur without dramatically affecting other industries or producers. Some exceptions to this rule should be noted, however. These exceptions occurred when the production technique consisted of a complex system in which individual components interacted. Under such circumstances, the components were subject to *structural constraints* that made dramatic developments more difficult to bring about. One example was open-field agriculture, in which different arable crops and livestock cultivation were all part of a *system* and were thus constrained in changing individual components by the need to fit in with the others. Another was the sailing ship, in which propelling, steering, navigation, defense, and maintenance all interacted to form a unit that was complex by the standards of the time. Because of the structural constraints associated with such systems, technologies in agriculture and shipping tended to develop gradually and slowly. After 1850, the complexity of technological systems increased. Examples of networks are found in the railroads, in the electricity, telegraph, telephone, and water supply systems, and in the need to supply spare parts and information to individuals. As Paul David and others have pointed out, in such systems certain components may resist change in other parts of the system because compatibility with existing techniques has to be satisfied.[2] Simultaneous with the growing complexity of technical systems after 1850, however, was a rise in inventors' ability to solve such problems. The ability of the free market to create a set of universal standards observed by different producers and to coordinate an intricate network of supply is impressive, but there were many lapses that underline the vulnerability of markets in coping with complex technological systems. Thus, for example, the world was split into 110 V versus 220 V users of electrical current, left-side and right-side drivers, narrow and broad gauge rails, and electric and diesel locomotives.

In the second half of the nineteenth century, mass production became an important feature of Western technology, yet its progress was neither inevitable nor ubiquitous. The overall picture may be even more untidy than David believes. In many industries the small firm clung tenaciously to life. In part, mass production guaranteed the survival of small firms because, as Sabel and Zeitlin (1985) have pointed out, much of the special-purpose machinery needed for mass production could not itself be mass produced, but catered to a small market that demanded flexibility and custom-made designs. In part,

2. An example from our own time is the failure to adopt high-definition television, in which compatibility problems have condemned Americans to inferior color television.

indivisibilities in equipment could be overcome by the pooling and sharing of equipment, by forming cooperatives, and by renting rather than purchasing expensive inputs. Such arrangements were often costly, and eventually gave rise to large-scale firms, but not before a long struggle. At times, technological progress favored the small firm: electricity brought elastically-supplied energy to every customer, and the bicycle and the automobile allowed the survival of small-scale production in transportation.

The later nineteenth century is sometimes described as the age of steel and chemicals. It was that, but it was much more. In what follows, I can provide only a brief and selective survey of what must be regarded as a technological avalanche.

STEEL

By 1850, the age of iron had become fully established. But for many uses, wrought iron was inferior to steel. The wear and tear on wrought-iron machine parts and rails made them expensive in use, and for many purposes, especially in machines and construction, wrought iron was insufficiently tenacious and elastic. The problem was not to make steel; the problem was to make *cheap* steel.

Henry Bessemer was a classic tinkerer, a professional inventor. His process of cheap steel followed a typical pattern: the inspiration came to him from the search for a solution to a different problem altogether (a high-speed spinning cannon shell). A sudden brilliant insight led to the basic steel conversion concept, but years of debugging were necessary before it was fully understood how and when the process worked, and why it sometimes failed. The Bessemer converter used the fact that the impurities in cast iron consisted mostly of carbon, and this carbon could be used as a fuel if air were blown through the molten metal. The interaction of the air's oxygen with the steel's carbon created intense heat, which kept the iron liquid. Thus, by adding the correct amount of carbon or by stopping the blowing at the right time, the desired mixture of iron and carbon could be created, the high temperature and turbulence of the molten mass ensuring an even mixture. At first, Bessemer steel was of very poor quality, but then a British steelmaker, Robert Mushet, discovered that the addition of *spiegeleisen,* an alloy of carbon, manganese, and iron, into the molten iron as a recarburizer solved the problem. The other drawback of Bessemer steel, as was soon discovered, was that phosphorus in the ores spoiled the quality of the steel, and thus the process was for a while confined to Swedish and Spanish ores. Cheap steel soon found many uses beyond its original spring

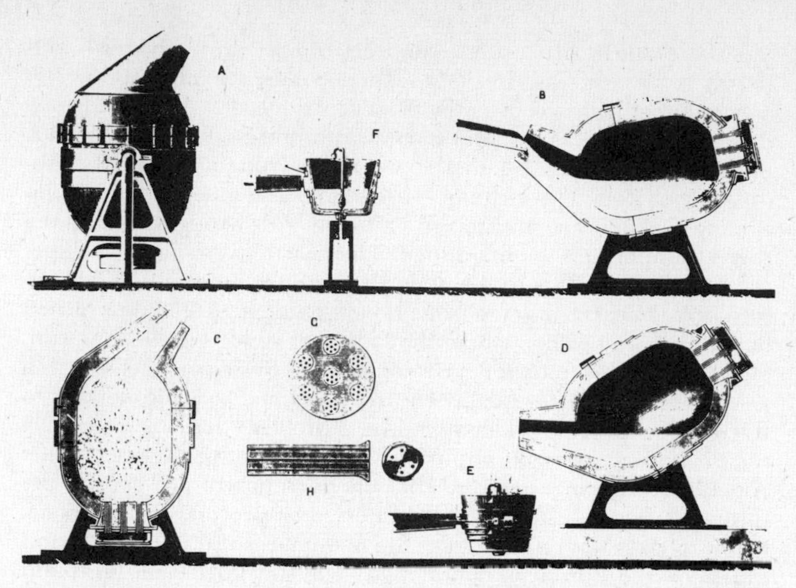

Figure 33. Drawing of an early Bessemer converter (1860). The airblast is introduced through the holes in the bottom. After the blast at (C), the steel is poured into the ladle (E).
Source: Sir Henry Bessemer. *An Autobiography,* Pl. XV. Fig. 43. London, Offices of Engineering, 1905.

and dagger demand; by 1880 buildings, ships, and railroad tracks were increasingly made out of steel.

Bessemer's invention occurred simultaneously with that of an American, William Kelly, who patented an identical process in June 1857, a year after Bessemer read his famous paper to the British Association for the Advancement of Science. Eventually Bessemer and Kelly came to an agreement to coordinate their patents. The almost uncanny simultaneity of the two inventions seems to lend support to Gilfillan's (1935) view that individual inventors do not matter much, and that inventions will come along when the time is ripe. Whether such a conclusion is warranted is unclear. What happened in this case, and in many others, was that a "problem" was defined jointly by a perceived market need and by the state of the art as defined by previous inventions and accumulation of knowledge. This "problem" led ingenious men to explore various paths leading to a solution. The two paths may not have been as independent as is generally supposed, and unless we can demonstrate that they were not in some subtle, indirect contact, through third- and fourth parties, the case for independence must remain suspect.[3]

A different path was taken by Continental metallurgists, who jointly developed the Siemens-Martin open-hearth process, based on the idea of cofusion, melting together low-carbon wrought iron and high-carbon cast iron. The technique used hot waste gases to preheat incoming fuel and air, and mixed cast iron with wrought iron in the correct proportions to obtain steel. The hearths were lined with special silica brick linings to maintain the high temperatures. The process allowed the use of scrap iron and low-grade fuels, and thus turned out to be more profitable than the Bessemer process in the long run. Open-hearth steel took longer to make than Bessemer steel, but as a result permitted better quality control. Bessemer steel also tended to fracture inexplicably under pressure, a problem that was eventually traced to small nitrogen impurities. In 1900 Andrew Carnegie, the American steel king, declared that the open-hearth process was the future of the industry. Like the Bessemer process, the Siemens-Martin process was unable to use the phosphorus-rich iron ores found widely on the European continent. Scientists and metallurgists did their best to resolve this bottleneck, but it fell to two British amateur inventors, Percy Gilchrist and Sidney Thomas, to hit upon the solution in 1878. By adding to their firebricks limestone that combined with the phosphorus to create a basic slag, they neutralized the harmful phosphorus. It seems safe to say that the German steel industry could never have developed as it did without this invention. Not only were the cost advantages huge, but the Germans (who adopted the "basic" process immediately) also managed to convert the phosphoric slag into a useful fertilizer. While the Bessemer and Siemens-Martin processes produced bulk steel at rapidly falling prices, high-qualitysteel continued for a long time to be produced in Sheffield using the old crucible technique (Tweedale, 1986).

Steel underwent the most dramatic changes of the iron industries, but its spectacular success after 1860 should not obscure important advances in other stages of the iron industry. In the blast furnaces, which smelted iron ore to produce pig iron, coke had long been the standard fuel by the 1850s. Most furnaces at that time were about 40–50 ft. high and heated the ore to about 600°F. Following the discovery of ore fields in the Cleveland district in northern Yorkshire, a set of improvements occurred that greatly increased the efficiency of the blast furnaces. Their height was gradually increased to 80 ft. and more; temperatures were raised to about 1000°F; waste

3. Cyril S. Smith (1981, p. 384) has argued that invisible communication in some form is behind nearly all cases of seemingly independent simultaneous inventions.

gases were recycled; and blowing engines were introduced. American inventors added a number of other improvements, such as "hard driving" (blowing large volumes of air at high pressure through the furnace); improved blowing engines; and direct casting of the pig iron into the steelworks (Allen, 1977, 1981; Temin, 1964).

CHEMICALS

In chemistry, Germans took the lead. Although Britain was still capable of achieving the occasional lucky masterstroke that opened a new area, the patient, systematic search for solutions by people with formal scientific and technical training suited the German traditions better. In 1840 Justus von Liebig, a chemistry professor at Giessen, published his *Organic Chemistry in Its Applications to Agriculture and Physiology*, which explained the importance of fertilizers and advocated the application of chemicals in agriculture. Other famed German chemists, such as Friedrich Wöhler, Robert Bunsen, Leopold Gmelin, August von Hofmann, and Friedrich Kekulé von Stradonitz, jointly created modern organic chemistry, without which the chemical industry of the second half of the nineteenth century would not have been possible. And yet it was an Englishman, William Perkin, who made the first major discovery in what was to become the modern chemical industry.[4] The 18-year-old Perkin searched for a chemical process to produce artificial quinine, an antimalarial drug that was much in demand because Europeans were making deep inroads into tropical areas at about this time. While pursuing this work, he accidentally discovered in 1856 aniline purple, or as it became known, mauveine, which replaced the natural dye mauve. Three years later a French chemist, Emanuel Verguin, discovered aniline red, or magenta, as it came to be known.

German chemists then began the search for other artificial dyes, and almost all additional successes in this area were scored by them. In the 1860s, Hofmann and Kekulé formulated the structure of the dyestuff's molecules. In 1869, after years of hard work, a group of German chemists synthesized alizarin, the red dye previously produced from madder roots, beating Perkin to the patent office by one day. The discovery of alizarin in Britain marked the end of a series of brilliant but unsystematic inventions, whereas in Germany it marked the beginning of a process in which the Germans established their hegemony in chemical discovery (Haber, 1958, p. 83). In 1874, re-

4. Perkin was trained by von Hofmann, who was teaching at the Royal College of Chemistry at the time, and his initial work was inspired and instigated by von Hofmann.

alizing that he could not compete with the Germans, Perkin sold his factory and devoted himself to pure research. The synthesis of indigo was carried out by Adolf von Baeyer in 1880, but the costs of indigotin remained high. Here Fortune favored the diligent: the accidental breaking of a thermometer in 1897 led chemists at *Badische Anilin und Soda Fabrik* (BASF) to the use of mercury sulphate as a catalyst, and within a decade synthetic indigo replaced the natural product (Holmyard, 1958b). Another area in which British and German chemists first competed, but that was eventually dominated by Germans, was sulphuric acid. In 1875 W. S. Squire of England and Clemens Winkler at BASF developed the catalytic-contact process for making sulphuric acid, but eventually the Germans came to dominate catalytic processes, which helped them become self-sufficient in ammonia, nitrates, and saltpeter in the twentieth century. The famous Haber process to make ammonia, developed by Fritz Haber in the early 1900s, and BASF chemist Carl Bosch's discovery of how to convert ammonia into nitric acid, made it possible for Germany to continue producing nitrates for explosives during World War I after its supplies of Chilean nitrates were cut off (Hohenberg, 1967, pp. 29–30).

Progress in the chemical industry during the nineteenth century remained an international collaborative effort. In 1847, the Italian Ascanio Sobrero discovered nitroglycerine, the most powerful explosive known at the time. Its instability caused numerous accidents, in one of which a Swedish manufacturer was killed. The victim's brother, Alfred Nobel, grimly set upon the task of controlling nitroglycerine, and discovered in 1866 that upon being mixed with diatomaceous earth nitroglycerine retained its full blasting power, yet could be detonated only by using a detonating cap. Dynamite, as the new compound was called, was used in the construction of tunnels, roads, oilwells, and quarries. If ever there was a labor-saving invention, this was it.

In the production of fertilizer, developments began to accelerate in the 1820s. Some of them were the result of the resource discoveries such as Peruvian *guano,* which was imported in large quantities to fertilize the fields of England. Others were by-products of industrial processes (Grantham, 1984, pp. 199, 211). A Dublin physician, James Murray, showed in 1835 that superphosphates could be made by treating phosphate rocks with sulphuric acid. The big breakthrough came, however, in 1840 with the publication of Justus von Liebig's famous book, which had been commissioned by the British Association for the Advancement of Sciences. Research proceeded in England, where John Bennet Lawes carried out pathbreaking work in his famous experimental agricultural station at Rothamsted, where he put into practice the chemical insights provided by von Liebig. In 1843 he established a superphosphates factory that used mineral phosphates. Yet Lawes' station remained isolated and the Germans soon took the lead. This was partly because Germany simply had better chemists. Because the physical and chemical processes in agriculture are far more complex than in manufacturing, better theoretical knowledge was required, and serendipity eventually ran into diminishing returns. In part, however, systematic research of the type needed here demanded that "its practitioners be shielded from demands for immediate practical results" (Grantham, 1984, p. 203). Private enterprise was unlikely to supply such patience, especially when the payoff was both distant and uncertain. In Germany, especially Saxony, state-supported institutions subsidized agricultural research and the results eventually led to vastly increased yields. Nitrogen fertilizers were produced from the *caliche* (natural sodium nitrate) mined in Chile. The third mineral crucial to plant growth was potassium, made from potash by burning wood. In 1870, forest-rich Canada was still the principal source of potash. By then, however, mineral deposits of potassium salts at Strassfurt in central Germany began to be exploited, the price of potash fell rapidly, and widespread application of this fertilizer began. By 1900 the Canadian asheries had disappeared (Miller, 1980).

In soda production, much work was done by engineers trying to improve the Leblanc process, which caused serious environmental pollution by spewing out clouds of hydrochloric acid gas and leaving a dark alkali ash known as *galligu.* In 1836, an English chemical manufacturer, William Gossage, constructed towers in which the gas was absorbed by falling water. The Leblanc process was ultimately replaced by carbonating towers that used ammonia to remove soda from salt, a process invented in 1861 by the Belgian Ernest Solvay. The ammonia process was much more efficient than the Leblanc process, not only because it produced few undesirable by-products, but also because it was able to recycle the ammonia and carbon dioxide it used. By the mid-1860s the process was perfected, and Solvay, like Perkin and Bessemer, became a very wealthy man.

Not all inventors in chemical engineering were so lucky. Charles Goodyear, the American tinkerer who in 1839 invented the vulcanization process of rubber that made widespread industrial use of rubber possible, died deeply in debt. Another American, John Wesley Hyatt, succeeded in creating the first synthetic plastic in 1869, which he called celluloid. Its economic importance was initially modest because of its flammability, and it was primarily used for combs, knife handles, piano keys, and baby rattles, but it was a harbinger of things to come.[5] The breakthrough in synthetic materials came only in 1907, when the Belgian-born American inventor Leo Baekeland discovered Bakelite. The reason for the long delay in the successful development of Bakelite was simply that neither chemical theory nor practice could cope with such a substance before (Bijker, 1987, p. 169). Yet Baekeland did not fully understand his own process, as the macromolecular chemical theories that explain synthetic materials were not developed until the 1920s. Once again, science and technology were moving ahead in leapfrogging fashion.

Perhaps the classic instance of a "free lunch" in which large gains in well-being were achieved at low cost was in the fine chemical industry, which after 1870 began to rationalize the hitherto chaotic industry of pharmaceutics. The use of anesthetics became widespread after Queen Victoria used chloroform when she gave birth to Prince Leopold in 1853. Disinfectants and antiseptics, particularly phenol and bromines, were produced in large quantities after Joseph Lister's discovery of the role of microbes in infection. One of the most remarkable inventions was salicylic acid. The medicinal properties of willow bark had been known since antiquity, and in 1838 it was shown that the active ingredient was salicylic acid. Because of the corrosive properties of the acid, it had unpleasant side effects, and in the 1850s the German firm of Bayer began to experiment with various derivatives. One of these was acetyl salicylic acid, which was synthesized and then forgotten. Instead, sodium salicylic was synthesized and sold as an analgesic. In 1899, a Bayer chemist by the name of Felix Hoffman on a hunch took the old compound off the shelf for a patient who could not tolerate the side effects of sodium salicylic acid. Immediately it became clear that the acetyl compound of salicylic acid, later known as aspirin, was a true wonder drug: effective, without serious negative side effects, and cheap to produce. Within months Bayer sent samples to 30,000 German doctors and the new drug was soon used universally (Krantz, 1974).

ELECTRICITY

Like chemistry, electricity was a field in which totally new knowledge was applied to solve economic problems. The economic potential of electricity had been suspected since the beginning of the nineteenth century. Humphrey Davy had demonstrated its lighting capabilities as early as 1808. Relying on the scientific discoveries of scientists

5. Contrary to some accounts, Hyatt did not succeed in producing celluloid billiard balls (Friedel, 1979, p. 52).

such as Hans Oersted and Joseph Henry, Michael Faraday invented the electric motor in 1821 and the dynamo in 1831. Yet there was still considerable uncertainty about the possibilities of using electricity.[6] Electric motors could not be made to work cheaply; as long as batteries remained the source of electric power, its costs were 20 times greater than those of steam engines (Passer, 1953, p. 212). From the mid-1840s on, the earlier enthusiasm about electricity as a source of cheap energy waned. Its first effective use was not in power transmission, but in the telegraph. The telegraph was associated with a string of inventors, the most important of whom were S. T. von Soemmering, a German, who demonstrated its capabilities in 1810; William Cooke, an Englishman, who patented a five-needle system to transmit messages (1837); and Samuel Morse, an American, who invented the code named after him that made the single-needle system feasible. The first successful submarine cable was laid by Thomas Crampton's Company between Dover and Calais in 1851, and became a technological triumph that lasted 37 years.

The telegraph, like the railroads, was a typical nineteenth-century invention in that it was a combination of separate technological inventions that had to be molded together. Just as the strength of a chain can never be greater than that of its weakest link, the efficiency and reliability of a system can never be greater than that of its weakest component. Long-distance telegraph required many subsequent inventions and improvements, which took decades to complete. Submarine cables were found to be a difficult technology to master. Signals were often weak and slow, and messages distorted. Worse, cables were subject at first to intolerable wear and tear. Of the 17,700 kilometers of cable laid before 1861, only 4,800 kilometers were operational in that year; the rest were lost. The transatlantic cable through which Queen Victoria and President James Buchanan exchanged messages in August 1858 ceased to work three months later. The techniques of insulating and armoring the cables properly had to be perfected, and the problem of capacitance (increasing distortion on long-distance cables) had to be overcome. Before the telegraph could become truly functional, the physics of transmission of electric impulses had to be understood. Physicists, and above all William Thomson (later Lord Kelvin), made fundamental contributions to the technology. Thomson invented a special galvanometer, and a technique of sending short reverse pulses immediately following the main pulse, to sharpen the signal (Headrick,

6. When a visitor to the Royal Institute asked Faraday what use his newly invented toy had, he reputedly answered; "What use is a newborn child?" Thirty years later, in 1862, the 70-year-old Faraday was taken on a tour of the English lighthouses and was able to see the beneficial effects of his "magnetic spark."

1990, pp. 215–218). In this close collaboration between science and technology, telegraphy clearly belonged to the second generation, in which science played a more prominent role.

The telegraph had an enormous impact on nineteenth-century society—possibly as great as that of the railroads. Its military and political value was vast, as was its effect in coordinating international financial and commodity markets. Unlike the railroad, it had no close substitutes, the closest being homing pigeons and semaphore. Semaphore systems were invented by the Frenchman Claude Chappe in 1793 but were largely monopolized by governments. Information had never before traveled faster than people. Moreover, because telegraph messages often had to cross international borders, they required something that few technological innovations had required before: international cooperation. Bilateral agreements and treaties were signed in the 1850s and early 1860s, and the International Telegraph Union was established in 1865. Clearly, technology can, under the right circumstances, create the institutional environment it needs (ibid., pp. 204–208).

Harnessing electricity as a prime means of transmitting and using energy was technically even more difficult than the development of the telegraph. Before electricity could be made to work, an efficient way had to be devised to generate electrical power using other sources of energy; devices to transform electricity back into kinetic power, light, or heat at the receiving end had to be created; and a way of transmitting current over large distances had to be developed. In addition, electricity came in two forms, alternating and direct current, and a decision had to be made regarding which of the two forms was to dominate.

Electric generators were crucial. Although Davy had shown as early as 1808 how electricity could drive an arc lamp, apart from lighthouses it was not widely used in lighting. Davy's arc light was driven by expensive batteries, and it was difficult to produce the vacuum inside the arc lamps. In 1860, the Italian Antonio Pacinotti built a dynamo with ring armature that could produce a steady current. Following the discovery in the mid-1860s of the principle of the self-excited generator by C. F. Varley and Werner von Siemens, the Belgian Z. T. Gramme built in 1870 a ring dynamo, which produced a steady continuous current without overheating. Gramme's machine substantially reduced the cost of alternating current. The vacuum problem was solved in 1865, when Hermann Sprengel designed a vacuum pump. Only then could the arc lamp be made practical. In 1876 a Russian inventor, Paul N. Jablochkoff, invented an improved arclamp (or "candle") that used alternating current. Subsequently, factories, streets, railway stations, and public places began to replace

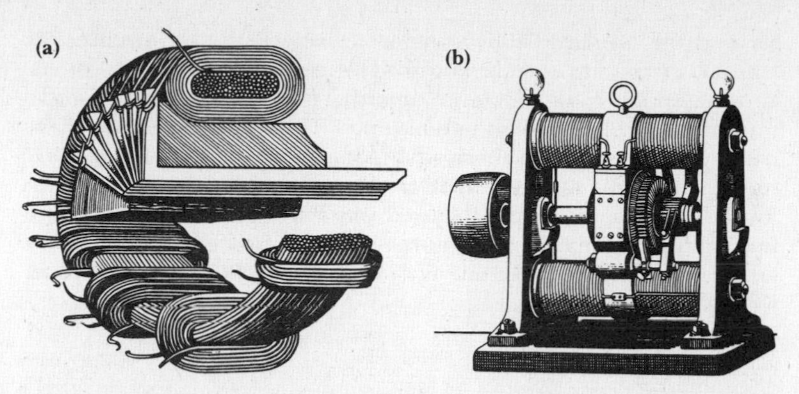

Figure 34. The first successful dynamos, built by Z.T. Gramme. On the left, typical ring armature, on the right, commercially made dynamo (1874). *Source:* (a) After J. Dredge (ed.). 'Electric Illumination,' Vol. 1, Fig. 125. London, Offices of Engineering, 1882. D. E. Woodall. (b) After S. P. Thompson. 'Dynamo-Electric Machinery' (3rd ed., enl. and rev.), Fig. 87. London, Spon, 1888. D. E. Woodall.

gaslight with arc light. In 1878, Charles F. Brush of Ohio invented a high-tension direct-current lamp, which by the mid-1880s had come to dominate arc lighting (Sharlin, 1967a). Inventors such as Thomas Edison and George Westinghouse realized that electricity was a technological network, a system of closely interconnected compatible inventions. In this regard it resembled gaslighting systems, but electricity was recognized to be a general system of energy transmission. Edison was particularly interested in systems of technology, and his ability to see the holistic picture and to coordinate the research efforts of others was as developed as his own technical ingenuity (T. P. Hughes, 1983, pp. 25–27).

The use of electricity expanded quickly in the 1870s. A miniature electric railway was displayed at the Berlin Exhibition in 1879; electric blankets and hotplates appeared at the industrial exhibition in Vienna in 1883; and electric streetcars were running in Frankfurt and Glasgow by 1884. The early 1880s saw the invention of the modern lightbulb by Joseph Swan in England and Thomas Edison in the United States. An electric polyphase motor using alternating current was built by the Croatian-born American Nikola Tesla in 1889, and improved subsequently by Westinghouse. Of equal importance was the transformer originally invented by the Frenchman Lucien Gaulard and his British partner John D. Gibbs and later improved by the American William Stanley, who worked for Westinghouse (T. P. Hughes, 1983, pp. 86–92). The polyphase motor and the Gaulard-Gibbs transformer solved the technical problems of alternating current and made it clearly preferable to direct current, which could not overcome the problem of uneconomical transmission. Led by Westinghouse and Tesla, the forces for alternating current defeated those advocating direct current, led by Edison. By 1890, the main technical problems had been solved; electricity had been tamed. What followed was a string of microinventions that increased reliability and durability and reduced cost. In 1900, an incandescent bulb cost one fifth what it had 20 years earlier and was twice as efficient.[7]

TRANSPORTATION

The development of steel, chemicals, and electricity required new scientific information for their perfection before they could become practical. Yet much technological progress in the years between 1830 and 1914 took the form of novel applications and refinements of existing knowledge. Moreover, the application of known techniques in new combinations often called for further inventions if the new idea was to become practical. Consider, for example, the railroad. The railroad was not a proper "invention" because it was in essence nothing but a combination of high-pressure engines with iron rails. The critical insight of combining the high-pressure steam engine with the principle of the railway, which minimized the friction between wheel and surface, originated from the genius of Richard Trevithick in 1804. For two decades, engineers struggled with the unfamiliar problems of high-pressure engines, the delicate balancing of heavy engines placed on iron rails, the driving-rod and crank mechanism connecting the piston to the wheels, the mechanics of suspension, the need to make stronger and more durable rails, and the design of efficient boilers. Once these problems were solved, thanks to such engineers as Timothy Hackworth, George Stephenson, and his son Robert, the railway became inexorably one of the most potent forces of the nineteenth century. A variety of competing types of locomotives was designed, but the famous Rainhill competition in October

7. Electricity found an unexpected use in electrochemistry, at this time largely confined to aluminum refining. The properties of aluminum had been known since the 1820s, and in 1855 the French chemist St. Claire Deville produced aluminum by electrolysis, selling it at about $200/kg, thus confining its use to jewelry and other esoteric applications. In 1886, a 22-year-old undergraduate at Oberlin College, Charles Martin Hall, perfected a refinement method using molten cryolite and dynamos instead of voltaic cells. Aluminum prices fell sharply. After K. J. Bayer invented a system to use bauxite and the age-hardening process was discovered by the German A. Wilm, aluminum became of major importance (Marie Boas Hall, 1976; Multhauf, 1967, pp. 477–79; Cyril Stanley Smith, 1967c, pp. 599–600).

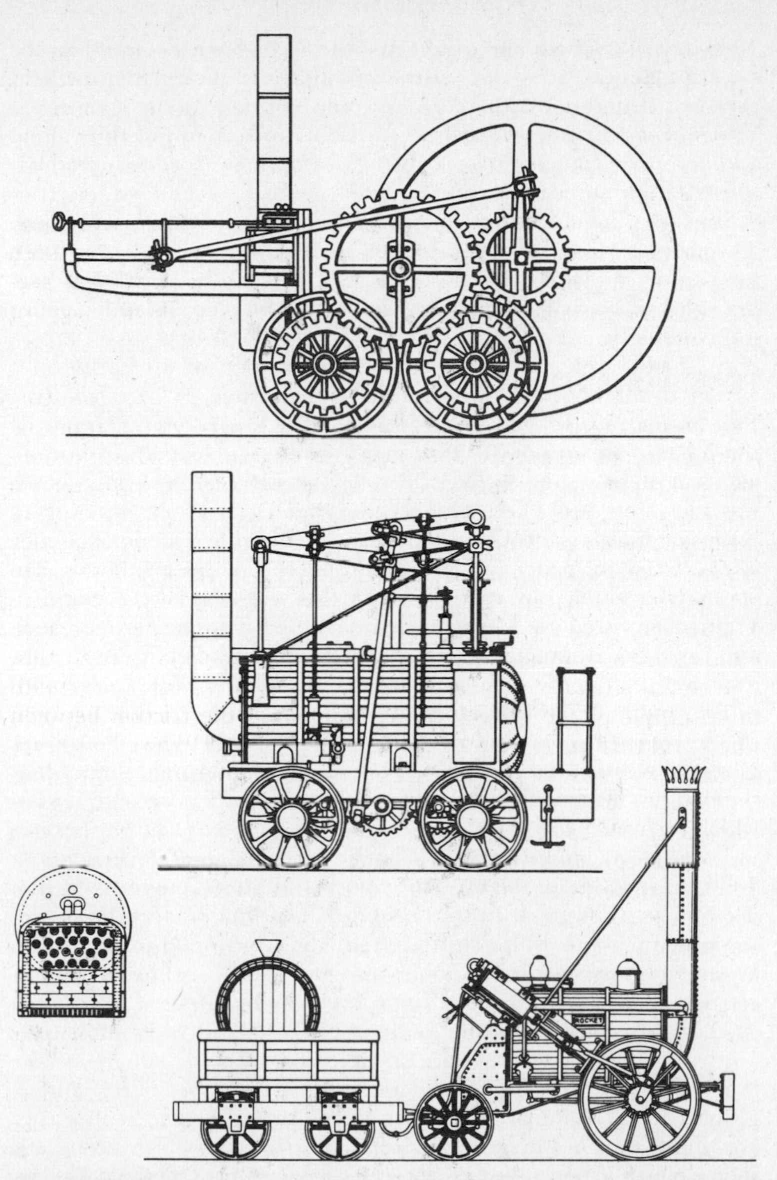

Figure 35. Early high-pressure engines: first Trevithick locomotive, 1803; William Hedley's "puffing Billy," used in collieries; Stephenson's *Rocket,* which won the Rainhill competition.

Source: Maurice Daumas. *A History of Technology and Invention,* Vol. III, Crown Publishers Inc., New York.

1829 determined without doubt that the Stephensons' *Rocket* had the superior design. It is clear, however, that the meteoric growth of railroads between 1825 and 1850 would not have occurred without subsequent refinements, such as horizontal placement of the engine and improved axles, brakes, springs, couplings, roadbed grading, tunnels, communications, and so on.

Similarly, the steamship contained few ideas that were truly new. The paddle wheel, which propelled the early steamboats, had been known in China for centuries. The first steamships were really sailing ships with auxiliary engines, with steam helping out only against unfavorable winds and tides. The *Savannah,* the first steamship to cross the Atlantic, used her engines for a total of 80 hours on a voyage of 30 days. It was not until 14 years later, in 1833, that the *Royal William* made the voyage from Quebec to Gravesend largely by the force of steam power. Two major inventions helped revolutionize steamships in the second third of the nineteenth century. One was the screw propeller. The screw propeller is an example of an invention that is impossible to associate with a single name. The idea of the screw propeller was proposed in 1753 by the mathematician Daniel Bernoulli, but early experiments did not succeed until a Frenchman, Frédéric Sauvage, demonstrated its capacities and took out a number of patents in the early 1830s. Propellers were further improved by the Swedish-born John Ericsson and by Francis Smith in England in 1838. The other invention crucial to the development of the steamship was the marine steam engine. High-pressure engines were difficult to use at sea, because sea water deposited salt at the bottom of the boiler, causing explosions. The surface condenser, which separated the water that cooled the condenser and the water in it, was developed in the 1830s and came into general use in about 1850, solved this problem. The compound steam engine was first used on ships in 1854. In the second half of the nineteenth century the construction of ships shifted gradually from wood to iron hulls. Wood had imposed an inevitable limit on the size of ships. The largest oak ships were no more than 250 ft. long, the average much smaller. Iron ships could be made in any size, and while most costs (and the ship's water resistance) increased with the square of the dimensions, carrying capacity increased with the cube. The great English engineer Isambard Brunel was one of the first to realize this. He designed the first large iron, propeller-driven transatlantic steamships, the *Great Western* and the *Great Britain.* The *Great Eastern,* launched in 1858, was almost 700 ft. long, the largest ship built in the nineteenth century (Headrick, 1981, p. 143). In the late 1870s steel began to replace iron as the main construction material. Here the transition was relatively quick: by 1891, over 80 percent of all ships under construction were made out of steel (Fletcher, 1910, p. 281). Larger ships meant lower shipping costs: fuel costs and crew size increased at a rate less than proportional to displacement. Thus technological change allowed the realization of economies of scale.

At first, the advantage of the steam engine was primarily speed and flexibility: in the beginning of the nineteenth century letter-writers in England could expect to wait up to two years for an answer to a letter to Calcutta. Ships often had to wait for weeks to sail the 80 miles up the Hooghly river from the Bay of Bengal to the city, and the monsoon winds precluded fast return journeys. By the 1840s the time of a one-way journey from London to Calcutta had fallen to six weeks, and by 1914 the voyage could be made in two weeks (Headrick, 1988, p. 20). As costs fell, steamships became a serious and eventually invincible competitor to the sailing ship.

Nonetheless, it took decades for the new shipbuilding technology to supplant the old (Harley, 1973), in part because the production of wooden sailing ships was also experiencing rapid progress (Daumas and Gille, 1979b, pp. 291–98; Davis, Gallman, and Hutchins, 1991). Between 1820 and 1860, sailing vessels were completely redesigned. Rigging and sails were revamped: the number of sails was increased and reefing (reducing the area of sails by folding or tying part of the sail) was introduced.

The newly designed sailing vessels were faster yet required fewer sailors. Ropes were supplanted by more durable metal cables, and wooden fittings were replaced by metal fittings. Ships became more elongated and streamlined, and sails became smaller and more numerous, culminating in the famous tea clippers built by American and British shipbuilders after 1850, which could attain speeds of up to 15 knots. The fastest clippers could sail from China to Britain in three months (Chatterton, 1909, p. 270). Steamships displaced sailing vessels on short routes, but sailing ships continued to dominate the long-distance voyages until the end of the nineteenth century. Harley (1971) has shown that this was the result of differences in costs reflecting the nature of the technology. The completion of canals connecting major oceans and the growing fuel efficiency that came with better marine steam engines brought about the demise of the sailing ship in the twentieth century.

Ships of either type grew bigger, more powerful, and faster at unprecedented rates. While the typical ship of 1815 was not much different from the typical ship of 1650, by 1890 both merchant ships and men-of-war had little in common with their predecessors of half a century earlier. The result was a sharp decline in transportation costs. In the first half of the nineteenth century freight rates fell by 0.88 percent a year, which reflected mostly improvements in sailing ships. The decline after 1850 accelerated to 1.5 percent a year, rates that are all the more impressive in view of persistently rising labor costs. Despite some organizational improvements, there can be little doubt that the decline in transatlantic freight rates was the result of technological improvements (Harley, 1988).

On some occasions, a technical solution looked simple, and it may seem surprising at first sight that it took so long before producers got it right. Throughout the entire nineteenth century, mechanics experimented with a device that would allow individuals to propel themselves rapidly while seated. A variety of velocipedes and "penny-farthing" types of bicycles were experimented with, largely for recreational purposes. Yet it was not until John K. Starley, a Coventry mechanic, built the Rover safety cycle in 1885 that the balanced position and easy steering of today's bicycles became feasible. The flat-link-chain drive of these bicycles, and the balanced frame due to two wheels of equal size contained little that would have been beyond the ability of, say a Henry Maudslay or a Richard Roberts. The case of the bicycle illustrates that neither purely technical factors nor purely economic factors, nor even a combination of the two, can fully account for technological change. The bicycle was a novelty in the deepest sense of the word; it did not replace an existing technique with a similar, more efficient one. The people who adopted the bicycle in the 1890s had previously walked or used public transportation. The optimal design of the bicycle was difficult because the attributes of the bicycle spanned a number of dimensions: speed, comfort, safety, elegance, and price were all considered and had to be traded off against each other. Long experimentation was necessary before the best type emerged (Pinch and Bijker, 1987).[8] In cases of a completely new product, learning-by-doing on the part of the consumer is as important as learning-by-doing on the part of producers. The bicycle became a means of mass transportation with incalculable effects on urban residential patterns. After a few years of further improvements, the design of the bicycle stabilized, and few further improvements were introduced after 1900. The bicycle, in some views, prepared the way for the automobile and the motorcycle, and supplied a form of cheap and democratic yet personal transportation. Hiram Maxim, the inventor of the Maxim machine gun (as well as an unsuccessful steam-driven aeroplane) pointed out

8. The pneumatic tire, on which the comfortable bicycle ride depends, had been patented in 1845 and tested successfully. Yet the needs of horse carriages were apparently insufficient for its widespread production, and the invention was soon forgotten until a Belfast veterinary surgeon, J. B. Dunlop, who was unhappy with the comfort of his ten-year-old son's tricycle ride, resurrected it in 1888.

that "the bicycle created a new demand that was beyond the ability of the railroad to supply. Then it came about that the bicycle could not satisfy the demand which it had created . . . the automobile was the answer" (cited in Cardwell, 1972, p. 199–200).

The classic case of a novel combination of known techniques laced with a number of important original contributions was the development of the automobile. The internal combustion engine was first suggested by Huygens in the seventeenth century. In 1824, Carnot had described the limitations of the steam engine as an energy source and pointed to heated air as the best potential means to generate motive power. Despite prolonged research efforts, it turned out to be difficult to employ steam power for carriages (Evans, 1981).[9] During the nineteenth century dozens of inventors, realizing the advantages of an internal combustion engine over steam tried their hand at the problem. A working model of a gas engine was first constructed by the Belgian Jean-Etienne Lenoir in 1859 and perfected in 1876, when a German traveling salesman, Nicolaus August Otto, built a gas engine using the four-stroke principle. Otto worked on the problem from 1860 on, after he read about Lenoir's machine in a newspaper. He was an inspired amateur without formal technical training. The problem was eventually solved with the four-stroke engine, but Otto initially saw this as a makeshift solution to the problem of achieving high enough compression. Only later was Otto's four-stroke engine, which is still the heart of the automobile engine, acclaimed as a brilliant breakthrough (Bryant, 1967, pp. 650–57). Interestingly, the four-stroke principle had been recognized earlier as the only way in which a Lenoir-type engine could work efficiently. Using Carnot's thermodynamic theory, a French engineer, Alphonse Beau de Rochas, specified in 1862 an Otto-type principle to describe an efficient internal combustion engine and obtained a patent for his idea, although it was never actually put into effect. Otto's invention, however, was unaffected by Beau de Rochas' theoretical in-

9. It is important not to underrate the steam engine as an alternative to internal combustion engines. Technological developments in the early twentieth century allowed the Stanley brothers to produce their famed steam-powered automobile, which in one view (McLaughlin, 1967) was technically and economically on a par with the internal combustion engine. The steam-powered automobile was quieter, produced a smoother ride, handled more easily, and could even be started within a few minutes if the pilot line was left on. One explanation why the steam-powered automobile lost out is because the high pressure it required produced serious maintenance difficulties (Rae, 1967c, p. 124). The Stanley Steamer was also handicapped by its need to replenish its water supply every 30 miles. These problems were compounded by public ordinances prohibiting water troughs on the road side, which were thought to spread foot and mouth disease among cattle.

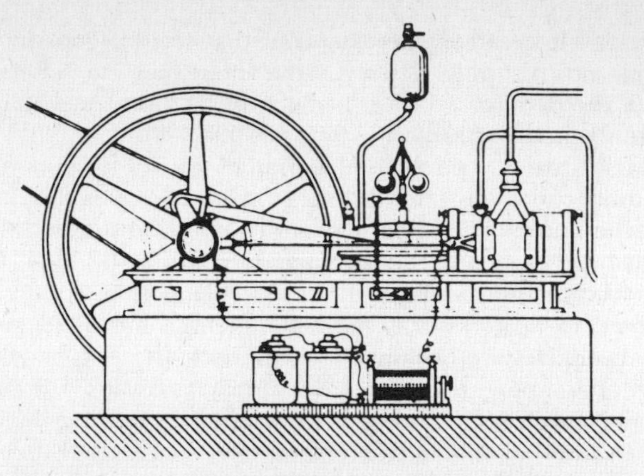

Figure 36. Gas engine, built by Lenoir in 1860.
Source: After J. W. French. "Modern Power Generators," Vol. 2, Fig. 320. London, Gresham Publishing Company, 1906. E. Norman.

sight.[10] The "silent Otto," as it became known (to distinguish it from a noisier and less successful earlier version), was a huge financial success. The advantage of the gas engine was not its silence, but that, unlike the steam engine, it could be turned on and off at short notice.

Otto's invention fits the mold of the classical inventions of the earlier Industrial Revolution years, and his career parallels that of James Watt in some important aspects. Both introduced a crucial improvement into an existing machine, transforming it from a comparative rarity into a dominant technique. Both were successful in large part due to the trust and business acumen of their partners, Matthew Boulton in the case of Watt, and Eugen Langen in the case of Otto. Both became wealthy thanks to patent protection (in Otto's case, despite the invalidation of one of his patents on the grounds that Beau de Rochas had preceded him). Both had contact with scientists, but were eventually inspired in their crucial breakthrough by sheer intuition.

Otto's gas engine was soon to adopt a new fuel. Somewhat earlier, in the 1860s, the process of crude oil refining using a method called cracking was developed. At that time the main interest was in lubricants, paraffins, and heavy oils, with petrol, or gasoline, considered

10. Otto apparently never read Beau de Rochas's pamphlet, and at first misunderstood the principles that made his invention work. As Bryant (1966) remarks, his invention was another example of how good engineering could be based on bad science.

a dangerously inflammable by-product. In 1885 two Germans, Gottlieb Daimler (who had previously worked for Otto) and Karl Benz, succeeded in building an Otto-type, four-stroke gasoline-burning engine, employing a primitive surface carburetor to mix the fuel with air. Benz's engine used an electrical induction coil powered by an accumulator, foreshadowing the modern spark plug. The Dunlop pneumatic tire, first made for bicycles, soon found application to the automobile. In 1893 Wilhelm Maybach, one of Daimler's employees, invented the modern float-feed carburetor. Other technical improvements added around 1900 included the radiator, the differential, the crank-starter, the steering wheel, and the pedal-brake control. The effect of the automobile and the bicycle on technology was similar to that of the mechanical clock five centuries earlier: mechanics involved in making and repairing the devices acquired the skills and the ideas to extend the principles involved to other areas.[11]

In steam technology, improvements in fuel efficiency and design continued, although progress encountered diminishing returns and fewer revolutionary advances were made. After 1850, scientists gradually understood the scientific principles that made the engine work. A German-French physicist, Henri Regnault, established the physical properties of steam in the early 1850s, and in 1859 the Scotsman William Rankine published his famous *Manual of the Steam Engine*, a practical guide for engineers. George Corliss, an American, made many improvements to the steam engine, culminating in the famous Centennial engine shown at the Philadelphia Centennial Exhibition in 1876, the largest and most powerful steam engine ever produced. Between 1860 and 1914 steam engines gradually became more efficient, cheaper, easier to maintain, and quieter, but they did not change radically. More revolutionary in concept was the steam turbine, invented by Charles Parsons in 1884, which returned to the idea of Hero's aeolipile. The problem Parsons and, working independently in Sweden, Gustav de Laval solved was that reciprocating compound steam engines could develop only limited speed.[12] For

11. As Rosenberg (1976, p. 24) has suggested, much of the experience accumulated in bicycle production was put to good use in the automobile industry. The best known case in which bicyclemakers went on to bigger things is, of course, that of the Wright brothers, builders of the first operational airplanes. Another example is Sterling Elliott from Watertown, Mass. who in 1902 invented the kingpin-joint steering wheel used in automobiles.

12. The turbines built by Parsons and de Laval were quite different in design. The history of the turbine serves as a good warning against easy generalizations about inventions emerging "when the time is ripe." Although Parsons and de Laval arrived at their inventions almost simultaneously, no fewer than 200 patents for gas and steam turbines had been taken out in Britain alone during the century before 1884. None of these inventions, apparently, was satisfactory.

electrical generation and ship propulsion much higher speeds were needed, and the turbine provided them: the prototype that Parsons built in 1884 ran at 18,000 rpm and had to be geared down. By 1900, the turbine had become a serious rival to regular steam engines. Parsons' engineering genius was heavily indebted to his scientific background, and his example should warn us against facile generalizations about scientific backwardness causing the decline of British technology in this period. If science owed much to the steam engine in the first half of the nineteenth century, it repaid its debt with interest in the second half (Hall, 1978, p. 99).

By the late nineteenth century, the steam engine was coming under attack from another corner: Rudolf Diesel built an internal combustion engine based on the idea that the temperature of air inside a combustion chamber could be raised sufficiently by compression to ignite the fuel, thus converting all of the energy from combustion into work. In 1897 he built the first engine that burned heavy liquid fuel, and after a decade of further development and improvements, Diesel engines began to challenge steam everywhere. The Diesel invention is paradigmatic of the Second Industrial Revolution. Like Watt and Trevithick, Diesel was a "rational" inventor, in search of efficiency above all else. He was not a tinkerer, however, but a trained German engineer, working with state-of-the-art scientific techniques. He started off searching for an engine incorporating the theoretical Carnot cycle, in which maximum efficiency is obtained by isothermal expansion so that no energy is wasted, and a cheap, crude fuel can be used to boot (originally Diesel used coal dust in his engines). Isothermal expansion turned out to be impossible, and the central feature of Diesel engines today has remained compression-induced combustion, which Diesel had at first considered to be incidental (Bryant, 1969).

The railroad, steamship, bicycle, and automobile all helped make transport cheaper, faster, and more reliable. The gains from trade made possible by these innovations constitute a clear link between what we have termed Schumpeterian and Smithian growth. Marshall ([1890] 1930, p. 674) postulated that lowering the cost of transport accounted for three quarters of the progress of manufacturers. How Marshall arrived at this proportion is not clear, and modern economic historians would probably contest it. But the effects of cheap transportation are profound and subtle once we recognize their positive feedback into technology (Szostak, 1986). With improved mobility, technology itself traveled easier: the minds of emigrants, machinery sold to distant countries, and technical books and journals all embodied the technological information carried from country to

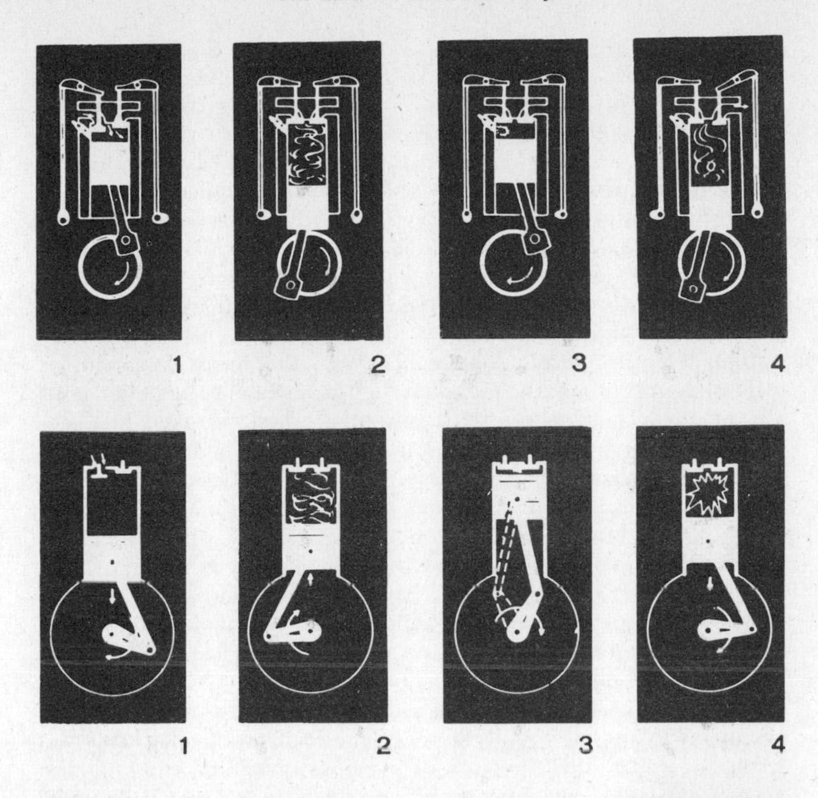

Figure 37. The two main types of internal combustion engines. On the top, the four stroke cycle invented by Otto. The bottom shows the five stages of the compression-ignited Diesel engine.
Source: Umberto Eco and G. B. Zorzoli, *The Picture History of Inventions.* © Gruppo Editoriale Fabbr: Bompiani Sonzongo Etas S.p.A.

country. More mobility also meant more international and interregional competition. Societies that had remained impervious to technological change, from Japan to Turkey, felt left behind and threatened as distance protected them less and less.

PRODUCTION ENGINEERING

From a purely economic point of view, it could be argued that the most important invention was not another chemical dye, a better engine, or even electricity, since, with the exception of steel, most of the inventions described had serviceable, albeit less efficient and more expensive substitutes. There is one innovation, however, for which social savings calculations from the vantage point of the twentieth century are certain to yield large gains. The so-called American System of manufacturing assembled complex products from mass-produced individual components. Modern manufacturing would be unthinkable without interchangeable parts. The term "American" is somewhat misleading: the idea that interchangeability had enormous advantages in production and maintenance had occurred to Europeans in the eighteenth century, and French and later British engineers fully realized its economic possibilities.[13] Moreover, what was regarded in the 1850s as the American System was not exactly interchangeability, but the application of high-quality, specialized machine tools to a sequence of operations, particularly in woodworking, as well as higher operating speeds and sequential movements of materials. As Ferguson (1981) has pointed out, mechanized mass production and interchangeable parts were not identical, and the former did not imply the latter.

The system of interchangeable parts was not an "invention." It was eventually to become a vastly superior mode of producing goods and services, facilitated by the work of previous inventors, especially the makers of accurate machine tools and cheap steel. To be truly interchangeable, the parts had to be identical, requiring high levels of accuracy and quality control in their manufacture. It is now realized that full interchangeability was more difficult to achieve than had

13. One could argue that moveable type was the first application of interchangeable parts. Christopher Polhem, the pioneering Swedish engineer, used interchangeable parts as early as 1720. In 1785 Thomas Jefferson, then ambassador to France, reported that a French gunsmith named Honoré Blanc was making musket locks entirely from interchangeable parts. Like many other French technological advances, Blanc's was buried by jealous competitors: pressure from gunsmiths forced the closure of his arsenal in 1797. For Britain, too, the textbook tale that Europe lagged behind the United States in the adoption of interchangeable parts is largely a myth. The Portsmouth blockmaking plant producing perfectly identical pulleys for ships already contained many of the elements of interchangeability by 1801. In 1832, the London mechanic Bryan Donkin built a continuous-web paper machine that astounded American visitors by the ease with which repairs could be made by replacing a defective part with a new part taken from the shelves. The engineer Joseph Whitworth was a tireless advocate of interchangeability from the 1840s (Musson, 1975a). Other British engineers who experimented with standardized parts were John Bodmer, James Nasmyth, and Daniel Gooch, a manufacturer of locomotives (Musson, 1981).

previously been believed. The use of interchangeable parts grew slowly after 1850, and recent research has shown that the American System was adopted far more haltingly and hesitantly than had hitherto been thought. Many American firms, such as McCormick, Singer, and Colt, owed their success to factors other than complete interchangeability (Hounshell, 1984; Howard, 1978). At first, goods made with interchangeable parts were more expensive and were adopted mostly by government armories, which considered quality more important than price (Howard, 1978; M. R. Smith, 1977). Only after the Civil War did U.S. manufacturing gradually adopt mass production methods, followed by Europe. First in firearms, then in clocks, pumps, locks, mechanical reapers, typewriters, sewing machines, and eventually engines and bicycles, interchangeable parts technology proved superior and replaced the skilled artisan working with chisel and file. Although in the long run true interchangeability was inexorable, its diffusion in Europe was slowed down by two factors: its inability to produce distinctive high-quality goods, which long kept consumers faithful to skilled artisans, and the resistance of labor, which realized that mass production would make its skills obsolete (Cooper, 1984).

Of related importance was the development of continuous-flow production, in which workers remained stationary while the tasks were moved to them. In this way, the employer could control the speed at which operations were performed and minimize the time wasted by workers between operations. The American Oliver Evans is generally credited with inventing this system at his famous grain mill in rural Delaware. A combination of Archimedes screws and endless belts moved the grain from task to task in fully automated fashion. Milling grain, or for that matter slaughtering cattle, as Siegfried Giedion (1948, p. 91n) has observed, involved not assembling but rather disassembling a product. Most manufacturing entailed assembling and combining, and the continuous-flow process was eventually adopted there as well, using endless belts that moved the materials and parts among workers. The first documented occurrence of assembly-line production was in a biscuit factory in Deptford, Britain in 1804. In 1839, J. G. Bodmer's machine tool workshop was organized on similar principles. Yet it was not until the last third of the nineteenth century that continuous-flow processes were adopted on a large scale, especially in the great stockyards of Chicago and Cincinnati. Henry Ford's automobile assembly plant combined the concept of interchangeable parts with that of continuous-flow processes, and it allowed him to mass-produce a complex product and yet keep the price low enough so that it could be sold as a people's vehicle. Giedion (ibid., p. 117) points out that Ford's great success was rooted in the fact that, unlike Oliver Evans, he came at the end of a long development of interchangeability and continuous-flow processes. Success depends not only on the ingenuity and energy of the inventor, but also on the willingness of contemporaries to accept the novelty.

AGRICULTURE AND FOOD PROCESSING

The standard of living of the population depended, above all, on food supply and nutrition. The new technologies of the nineteenth century affected food supplies through production, distribution, preservation, and eventually preparation. In agriculture, the adoption of the new husbandry based on fodder crops and stall-fed livestock continued apace, though in France and in most of eastern Europe progress was slow. New implements and tools appeared on the scene, but here the traditional obstacles to technological progress in agriculture retarded growth: inventions that were useful in some environments failed elsewhere. The mechanical reaper, invented by the American Cyrus McCormick in 1832 for the flat, wide-open midwestern fields, had to be modified extensively before it was suitable for Europe (David, 1975, pp. 233–75). The invention of barbed wire in 1868 by Michael Kelly of DeKalb, Illinois was of great importance in delineating property rights in agriculture and must be counted as one of those simple but crucial inventions that seem so obvious once they exist.[14] Of the new crops adopted after 1830 or so, the sugar beet deserves mention. A French engineer, Benjamin Delesert, discovered the beet sugar refinement process in 1812, but only after 1840 did beet sugar become a serious threat to cane sugar.

Agricultural productivity owed much to the extended use of fertilizers. Farmers learned to use nitrates, potassium, and phosphates produced by the chemical industries. In addition, *guano* (bird droppings) was imported in large quantities from Peru. In the United States, the large stockyards produced fertilizers made from animal bones combined with sulphuric acid. The productivity gains in European agriculture are hard to imagine without the gradual switch from natural fertilizer, produced mostly in loco by farm animals, to commercially produced chemical fertilizers. Fertilizers were not the

14. Cardwell (1968, p. 114) cites barbed wire as an example of an invention that could have been made almost any time, but simply did not occur to anyone before Kelly. Surely demand cannot be the explanation here: the British enclosures required the separation of plots of land by means of expensive hedges and fences. Steel wire, of course, was cheaper in 1870 than it had been earlier, but it had been used in mail armor in the Middle Ages. Like the wheelbarrow and the functional button, then, it must be regarded as an example of a purely empirical invention whose timing we cannot explain.

only scientific success in farming: the use of fungicides, such as Bordeaux mixture, invented in 1885 by the French botanist M. Millardet, helped conquer the dreaded potato blight that had devastated Ireland forty years earlier.

Technological progress outside agriculture affected food supplies in many ways. Steel implements, drainage- and irrigation pipes, steam-operated threshers, seed drills, and mechanical reapers slowly but certainly improved productivity and expanded the supply of food and raw materials. Yet here more than anywhere else the old resisted the new, and modern tools and techniques continued to coexist with manual operations that had not changed in centuries. Mechanizing agriculture involved some major technical difficulties. Much work in agriculture, such as weeding, picking, and milking, was carried out by movements of the human fingers, as opposed to the sweeping or beating motions of the human arm. We have already seen how hard it was to substitute technology for human fingers in the spinning of yarn; much of agriculture was subject to the same kind of difficulties.

A second reason for the slowness of the mechanization of agriculture (compared to manufacturing) was the lack of power substitutes. In most industrial processes, the act of manufacturing can occur at the site of the power source. The utilization of more efficient energy sources was thus rather simple. In agriculture, the power sources had to be brought to the production site (i.e., the land) for most activities, and thus plowing, harrowing, reaping, raking, and binding remained dependent on draft animals long after manufacturing and transportation had adopted the steam engine.[15] Only when the work could be carried out near the power source did mechanization come relatively early: the threshing machine built in 1784 by the Scotsman Andrew Meikle spread quickly, as did the winnowing machine built in 1777 by a London mechanic, James Sharp. These machines were attached to steam engines in the first half of the nineteenth century, but they remained something of an exception. The internal combustion engine solved all that, and by the eve of World War I, tractors and combines were being introduced on both sides of the Atlantic. In 1880, it still took 20 man-hours on average to harvest an acre of wheat in the United States; by 1935 this figure had fallen to 6.1.

Of special interest to the historian interested in economic welfare is the development of food preparation and preservation. Much human suffering has been caused over the ages by nutritional deficien-

15. Attempts to use steam power on the fields produced both John Fowler's cable plowing in Britain (1850) and steam tractors in the United States around 1830. Except for stationary threshing machines, however, steam power was not a lasting success in agriculture.

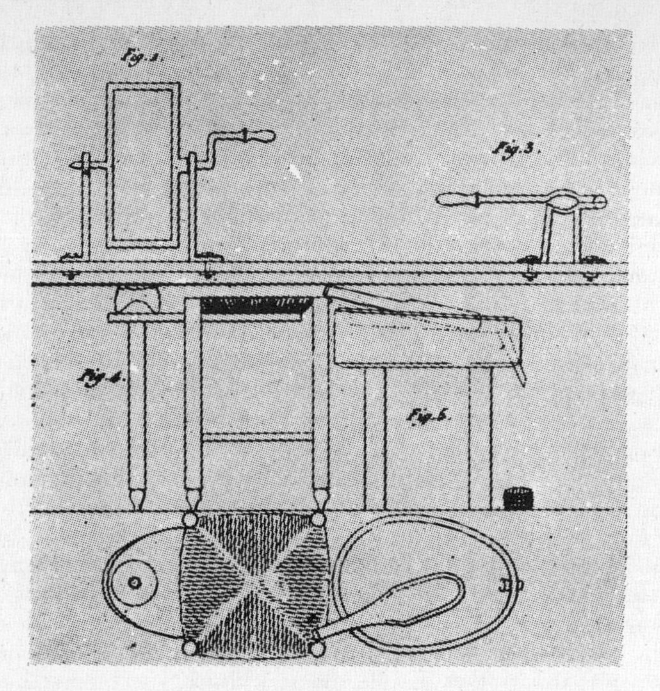

Figure 38. Early food canning. On the left the apparatus used by Appert, on the right a can of fruit dating from 1860. *Source:* Archibald Clow and Nan L. Clow, *The Chemical Revolution* Batchworth Press.

cies and by the unwitting consumption of contaminated foods. The idea of preserving food by cooking followed by vacuum sealing was hit upon by Nicolas Appert in 1795. Appert originally used glassware to store preserved foods, but in 1812 an Englishman named Peter Durand suggested using tin-plated cans, which were soon found to be superior. By 1814, Bryan Donkin was supplying canned soups and meats to the Royal Navy. Canning, too, was a good example of technology leading science, because only after Louis Pasteur's pathbreaking discoveries was it understood why canning worked, and not until the end of the century did it become clear that the optimal cooking temperature was about 240°F. Canned food played an important role in provisioning the armies in the American Civil War, and led to vastly increased consumption of vegetables, fruit, and meat in the rapidly growing cities.

Other food preservatives were also coming into use. Gail Borden invented milk powder in the 1850s and helped win the Civil War for the Union and a fortune for himself. By the end of the century his dehydration idea was also successfully applied to eggs and soups. The centrifugal cream separator, invented by Gustav de Laval in 1877, soon became the cornerstone of the dairy cooperatives in Denmark, the Netherlands, and Ireland. Pasteur himself showed how to sterilize bottled cow milk. Cooling was an alternative form of preservation. In the eighteenth century, ice was preserved in special ice-houses, and an international market in ice emerged in the early nineteenth century, though in the warm seasons the price of ice was so high that only the rich could afford it. Mechanical refrigeration was gradually developed and improved upon between 1834 (when the first patent for the manufacture of ice was issued in Britain) and 1861 (when the first frozen beef plant was set up in Sydney, Australia). By 1870, beef transported from the United States to England was preserved by chilling (29–30°F). The efficient method of preserving beef, however, was by deep freezing at about 14°F. In 1876 the French engineer Charles Tellier built the first refrigerated ship, the *Frigorifique,* which sailed from Buenos Aires to France with a load of frozen beef. By the 1880s, beef, mutton, and lamb from South America and Australia were supplying European dinner tables. Farming in Europe suffered from this competition, but the consumer, the ultimate and final arbiter of all questions of economic progress, benefitted greatly. Technological changes reduced the price of food in general to the point where after 1870 in many countries farmers, rather than consumers, turned to their governments for help. The decline of the price of proteins relative to carbohydrates helped not only to augment, but to improve European diets.

OTHER SECTORS

Two industries that played a central role in the early stages of the Industrial Revolution experienced comparatively slow development after 1830, yet continued to score solid productivity gains. In mining, a major technical difficulty was overcome when the advantages of compressed air were recognized. Between 1849 and 1856 pneumatic mining tools such as circular cutters for underground hewing were devised. Yet not until at least the end of the century did power tools supersede the pick and the shovel.

In textiles, one major innovation stands out: the sewing machine. Apparel making had lagged behind the rest of the textile industry during the early stages of the Industrial Revolution despite an international search for a machine that would replace the motion of the human hand in the stitching process. In the sixty years before 1846, seventeen different machines capable of stitching were invented in the United States, Britain, France, Austria, and Germany (Thomson, 1984, p. 249).[16] These machines were at first technically unworkable, but after 1830 a solution began to appear on the horizon. Elias Howe, an American, is usually credited with the invention, but he merely perfected one crucial feature, the lock stitch, patented in 1846, and made his machine of metal parts. The important insight embodied in the sewing machine was that it should not mimic sewing motions of the human hand, but rather use a double thread, with the eye of the needle far down, forming a stitch by intertwining the two threads. The man who deserves the most credit for perfecting the sewing machine is Isaac Merritt Singer, who powered his machine with a foot treadle. A conservative estimate of the increase in productivity resulting from the sewing machine puts it at 500 percent (Schmiechen, 1984, p. 26). It was later adapted to make shoes (the McKay shoe-sewing machine dates from 1861) and carpets (1880). Annual production of sewing machines went from 2,200 in 1853 to half a million in 1870. Unlike many other inventions in textiles, the sewing machine did not lead to factories, as it did not require a centralized power source. Instead, it kept struggling domestic workers (mostly women) occupied, and created a system of notorious sweatshops. The sewing machine was slow in adopting interchangeable parts; the Singer Company was successful primarily because of brilliant marketing, but still used skilled fitters long after its competitors had switched to fully interchangeable parts (Hounshell, 1984, pp. 67–123).

In the rest of the textile industry, the period after 1850 completed what had been begun earlier. The combing of wool, which had long defied mechanization, was further improved by the introduction of the Lister-Donisthorpe nip machine in the 1850s, reviving the fortunes of the Yorkshire worsted industries. The Heilmann combing machine was used widely on the Continent, especially in Alsace. In spinning, the throstle that had dominated the scene until the 1860s was slowly superseded by ring spinning, invented in 1828 by a Rhode Island mechanic, John Thorp. Unlike mule spinning, in which the twist was imparted by the combined action of rollers and the revolving spindles, ring spinning involved twisting the yarn by the circular movement of a clip on the rapidly turning ring, called the *traveler*. The traveler guided the thread and ensured its winding on the spindle. Ring spinning was continuous rather than intermittent, and re-

16. The inventor with the best claim to the sewing machine is the Frenchman Barthélemy Thimonnier, who devised a working model in 1830 using a chain-stitch system that made it a combination sewing- and embroidery machine. Thimonnier ran into violent opposition from tailors. His army uniform factory in Paris, in which he had the new devices installed, was raided twice in the 1840s, and the machines were destroyed. Like many other French inventors, he died a poor and bitter man.

quired less skill and strength than mule spinning. It produced a slightly inferior product, however, especially for fine yarns. Ring spinning did not spread widely until the second half of the nineteenth century, but then it conquered the U.S. textile industry rapidly. Lancashire in Britain remained loyal to mules, however, a fact that has long intrigued economic historians, as it seemingly indicates a reluctance on the part of the British industry to adopt a superior technology.[17] In weaving, the power loom continued to replace handloom weavers after 1850, but automation came to weaving only after J. H. Northrop built the first automatic loom in 1894. Within twenty years it did to the U.S. weaving industry what Roberts' self-actor had done to British spinning in 1830. By this time, however, Britain's textile industries had lost their position at the cutting edge of technology, and their adoption of the Northrop loom was slow.

Mechanization in the textile industries during the second half of the nineteenth century had the character of a mopping-up operation. Industries that had been skipped for one reason or another by the Industrial Revolution were now caught up in its maelstrom. A good example is the hosiery industry. Lee's knitting frame, invented in 1589, remained at the heart of stocking-frame technology until late into the nineteenth century. The only major improvement introduced before then was an attachment for knitting ribbed fabrics, invented by Jedediah Strutt in 1758, who later became Arkwright's partner. It is often maintained that technological progress in frame-knitting was impeded by low wages, as workers had essentially no alternative employment (Wells, 1958, p. 598). Another alleged cause for the lateness of mechanization was the riotousness of the Nottingham knitters, which may have scared off potential innovators. Such factors, however, were also present in other industries, where they seem to have had little effect in impeding progress. The main problem in applying power to the knitting industry was purely technical: it was difficult to adapt the fine mechanism that varied the tube width, on which the quality of the stockings depended, to more powerful machinery. When the technical problems were finally solved by Matthew Townsend in the late 1840s and William Cotton in the early 1860s, hosiery mechanized just as fast as any other industry.

Finally, I turn briefly to what may best be called information-processing technology, though the term rings vaguely anachronistic. Following the telegraph in the 1830s, a "modern" pattern starts to emerge, in which practice followed theory. Thus, Hermann Helmholz, a German physicist, experimented with the reproduction of

17. See Sandberg (1974) and Lazonick (1981) for different interpretations of the diffusion of ring spinning.

sound, which inspired a Scottish-born speech therapist and teacher at Boston University named Alexander Graham Bell to work on what became the telephone (1876). Supplementary inventions, such as the switchboard (1878) and the loading coil (1899), made the telephone one of the most successful inventions of all time. Wireless telegraphy is one of the best examples of the new order of things, in which science led technology rather than the other way around (Aitken, 1976). The principle of telegraphy, as yet unsuspected, was implicit in the theory of electromagnetic waves proposed on purely theoretical grounds by James Clerk Maxwell in 1865. The electromagnetic waves suggested by Maxwell were finally demonstrated to exist by a set of brilliant experiments conducted by Heinrich Hertz in 1888. The Englishman Oliver Lodge and the Italian Guglielmo Marconi combined the theories of these ivory tower theorists into wireless telegraphy in the mid-1890s, and in 1906 Lee DeForest and R. A. Fessenden showed how wireless radio could transmit not only Morse signals but sound waves as well.

Little science was necessary for another invention that had an enormous effect on information technology: the typewriter. The idea of the typewriter is conceptually obvious, but a number of minor technical bugs, such as bars clashing when two letters were typed very closely together, bedeviled its perfection. These problems were finally solved by Christopher L. Sholes of Milwaukee, reputedly the 52nd man to invent the typewriter. Sholes sold his patent to the Remington Company in 1874, and a small revolution in the office began. The technical problems in the printing industry were more complex. Typesetting had always been laborious and slow work, and the need for improvement was becoming acute as literacy rose and the thirst for information grew in the late eighteenth- and nineteenth centuries. The use of cylinders in printing was pioneered in cotton printing around 1780. Cylindrical impression and inking were applied successfully to newspaper and book printing in 1812 by a German immigrant in London, Friedrich Koenig. The first rotary press was built in Philadelphia in 1846. A horizontal cylinder contained the printed material, and rotated in contact with smaller cylinders, each of which corresponded to a page with automatic grippers guiding the pages from cylinder to cylinder. This machine, originally conceived by Robert Hoe, found its way to Europe, where many leading newspapers adopted it. It was fast but labor intensive. Typecasting, equally revolutionary in that it recast the type each time anew using an automatic process, was perfected in the United States in 1838; by 1851 it had spread all over Europe. An alternative technique, invented by Henry Bessemer (of steel fame), was the pianotype, where the operator worked at a keyboard. For a while, a con-

Figure 39. Linotype machine, invented by Mergenthaler. The keyboard controlled bronze letters that were set on a line, used as a mold on which molten type-metal was poured and then used as type.
Source: Linotype, model Alfa-I (Linotype S.P.A., Milan).

fusing multitude of automatic typesetting techniques were in operation at the same time. Between 1886 and 1890, a German immigrant to the United States, Ottmar Mergenthaler, invented the linotype machine, which cast and set a whole line at a time using a keyboard controlling hundreds of matrices, from which the letter molds were made. Linotype machines were primarily used for newspapers; for books, a related machine, the monotype typesetter, was developed in the 1890s. With the increase in demand for paper, new raw materials became necessary and, after much experimentation, the use of wood pulp was perfected in about 1873.

Pictures as well as written words underwent changes. In 1798 Alois Senefelder, a Polish typographer, invented lithography to reproduce drawings. Photography developed gradually in the nineteenth century, exploiting the well-known property of silver salts to blacken under the influence of light. In the 1830s, two Frenchmen, Joseph N. Niépce and Louis Daguerre, discovered how to make the pictures permanent, and the Englishman W. H. Fox Talbot succeeded in taking photographs on paper (patented in 1841), so that many copies could be printed. In 1888 George Eastman introduced his Kodak camera, adding one more item to the growing list of consumer durables becoming available to better-off people in Western nations.

* * *

Technology in the twentieth century has developed so rapidly and become so complex that no justice can be done to it here. The trend toward a more "scientific" approach to technology continued, and many developments would not have been possible without the advances in mathematics, physics, chemistry, and biology that occurred after 1870. Three hundred years after Francis Bacon, his dream of advancing technology and material welfare by constantly increasing their scientific basis has become reality. Yet the nonscientific taproots of invention—serendipity, luck, and inspiration—have not disappeared, and probably never will. The lonely individual inventor will not be superseded entirely by the corporate research team, any more than the brilliant general will be replaced by a war-game-playing computer. Technological progress is more efficient today in the sense that fewer false turns are taken, and blind alleys are easier to avoid. But the twentieth century still runs into two dilemmas common to all ages. The first is rooted in the fact that some devices can be made to work long before it is understood why or how they do. Especially in medicine and biological technology, such empiricist methodology ("try every bottle on the shelf") is still rampant, and explains, for instance, the painfully slow pace of progress in cancer therapy. The second may be called the Leonardo problem: gadgets and devices can be conceived that are known to be possible, but cannot be built efficiently because supporting technologies are lacking. Energy generation without long-term environmental damage, safe and effective contraception, insect control, and superfast long-distance travel are among the technological bottlenecks of this type of our own age.

The difference between the nineteenth and the twentieth centuries is the technological optimism of our age. In spite of immense progress, the nineteenth century was an age in which technology was believed to be constrained and ultimately incapable of lifting mankind out of poverty. The sources of pessimism were varied. Most economists believed with Ricardo that the standard of living would eventually be set by the minimum of subsistence, or some other ceiling governed by demographic factors. As late as 1890, Alfred Marshall's *Principles* epitomized economics as the study of small, continuous changes rather than abrupt breakthroughs. Despite a century and a half of innovations, many of which were abrupt and revolutionary enough, Marshall still believed with Leibniz that Nature does not make leaps. Economists were not alone in this. Physicists had discovered the laws of thermodynamics, and with them the limitations on energy generation. Geologists began to draw attention to the limits of the earth's resources, leading some economists, such as W. S. Jevons, into fallacious predictions of an imminent energy crisis. In a related context, the theory of evolution suggested that long-term change was typically slow and gradual. The twentieth century, on the other hand, has understood that technology is limitless and can advance in leaps and bounds, and that only society's proclivity to destroy the conditions for its growth limits progress. Twenty years after Marshall, Schumpeter wrote about "spontaneous and discontinuous changes," while a new physics, in which critical masses and quantum leaps occupied a central position, was emerging.

By 1914 the Western world had been growing for many decades, and the gap between it and the rest of the world had become huge. The sources of its economic growth had been the triad of gains from trade, capital accumulation, and technology. The gains from trade on all levels were still rising, but as transportation and information networks became more efficient, the marginal effects of trade expansion declined. As the costs of trading (such as transportation, communication, and insurance) decline more and more, the additional gains in terms of income per unit of cost saved fall. In the (purely hypothetical) limit at which these costs have fallen to zero, no further gains from trade can be attained. Moreover, Smithian growth is always vulnerable to political events. In the late 1870s, growing protectionism partially offset the gains achieved by the railroad and the steamship; in August 1914 it all fell apart. Capital widening raised incomes, but by 1900 capital from the technologically progressive countries (Britain and France) was flowing out to newly settled areas in North America and Australia, or to comparatively less developed European economies, suggesting either strongly diminishing returns, distorted and inefficient capital markets, or both. In any case, capital accumulation on its own was an equally weak reed for long-term economic change. Technological progress alone can support sustained growth, because it alone has not run into diminishing returns. And yet technology, in its own way, was a mixed blessing. World War I highlighted the dangers of the Second Industrial Revolution. It was a war of steel, chemicals, and internal combustion engines. The inventions that had the power to bring unlimited prosperity could also bring unlimited misery.

Technological change, then, accounted for *sustained* growth. It was not caused by economic growth, it caused it. It had no substitutes. Economic historians have been fond of judging particular innovations in terms of their forward linkages, that is, the savings in cost in using a new technique compared to the next best alternative. Some innovations, such as the railroad or the steamship or ring spinning, are put in their rightful modest place this way. The printing press, Cort's puddling-and-rolling process, and chemical fertilizers probably look more important. But taken ensemble, the plethora of innovations that emerged between 1750 and 1914, whether narrowly technical or organizational, obey the "indispensability axiom." Had there been no technological change, other forms of economic progress would eventually have ground to a halt, and Europe would have ended up like the Roman or Chinese Empires. The question is, why didn't it?

PART THREE

ANALYSIS AND COMPARISONS

CHAPTER SEVEN

Understanding Technological Progress

Why does technological change occur in some societies and not in others? In what follows, I concentrate primarily on the supply side of technological change: whence all those free lunches? Statements such as "necessity is the mother of invention" are clearly nonsense in this context. Human appetites being what they are, necessity is always there; the ability to satisfy it is not. In fact, it would be closer to the truth to say that "invention is the mother of necessity," in that new technological possibilities often give rise to hitherto unrecognized desires.[1] The "demand" for technology is a derived demand, that is, it depends ultimately on the demand for the goods and services that technology helps produce; there is little or no demand for technology for its own sake. Hence the structure of preferences influences the direction of the search for new technical knowledge, just as relative factor prices (the cost of labor relative to that of capital or raw materials) may affect its factor-saving bias. But did demand determine the overall intensity of the effort, its success rate, the implementation of improvements, and thus the technological creativity of a society?

If technical information is viewed as just another input into the

1. White (1978, p. 222) points out that this fallacious aphorism can be traced to the twelfth century and cites numerous counterexamples. Similarly, Cipolla (1980, p. 181) states that "necessity explains nothing; the crucial question is why some groups respond in a particular way to needs or wants which in some other group remain unfulfilled." Veblen (1914, pp. 314–17) heaped scorn on the aphorism as a "fragment of uncritical rationalism" and insisted sarcastically that invention was "always and everywhere" the mother of necessity. Schumpeter (1939, Vol. I, p. 85n) pointed out that for a given innovation to occur, some kind of need must exist, but that such a need rarely determines what kind of solution will satisfy it, and may in fact go unsatisfied for an indefinite period of time.

production process, for which producers equate on the margin the efforts involved in producing it and the benefits accruing from it, autonomous shifts in demand would increase inventive activity and thus technological change. If that were how technological change occurred, technology would be analogous to labor: if demand for a good goes up, producers increase its production by hiring more workers. The question is, of course, whether technology can be "hired" like labor or capital. Over most of human history, technological change did not take place, as it does today, in specialized research laboratories paid for by research and development budgets and following strategies mapped out by corporate planners well-informed by marketing analysts. Technological change occurred mostly through new ideas and suggestions occurring if not randomly, then certainly in a highly unpredictable fashion. Demand conditions may have affected the rate at which these ideas occurred, and may have focused them in a particular direction, but they did not determine whether a society would be technologically creative or not. If technological change was a very cheap lunch, with the benefits typically vastly exceeding the costs of invention, then further increases in demand may not have had much effect. After all, if desire for more goods and services was insatiable, and if in the cases of macroinventions the social benefits typically exceeded the social costs, the "excess demand" for technology was always a fact of life. Thus changes on the margin in demand would have had little effect on the pace of innovation. This is not to say that in the past inventors and inventive activity did not respond to rewards, but rather that these rewards were not necessarily closely correlated with changes in demand. In short, demand factors were not negligible, but the intuition that there was a certain symmetry between demand and supply in the generation of technological progress should be resisted.[2]

Moreover, by focusing on demand factors in trying to explain the overall *level* of technological creativity, rather than the *direction* of innovative activity, there is a danger of circular reasoning. Demand was not an exogenous variable. It is always and everywhere constrained by income. Income, of course, could rise for a variety of reasons, including population growth, trade and specialization, or natural windfalls, such as improvements in climate. But since medieval times, technological change itself has been one of the chief fac-

2. In this regard, too, technological progress resembles the geographical discoveries. Adam Smith's ([1776] 1976, book II, p. 68) wise words on the discoveries apply to technological progress as well: "The establishment of European colonies in America and the West Indies arose from no necessity; and though the utility which has resulted from them has been very great . . . it was not understood at their first establishment and was not the motive of that establishment."

tors determining incomes. By 1700, it appears, Europe was already richer than all non-European economies. Technological progress was largely responsible for this gap. It is, then, inappropriate to cite this higher income as a demand factor that encouraged the search for more new technologies. In short, technology determined real income, not the other way around, though feedbacks and other complications did occur.[3]

Preferences do matter, of course. Many costs associated with the adoption of a novel technique must have seemed high to contemporaries, even if they do not appear so in retrospect. Sustained innovation requires a set of individuals willing to absorb large risks, sometimes to wait many years for the payoff (if any). It often demands an enormous mental and physical effort on the part of the pioneers. Risk aversion, leisure preference, and time preference are thus of major importance in determining the rate of innovation of a particular society. However, the costs of the technological "lunch" borne by the inventors and developers are only a part of the total social cost. In some cases, technological changes had powerful environmental consequences such as pollution and deforestation. Not all such changes need be destructive, but insofar as they imply changing nature from its virgin state, they usually involve some cost. Beyond that are the social costs incurred by labor: the obsolescence of skills, the forced migration of workers, the changing organization of production, and in some cases technological unemployment. Economists tend to believe that most of these costs are at most temporary phenomena. Yet technological progress is hardly ever Pareto superior, that is, an improvement for everyone affected: there are losers in the process, and while the gainers could compensate them, it is only rarely that they do. The stronger the aversion to the disruption of the existing economic order, the less likely it is that an economy would provide a climate favorable to technological progress.

The total cost of adopting a new technique thus consists of two parts, the private costs paid by the inventors, and the social costs paid by society as a whole. Social costs usually exceeded the private costs, though the latter were not always negligible. Invention usually involved problems that were exceedingly difficult. Hindsight creates a serious distortion in this respect; the technical problems invariably seem easier ex post than ex ante. There was nothing natural and inevitable about technological progress. Most societies experienced some of it, but none of them, except the West, was able to turn it into a sustained and almost self-perpetuating mechanism of contin-

3. For a detailed argument on this issue in the context of the Industrial Revolution, see Mokyr (1985).

uous expansion. Technologically creative societies in the past, as in the present, can be defined as those that generated innovations whose benefits dwarfed the costs of invention and development and thus created a free lunch.

Once an invention was perfected, its benefits may have dwarfed the costs of research and development, but the research was usually long and tedious, and the end result dependent on a combination of luck, brilliance, and perseverance. Moreover, looking for the costs incurred by known (and therefore successful) inventors involves a bias, because for every successful inventor, there were hundreds of failures, whose efforts may not have been less valiant than those of their more fortunate colleagues. In an ex post sense these failed inventors were dispensable. But in an ex ante sense, all major and minor inventors operated in complete uncertainty about whether their projects would succeed, and thus all were part and parcel of technological creativity. The last century has seen an improvement in this regard: the greatest invention of the nineteenth century, the cliché goes, was how to invent. Theory, careful measurement, and accurate instruments were science's gift to technology, and the importance of serendipity has declined.

There is a desire for stability that exists in every society, though its intensity varies. The sentiment "if it ain't broke, don't fix it" is the archenemy of technological progress. Technological conservatism refers to the tendency to adopt a certain technique only because it happened to be used in a previous period. Economists have traditionally dismissed such conservatism as irrational, costly, and thus nonviable. But recent thinking, much of it summarized in Kuran (1988), has revealed that economic theory can and must deal with conservatism. Kuran distinguishes between personal and collective conservatism. The most interesting explanation of personal conservatism is what is known as bounded rationality, the inability of individuals to process and manipulate large amounts of information, thus making it difficult for them to re-evaluate alternatives. Bounded rationality seems particularly pertinent to technological progress, because changes in technique usually involve the evaluation of rather complex information. This does not mean, of course, that techniques are completely fixed, but that relatively small changes involving choices between alternatives that are only marginally different would be biased toward the status quo. On the collective level it is possible for a society to be conservative even when individuals are not. This happens when well-organized interest groups in a society have a stake in maintaining the status quo in the face of a superior alternative. I shall return to this issue later in this chapter. Technological conservatism produces an economic inertia that seems an apt description of the history of most societies that ever existed. The pervasive influence of the status quo, tradition, custom, routine, and adherence to precedent were powerful obstacles to innovation and advances. When technological progress occurred, it did so in the face of considerable odds.

Technological change is a game against nature rather than against other players, what von Neumann and Morgenstern have called a "Crusoe game." Invention occurs at the level of the individual, and we should address the factors that determine individual creativity.[4] Individuals, however, do not live in a vacuum. What makes them implement, improve, and adapt new technologies, or just devise small improvements in the way they carry out their daily work depends on the institutions and the attitudes around them. It is exactly at this level that technological change is transformed from invention, a game against nature, to innovation, a complex, positive-sum game with many players and very incomplete information.[5] As C. S. Lewis pointed out, "Man's power over Nature often turns out to be a power exerted by some men over other men with Nature as its instrument" (cited by Pacey, 1986, p. 12). Wealth, however, not power, is the main object of nonmilitary technology.

The physical and social environment is important in determining the actions of individuals, although it is not solely responsible for the outcome. I shall try to show which factors on the aggregate level help determine the propensity of a member of a society to invent and which factors make others want to adopt his or her inventions. Before that, however, it is worth examining some factors that operated, consciously or otherwise, on the minds and actions of individuals in their lonely struggles with the laws of physics, chemistry, and biology.

LIFE EXPECTANCY

Boulding (1983) asserts that life expectancy is an important factor in determining the level of technological progress. People who live very short lives have little time or incentive to generate new knowledge.

4. As one historian of technology (White, 1968, p. 105) has noted, "a group can conceive nothing which is not first conceived by a person."

5. An apt description of the difference between the two types of games was provided by Rudolf Diesel, who distinguished two phases in technological progress: the conception and carrying out of the idea, which is a happy period of creative mental work in which natural obstacles are overcome, and the introduction of the invention (which we would call today innovation), which is "a struggle against stupidity and envy, apathy and evil, secret opposition and open conflict of interests, the horrible period of struggle with man, a martyrdom even if success ensues" (cited in Klemm, 1964, p. 346).

In earlier times such new knowledge depended more often than not on trial-and-error processes that were very time consuming, and thus less likely to be engaged in when life expectancy was short. Short (and uncertain) life expectancies also made it less attractive to save in order to buy or construct the capital goods that embodied technological progress. A short life expectancy implies a high subjective rate of discount, that is, a low level of patience and willingness to defer gratification, and almost all technological progress requires an element of abstinence. Closer examination of this reasoning raises some doubt. Life expectancy is an average concept. The thirty years or thereabout, of life expectancy associated with backward societies is an average of the many people who die in the first years of their life and the remaining whose lives were not substantially shorter than those in the West in the beginning of this century.[6] Moreover, there is little evidence of a rise in life expectancy before 1750 in Europe, and yet technological change was clearly accelerating long before then. Perhaps the effect predicted by the theory was there, but was so small that it is difficult to observe empirically.

NUTRITION

A more likely determinant of technological creativity is the level of nutrition of a society. It is well known that "hidden hunger," i.e., significant long-term shortfalls in net food intake, does not terminate life or even impair health, but does reduce the level of energy at which the individual operates to create a lethargic and supine personality. Thus, the output of energy is adjusted to the intake, and what is often described as "laziness" is really the result and not the cause of poverty and malnutrition. It is reasonable to expect that in societies in which hidden hunger is rampant, initiative and ambition, necessary for economic progress will be harder to find.

A particularly vicious form of hidden hunger is infant protein deficiency syndrome (IPDS). Levels of protein that are significantly below the minimum required for growth in the first 18 to 24 months of life will permanently cripple mental development. Demographic historians agree that differences in nutritional status led to substantial differences in height, biological maturation, ageing processes, resistance to disease, fecundity, and so forth. There is good reason to suspect that the nervous system and the brain were not immune from these influences. If nutritional deficiency can reduce the average realized intelligence of a population by a significant proportion, its ef-

6. Some medieval inventors and scientists whose ages at death are known: Johann Gutenberg lived to 70; Roger Bacon to 72, Albertus Magnus to 87, Gerard of Cremona to 73, and the Arab chemist al-Jabir to about 94.

fects on technological progress are likely to have been important. To break out of a predetermined work routine, particularly if the previously set technology has been in use for a long time, requires energy and initiative as well as the ability to think clearly on a purely intuitive level. To assess whether an idea works, the innovator needs some notional distinction between a conditional probability and an unconditional probability; that is to say, he or she must be able to evaluate the impact of the change while holding other things constant. This kind of thinking requires a mental experiment that depends on the ability to reason. Brain damage caused by inadequate nutrition irreparably impairs this ability.[7] Thus it may have come to pass that some societies had a larger supply of intelligence per capita than others not because of any biological differences between races, but because of poverty and ignorance about the nutritional needs of infants.

Did a higher average intelligence matter? The point is debatable. Clearly, if IPDS was important in the past, it did not prevent the emergence of exceptionally creative and intelligent individuals altogether. The effects of malnutrition were concentrated primarily among the lower ranks of society—slaves, peasants, and laborers. If technological change, and especially the implementation of new devices in daily routines, required a cooperative effort of management and labor, a more adaptable and intelligent work force may well have helped determine the outcome. Technological progress is not homogenous in this respect. In agriculture, mining, and shipping, most improvements would be aided by a higher quality work force. In manufacturing and services, especially in modern factories and mass production services, such as fast-food chains, discipline, docility, and a willingness to accept mind-deadening routines may be more important. It is possible, as Wright (1987, p. 326) has recently remarked, to devise technical systems that are designed by geniuses to be run by idiots. In the more distant past, before the age of mass production and uniform components, such systems appear less plausible. The age of the peasant and the artisan required intelligence and dexterity by the on-the-spot operators of any technique.

WILLINGNESS TO BEAR RISKS

The structure of preferences regarding risk also affects a society's ability to produce innovative individuals. Invention and innovation almost always involve some willingness to bear risk. Changing a known

7. For a complete analysis of the protein content of infants' diets in the European past and its likely effect on mental development, and through it on technological change, see Williams (1988).

and trusted production method, even in marginal ways, involves something of a gamble. In the past, when social safety nets were imperfect or nonexistent, the dangers of failing in a business adventure involving a new technique were far greater than they are today. In agriculture, experimenting with new techniques or crops might well have entailed a real risk of starvation. Moreover, even in cases of success the benefits of invention were limited in an age before mass production. Careers like those of nineteenth- and twentieth-century inventors—such as George Westinghouse, Charles Hall, or Edwin Land, who turned brilliant inventions into personal fortunes by building corporate empires based on mass production—were simply not available to medieval and early modern inventors. With the expected gains relatively small and the fallback positions of inventors more precarious, it is clear that invention and innovation were more risky in the past than in more recent years.

Economists believe that most individuals are risk averse, but little is known about the determinants of the degree of risk aversion. Some factors that determine the willingness to absorb risk are determined at the level of the economy, such as the existence of institutions that make diversification possible. Underlying the willingness to absorb risk is the shape (curvature) of the utility function, which is in part determined at the level of the individual. Household structure could be of importance here. Heads of extended families will be more cautious, all other things equal, than heads of nuclear families, since more people depend on them. Owners of dynastic family firms regarded themselves as caretakers of businesses that transcended their own lives, and thus may have been more cautious than managers of shorter-lived enterprises. Another possible factor was changes in income distribution, as individuals whose position in society is threatened are more willing to take risks (Brenner, 1983). The opportunities for diversification that large business units make possible could be of importance here. Yet as long as no good theory exists as to the determinants of risk aversion, we can do little more than speculate.

Willingness to bear risk is but one factor relating the intensity of innovation to uncertainty. In an interesting passage, Schumpeter (1950, p. 74) pointed to an important principle underlying innovative activity under uncertainty. If individuals consistently overrate their chances of success, their behavior may appear risk loving and society could enjoy more technological change relative to a situation in which individuals assess their chances correctly. Such an optimistic bias could offset to some extent the inherent bias to *under*produce technological change (because the social benefits typically exceed the private benefits). A large number of important inventors died in obscurity and poverty, indicating that the private returns to a socially useful invention were low ex post, but that the effort was carried out anyway because inventors overestimated the private payoff (Nye, 1991).

GEOGRAPHICAL ENVIRONMENT

Once we get beyond the level of the individual's creativity and start to examine some of the social factors in technological change, matters become considerably more complex. The impact of the environment, physical and cultural, on the amount of technological creativity in a given society is a matter of much controversy, in part because the concept of causality is not always made quite clear. If a flower grows in a garden can we "explain" the existence of the plant in any simple monocausal framework? A favorable environment consists of sunshine, water, soil chemistry, and the absence of pests, hail, and destructive children. Yet the real "cause" of the plant's existence is the gardener who sowed the plant. Few of the environmental factors are either strictly necessary or sufficient, and they are permissive rather than causal in a direct sense. In some cases the economic and social conditions function as focusing devices, directing and channeling the creative potential that is already there. All this has led to some confusion, which I shall try to unscramble by examining some key elements in the geographic, social, and economic environment in which technological change occurs.

The relationship between the physical environment and technological change is subtle, though at times the connection has appeared relatively straightforward (Rosenberg, 1976, essays 12–14). For instance, the diffusion of watermill technology that changed the economies of Europe north of the Alps in the early Middle Ages has an ostensible explanation in different climates (Strayer, 1980). Rainy northern Europe has more usable sites, and it may seem natural that when the center of gravity shifted northward in the early Middle Ages, water mills were employed more intensively and, through a learning-by-doing process, became more sophisticated. But such an explanation does not withstand scrutiny. Rainy England soon counted thousands of water mills, whereas rainy Ireland did not. Northern Italy and large parts of Gaul had many appropriate sites for water mills, of which only a fraction were utilized by the Romans. Roman hydraulic engineering was sufficient to adapt and utilize sites even in places where the flow of water was slow or irregular (Reynolds, 1983). Yet there is little evidence that the Romans made wide use of their hydraulic expertise to improve sites for mills. Water mills would not be of much use in the dry regions of central Asia or North Africa, which is why these areas developed a good substitute, the windmill. But why, then, did the Asian windmill fail to do for Eastern economies what water mills did for western Europe? Similarly, countries that had abundant coal deposits would become natural leaders in coal use and steam-engine technology. Learning-by-doing or learning-by-using models imply that the likelihood that a technology will be improved rises with the intensity of its use. However, not all technological change can be explained by learning models. It could just as easily be maintained that innovations were stimulated by the search for substitutes, rather than complements. Economies that had no coal would constantly be under pressure to develop more fuel-efficient techniques, or engines that used alternative sources of energy, or to produce goods requiring low energy intensity.

An argument supporting the alleged link between resource availability and use and technological change is provided by Wrigley (1987, p. 15), who argues that productivity soared during the Industrial Revolution because large quantities of energy (that is, coal) were put at the disposal of British workers. Per capita consumption of coal tripled between 1775 and 1830, and thus it may be concluded that Britain's good fortune in being located on top of a mountain of coal made the Industrial Revolution possible. After all, in the absence of coal, the timber requirements of a nation such as Britain in the early nineteenth century would have become massive (ibid., p. 79). What Wrigley does not stress sufficiently is that causality may have run not from resources to technology but rather in the opposite direction. Coal consumption increased because technological change increased its efficiency in use. Coal had been available for centuries and used in a variety of industrial processes. During the Industrial Revolution, the techniques used in coal mining itself changed little relative to those that used coal. Improvements in steam engines were, as we have seen, paralleled by advances in waterpower, and coal may have been less crucial than Wrigley has suggested. In the absence of coal, the ingenuity applied to using it would have been directed toward replacing it. Coal-poor regions, such as Switzerland, Northern Ireland, and Catalonia were able to generate substitutes or develop low-energy product lines.

The history of the adoption of coke smelting in Britain illustrates the difficulties associated with an interpretation in which causality runs from resources to technology. Traditional accounts, such as Ashton's (1924, pp. 8–9) or Clow and Clow's (1956), related the adoption of coke smelting to the rise in the price of charcoal caused by deforestation. Flinn (1959; 1978) has shown, however, that evidence on these prices does not confirm this view. More important, however, is the fact that attempts to substitute coal for timber as a fuel were made in a wide range of processes going back as far as the sixteenth century. In the sixteenth century, as we have seen, this substitution succeeded in a range of commodities: from soap boiling to brewing, fossil fuels were substituted for charcoal long before coking was used in iron smelting. The reason was that the labor costs involved in producing a coal-based fuel were in general lower per calorie than in preparing charcoal. Realizing this, ironmasters experimented with coal as a fuel in smelting furnaces from the mid-sixteenth century. Abraham Darby differed from the many nameless and forgotten other inventors who tried to solve this problem only in that he was successful, in part due to his experience in malting (which had been using coke) and in part because of luck.

Why do searches for substitutes succeed in some cases and fail in others? One of the biggest advantages that Europe had was its large supply of large domesticated animals, which were entirely lacking in pre-Columbian America and Africa, and scarce in most parts of Asia.[8] This scarcity may have had deep historical roots: African and East Asian adults suffer from lactase deficiency and cannot digest large quantities of fresh milk (although they can digest milk in the form of cheese or butter). Yet the lack of draft animals did not stop the Amerindians from inventing the wheel; they used it for toys, and then turned their attention elsewhere (Crosby, 1986, p. 43). It apparently never occurred to them to use wheels in devices such as pulleys, gears, and waterwheels. Can we blame the absence of animals for this? Man-pulled or -pushed vehicles existed throughout the Eastern Hemisphere—the principle of the wheel transcends the availability of a specific form of motive power. Where draft animals were scarce, wheels were adapted to humans, and wheelbarrows and rickshaws appear. Moreover, if a society, for some reason, had no access to horses, camels, or oxen, what better incentive could there be to invent techniques that got around this constraint, creating alternative sources of energy (such as windmills), fertilizer, and proteins? Yet pre-Columbian America had little success in generating these technologies, or perhaps it never tried hard. Similarly, deforestation and expensive timber did not stimulate coal mines in China, which British reports found to be primitive beyond belief in the 1870s (Brown and Wright, 1981, pp. 66–67).

Two theories thus coexist regarding the connection between natural resources and technological progress. One holds that abundant natural resources encourage complementary innovations, the other that the scarcity of natural resources, as a result of depletion or pop-

8. Harris (1977, p. 42), goes so far as to state flatly that the absence of animals in pre-Colombian America "explains why it was that Columbus 'discovered' America and Powhatan did not 'discover' Europe."

ulation growth, stimulates the search for substitutes. Neither view implies that natural resources (or their absence) are a necessary or sufficient condition for technological creativity to emerge, but both suggest strong correlations—of opposite signs.[9] The two theories do not necessarily contradict each other, but they do suggest a misspecified model. History displays a wide range of results from the interaction of environment and the economy, and correlation was rarely causation.[10] The true exogenous variable in the model, whatever it is, makes the economy technologically creative. Given that creativity, resource endowment will be one of the factors that determines the direction of the search for innovations, operating, in Rosenberg's (1976) term, as a focusing device. Resource scarcities, like demand, are a steering mechanism, not a *primum movens*, of technological progress; it is the engine that makes the car move, not the steering wheel.

PATH DEPENDENCY

A more persuasive version of the natural resources argument focuses on spillover effects from previous technologies, which were themselves environmentally determined, thus combining geography and history. A good example of this kind of process is the spillover from mining activities. Mining, of course, was located wherever some desirable mineral resource was found. But miners had to struggle with water, and hence they worked on developing better pumps, leading to more accurate boring machines and other tools, which eventually helped to develop steam- and modern waterpower. Mining required knowledge of metallurgy, chemistry, mechanics, and civil engineering; the convergence of so many different branches of knowledge—empirical and unscientific by our standards as they were before 1800—could not but lead to further technological progress. Agricola, Ercker, Newcomen, Biringuccio, Polhem, Watt, and Stephenson were among the great technological minds whose back-

9. This paradox has baffled many of the best scholars writing on the economic history of technology. Usher stated that the scarcity of labor in the sparsely settled Europe of the early Middle Ages was favorable to "certain kinds" of innovation, presumably labor saving. Yet on the next page he argues that the abundance of livestock in northern Europe, relative to its scarcity in the Mediterranean, encouraged the development of technologies that used draft animals in this period (Usher, 1954, pp. 182–83). For some similarly inconclusive reflections on this problem and examples of these arguments, see Rosenberg (1976, pp. 122–23 and n. 51).

10. For example, eastern Europe produced flax, whereas Britain had easy access to cotton. Flax was difficult to mechanize, whereas cotton was easy. Yet the difference in the rates of innovation in Britain and Russia between 1750 and 1830 can hardly be attributed to that difference alone.

grounds were in mining technology. A similar phenomenon can be discerned in late medieval Holland, where location determined an affinity with shipping. Starting off as a nation of fishermen, the Dutch learned one thing from another: shipbuilding led to rope- and sail-making, to the use of wind-driven sawmills, and to the development of provisioning industries. Its maritime sector created the technologically progressive sectors that depended on it.

This explanation of technological progress may appear trivial to some and false to others. The view that technological change depends primarily on its own past is known as *path dependency* (David, 1988). A variety of mechanisms for this autocorrelation can be proposed. One of them, due to David (1975), is that technological change tends to be "local," that is, learning occurs primarily around techniques in use, and thus more advanced economies will learn more about advanced techniques and stay at the cutting edge of progress.[11] Thus, societies that have been creative in the past would have a head start on others. Another rationalization of why technological progress depends on past progress is that it creates imbalances and "bottlenecks" in related and complementary processes, which then stimulate further search for innovations. This view, associated with Landes (1969), has been further refined and formulated by Rosenberg (1976), who pointed out that such complementarities operate as "focusing" devices in generating further progress. These theories are plausible, but some unresolved questions remain. One of them, emphasized by Parker (1984, p. 38) is that such a mechanism works only if there are immobilities in the allocation of resources. Otherwise any bottleneck could be readily resolved by reallocating labor and capital to industries in which technology had not been improved. Another difficulty is that it assumes that the supply of new technology is sufficiently elastic for additional search for technical solutions actually to yield the desired results. Any focusing device will be useless unless there actually is some real object to gaze at through the lenses.

For our present purpose, the importance of path dependency for technological change is that some paths lead to more and more progress, while others lead to dead ends. Because it is not possible to predict in advance which paths lead where, what appears in retrospect as technological creativity may have been the result of a past

11. A special case of this type of technological change is what Persson (1988, pp. 7–13) has called "endogenous" technological change, by which he means technological change that is more or less automatic since it is a by-product of trial and error, experience, and what he calls "economies of practice." He does not suggest why these technological sequences occur when and where they do, or what role other forms of technological creativity play.

lucky choice, whereas technological backwardness may simply be the price for having bet on a slow horse. In some cases a technological choice that initially appeared superior eventually led to stagnation. One example is the adoption of the potato as the staple crop in Irish agriculture in the eighteenth century. An acre planted with potatoes produced more than three times the calories produced by an acre of grain, and thus the adoption of potatoes appeared an obvious case of productivity growth. Since the eighteenth century, however, the increase in grain yields has been much larger than that of potato yields, so that areas that specialized in potatoes appear to be stagnant. A second example is the invention, some time between 500 B.C. and 100 B.C., of the camel saddle. As a direct result, the camel gradually came to replace wheeled transport in large parts of the Middle East and North Africa (Bulliet, 1975; Hill, 1984b). Although the principle of the wheel remained in use for industrial purposes other than transportation in these economies, the wheel in the ordinary sense of the word disappeared. The localized learning-by-doing effects in transportation technologies using camels were clearly less impressive than in wheeled transport.[12] Camels conserved resources by obviating road construction, but they did not inspire railroads.

In any event, these sequences do not provide a wholly convincing account of technological advance. It is misleading to think that nothing leads to technological progress like technological progress. Examples of successful innovation gradually petering out—as well as sudden "catch-up" phenomena in societies that had previously been left behind (such as post-1850 Germany)—show the inadequacy of these models. Why did the mining regions in southern Germany and England develop far greater technological spillover effects than the mining regions of Sweden, Hungarian Slovakia, or northern Spain? Why were the Dutch so much more adept at sea than, say, the Irish? And why did Portugal after 1500 fail to develop spillover effects similar to those of the Netherlands? To use Lynn White's metaphor, nature-based effects like this can open doors, but they cannot force an economy to walk through them. In symmetric fashion, if a door appears closed because of a previous technological choice, it is always possible to smash the door and cross over to a different technology. Neither nature nor history can lock a society forever in a dead-end technology.

The path-dependent nature of technological change, in which its

12. Bulliet (1975, pp. 222–23) has argued that the camel led to the emergence of "societies without wheels," (at least in transportation), which eventually led to an "unconscious prejudice against wheeled vehicles." Thus the absence of wheelbarrows on construction sites in modern Teheran, and the slowness with which Ottoman armies moved mobile field artillery.

course is explained mainly by its past, can be extended, though caution must be used in applying these models. The links with the past must be specified rather than assumed. Regions that specialized in clock- and instrumentmaking may have had a significant edge in technological innovation because clock- and instrumentmakers had to be craftsmen of superior ability with considerable knowledge of materials and mechanics. It is therefore not surprising to see many engineers and inventors who were trained as clock- and instrumentmakers during the British Industrial Revolution. Sometimes the prominence of a few great men created a "school" that became self-sustaining and eventually led to overlapping generations of engineers and inventors. The British machine-tool industry in the late eighteenth- and early nineteenth century and the German chemical industry in the nineteenth century owed part of their success to such traditions. Yet many traditions were extinguished or led to dead ends. Clearly, then, technology requires more than its own past to be explained.

LABOR COSTS

A related hypothesis maintains that high wages and labor scarcity stimulated technological creativity. This hypothesis, known as the Habakkuk thesis after H. J. Habakkuk (1962), maintains that high wages in the United States in the nineteenth century spurred technological progress, and in particular brought about the American System of manufacturing based on interchangeable parts. Much of this literature lies outside the scope of this book, but a few notes on the relationship between high labor costs and technological progress are apposite. The idea that high wages were necessary to encourage technological creativity is based on the false perception that technological progress was first and foremost a process of choice between more or less equivalent alternatives, and that these choices depended on factor prices. Some cases in which such choices had to be made doubtless took place, but they do not describe its essence. High wages would, according to the Habakkuk thesis, stimulate labor-saving inventions. There is no persuasive evidence, however, that technological progress has been predominantly labor saving in the past (von Tunzelmann, 1981, p. 158). MacLeod (1988, pp. 158–181) has analyzed the motives of eighteenth-century English patentees and shown that they overwhelmingly declared the goal of their innovation to save capital or improve product quality, whereas only 3.7 percent declared "labor saving" to be the purpose of the invention. To be sure, the desire to save labor intensified in the eighteenth century, but what was meant thereby was often different from the processes envisaged by Habakkuk. Instead, inventors wished to substitute away from labor to reduce the bargaining power of unions or because they believed their workers to be untrustworthy.

Moreover, biased technological change is a relative concept. A labor-saving innovation means that after the innovation is implemented the capital–labor ratio rises. But in the majority of cases, the absolute amounts of both capital and labor needed to produce one unit of output decreased, even if that of labor has decreased more. The producer, whether independent craftsman or mass manufacturer, will try to reduce costs by as much as possible, regardless of whether the saving is in labor or any other input. Labor costs are still costs, whether labor is cheap or not. Changes that reduce labor requirements, even in low-wage economies, will increase profits and thus be adopted. Furthermore, inventions that produced goods that were easier to maintain and use, better looking, and more durable; that reduced the drudgery and effort of the workers; and that reduced the use of fuel, raw materials, water, and wear-and-tear of tools and equipment were attractive to low-wage and high-wage economies alike. Labor-saving inventions seem to have occurred as often in periods of stable or slowly falling real wages (such as the last third of the eighteenth century) as in periods of rising wages.

To be sure, there are some cases in which the evidence suggests that high wages or labor scarcity due to high mortality or strikes stimulated the search for labor-saving technological progress.[13] The flip side of this argument is that cheap and abundant labor slowed down mechanization (Samuel, 1977, p. 47). When mechanization required the purchase of massive and expensive capital goods in which the new technology was embodied, the relative prices of capital and labor may well have been a consideration, at least in the short run. But such cases were rare before 1870, and some putative examples of this phenomenon do not stand up when examined closely. For instance, the slowness of technological progress in the British coal gas manufactures in the 1870s and 1880s has been attributed to cheap labor, and when wages in the industry started to rise in the 1880s, labor-saving innovations were allegedly rapidly introduced. A recent paper (Matthews, 1987) persuasively refutes this theory, and shows that new techniques in gas production (such as new sources of power

13. The idea that labor conflict could stimulate invention was pointed out already by Marx. The most famous case is the invention of Roberts's self-actor, which was explicitly aimed at relieving the plight of Manchester cotton manufacturers plagued by a strike of skilled mule operators. Other examples allegedly attributable to in Marx's words, "supplying capital with weapons against the revolt of the working class" are cylinder calico printing, the McCormick reaper, wool combing, and Richard Roberts's punching machine (Rosenberg, 1976, pp. 118–19; Bruland, 1982).

and gravity stoking) became available only after 1885, and probably would have been adopted even had wages remained unchanged. Similarly, it has been suggested that the *vallus,* the Roman harvesting machine, failed to spread because it was labor-saving and there was no labor shortage in most of the Roman empire (Pleket, 1967). K. D. White (1969) has pointed out, however, that the texts referring to the *vallus* imply a savings of both manpower and time. More important, perhaps, is that the absence of a labor shortage is not the same as costless labor. As long as labor costs anything, a machine that saved it could increase efficiency.

SCIENCE AND TECHNOLOGY

Was the supply of scientific ideas and knowledge ever a binding constraint on the creation of new technology? Rostow (1975) regards the Scientific Revolution of the seventeenth century as the crucial difference between West and East. Before that, "science did not teach those with access to or control over resources that the physical world could be understood in ways that permitted it systematically to be transformed to their advantage" (p. 31). There was always what could be called a body of metatechnological knowledge, from which technology drew its inspiration, consciously or subconsciously. The classical distinction made by Francis Bacon between inventions that depend on a state of knowledge (scientific or other) and those that could have been made at almost any time is relevant here. Before 1850, it seems, most important inventions, from the verge-and-foliot escapement mechanism to the cotton gin, were of the empirical kind, and the role of metatechnological knowledge was marginal. But appearances can be misleading in some instances: Galileo's theory of machines, as we have seen, was of great importance to all subsequent developments in machine design. Even that most empirical and pragmatic of all inventors, John Smeaton, was inspired by the experimental techniques developed by Newton, in which the partial effects of changes were measured by varying one component while keeping all others constant. Pacey (1975, p. 137) has argued that the Scientific Revolution of the seventeenth century taught engineers "the method of detail," that is, to analyze problems by breaking them down into their component parts, which may be easier to analyze than the problem taken as a whole. Jacob (1988, p. 208) claims that the importance of eighteenth-century science was in teaching manufacturers and merchants to "think mechanically," that is, in terms of understandable and controllable physical processes.

A practical distinction between science and technology is not easy to draw. Gille (1978, p. 1112), who has examined the nexus closely, suggests a distinction based on purpose: science aims at comprehension, whereas technology aims at utilization. Although Gille feels that science in the West did not start to play an important role until the Renaissance, he seems to agree with Rostow that its decline could lead to the blockage of further progress. Yet his own survey suggests that correlations between scientific advances and technological progress could as likely involve technology helping science as the other way around. Moreover, any correlation between the two could well be spurious in that both science and technology may well be functions of other social factors. Without denying the importance that scientific insights and practices and some individual scientists had on occasion on the progress of technology during the baroque period and early Industrial Revolution, we have seen that what we call today "scientific knowledge" was rarely a binding constraint on technological progress before 1850. The relationship between scientific and technological progress, as Otto Mayr (1976) has pointed out, has confounded the best minds in both fields. Indeed, an eminent historian of science (Kuhn, 1969, p. 428) has made the startling assertion that there was a negative correlation between the two, because the social conditions that promote science are antithetic to technology and vice versa.

As the work of Musson and Robinson (1969) has demonstrated, British inventors and manufacturers during the Industrial Revolution were in constant contact with scientists. Engineers and mechanics such as Smeaton, Watt, Trevithick, and Stephenson learned from scientists a rational faith in the orderliness of natural phenomena and physical processes; an appreciation of the importance of accuracy in measurement and controls in experiment; the logical difference between cause and correlation; and a respect for quantification and mathematics. In many cases the new knowledge developed by scientists did filter down, and in subtle ways affected the questions asked and the solutions sought. For example, steampower would have been unimaginable without the insights of Torricelli and Guericke into atmospheric pressure; the engineering of Smeaton and Maudslay would hardly have been possible without Galileo's theory of mechanics. It was certainly the case that before 1800 there was not much in the physical sciences that would be of direct use to these engineers, but scientific method and individual insights proved both an inspiration and a guide leading to technological breakthroughs.[14]

14. Theory had been a somewhat unreliable source of inspiration in the First Industrial Revolution. In power technology, for example, the French mathematician Antoine Parent calculated that the maximum useful effect of a waterwheel was only 4/27th of the natural force of the stream, and that the optimal speed of the waterwheel was 1/3 that of the stream. These calculations were widely accepted, although they

In some cases, clever experiments revealed truths despite incomplete or even erroneous scientific principles. Berthollet, the inventor of chlorine bleaching, thought that chlorine was a compound, not an element. Hot-air balloons were believed by their inventors to be lifted by a lighter-than-air substance given off by the heat. But would the hot-air balloon have been invented without the newly emerging physics of Cavendish, or Giovanni Alfonso Borelli's demonstration in 1680 that it was physically impossible for humans to fly like birds? Metallurgy presents another example of how mistaken scientific principles could lead to correct insights. The nature of steel had eluded the best minds of European science for centuries. Phlogiston theory, the reigning paradigm of eighteenth-century physics, greatly stimulated interest in the composition of useful materials. Its proposition that steel was a pure form of iron with phlogiston added to it did not stop the Frenchman René Réaumur from publishing, in 1722, a famous book in which he postulated that steel was an intermediate product between cast iron and wrought iron. The Swede Tobern Bergman, another convinced phlogistonist, discovered in 1781 that the difference between wrought iron, steel, and cast iron was in the quantities of "plumbago" contained in them. It was soon discovered that plumbago was merely charcoal, that is, pure carbon (Cyril S. Smith, 1981, p. 35–44). By the 1820s the role of carbon in iron and steel was well understood, and without it the advances in steelmaking are hard to imagine.

It is widely agreed that after 1850, the role of science became more important in generating technological progress. But the imaginative, original, energetic, bold, but basically serendipitous, untrained, and unsystematic mind of the eighteenth-century inventor did not disappear after 1850.[15] In fact, this kind of inventor remains important in the twentieth century. As a general rule, it seems likely that in the past 150 years the majority of important inventions, from steel converters to cancer chemotherapy, from food canning to aspartame,

were incorrect and did not square with empirical observations. In wind power, experiment and theory were applied in the 1780s to determine the optimal size and shape of the sails. It turned out that the concave shape and warped surface of Dutch windmills, in use for centuries, was in fact optimal.

15. The classic defense of the individual inventor is in Jewkes, Sawers, and Stillerman (1969), who insist that "in many fields of knowledge, discovery is still a matter of scouting about on the surface of things where imagination and acute observation, supported only by simple technical aids, are likely to bring rich rewards" (ibid., p. 169). It is also increasingly recognized that the innovation and diffusion stages of technological progress are rarely left to individuals. Although Langrish et al.'s (1972, p. 14) statement that " 'the individual innovator' is almost a contradiction in terms" is certainly exaggerated, it does reflect an important fact about technological progress in our time.

have been used long before people understood *why* they worked, and systematic research in these areas was thus limited to ordered trial-and-error operations. The proportion of such inventions is declining, but it remains high today.

The function of science after 1850 was as much to show what could *not* work as to show what *could.* In 1853 the versatile inventor John Ericsson produced a "caloric" engine based on regeneration of the heat produced. The machine was a failure because its inventor failed to realize the basic thermodynamic principle that an engine cannot use the same heat over and over again because the heat is converted into energy (Bryant, 1973). Even the best tinkerer inventors felt increasingly the need to collaborate with persons with a systematic training. Thomas Edison, for example, hired the mathematician Francis R. Upton and the chemist Reginald Fessenden (who later made important contributions to the understanding of the modulation of radio waves) to translate his ideas into rigorous form. Inspired outsiders still played a crucial role in the invention process, and their openness of mind was often essential in the initial breakthrough. In the "development" stage of the process, however, scientific training and systematic work proved increasingly necessary. Inventors who wished to remain active had to become experts themselves or give up (Hounshell, 1975).

RELIGION

Of particular interest in explaining technological change over the long term is the matter of religion. Of course, religion, especially in the past, was socially determined. Indeed, there have been relatively few episodes in history in which individuals could exercise much choice over their religious beliefs. Once set, these beliefs were a key variable on the microlevel in setting individuals' frames of mind regarding technological change. The literature on the relationship between religion and economic change (or "capitalism") is large and confusing. Here I wish to focus exclusively on its technological aspects.[16] Two important warnings should be stated. First, few, if any, religions are totally opposed to all innovation. The differences were differences of degree, but they were cumulative and in the long term could explain large divergences. Second, religions are themselves to some extent social-choice variables, at least at some critical junctures. Religions might have been chosen and adapted to reflect changing

16. Much of what follows is influenced by the pioneering writings of Lynn White, especially (1968) and (1978), and to the survey of the problem of technological change in Landes (1969, ch. I).

preferences and circumstances. Every society, so to speak, gets the religion it deserves. Yet the historical record suggests that religion was rarely entirely endogenous, and the predominant religion was often determined by military and political events. What people really believed, moreover, was not identical with the dogma of the official churches, and neither economists nor other social scientists have had much success in explaining it.

Precisely because the act of invention is a game against nature, what matters above all for invention is whether beliefs enhance the propensity to change production methods, that is, the willingness to challenge and manipulate the physical environment, what Landes has called the "Faustian ethic," the sense of the mastery over nature and things. Not all technological change is strictly of that type; coinage, new methods in accounting, or improved monitoring of workers would fall outside that definition. But agriculture, manufacturing, mining, hunting, and transportation require an intervention through which the innovator alters nature to suit his or her ends. The willingness to adjust ecology in this way depends on the attitude of the innovator toward the physical world, i.e., religion. For modern Western man and woman, whose view of nature (if conscious at all) is that it is there to be manipulated and enjoyed, it seems perhaps commonplace that the binding constraint on technology is knowledge (i.e., the ability to use nature). But this frame of mind, as we shall see, is exceptional in human history, and is mainly derived from the anthropocentric philosophies of Judeo-Christian religions. Mental changes are therefore a natural candidate for the explanation of the technological take-off in medieval Europe, a subject to which we shall return.

Economists have traditionally been leery of *mentalités* as a factor in long-term economic development. In the budding literature on the economic rise of the West, such factors have been ignored or curtly dismissed.[17] In part, economists' hostility to religious factors stems from the incompleteness of such theories. Attitudes were a matter of degree, not of absolutes. Non-Western societies have in some cases altered their environment drastically, and inflicted ecological disasters upon themselves (Jones, 1988, pp. 59–61). All economic production involves some manipulation of the physical environment, and all societies have had to play games against nature. Religion affected material attitudes in odd ways: some Judeo-Christian religions, first

17. North and Thomas (1973) and North (1981) ignore attitudes altogether, as does Hicks (1969). Jones (1981) dismisses them, while Rosenberg and Birdzell (1986) devote one footnote to White's thesis regarding the relationship between religion and technology.

and foremost Judaism itself, have shown little affinity for technological change, even if they shared the view that God created man as the center of the universe. Greek Orthodoxy and Islam, and at times Western Christianity as well, have shown tendencies toward mysticism and reactionary dogma that thwarted innovation. Moreover, non-Christians, especially the Chinese, have made major contributions to technology. Qualifications and doubts thus abound. Yet even if religion was far from being the sole or even the primary factor, it remains one of the most intriguing forces in trying to explain the European miracle. Technological change as defined here—a sustained, cost-reducing, output-augmenting change in knowledge—took off in Europe after Christianity had taken firm root and has remained for over twelve centuries an economic force of unparalleled and apparently inexhaustible potency.

Religion, economy, and society were often intertwined in ways that defy easy generalization, but that had an evident effect on technological creativity. Indian Brahmanism was created by Aryan conquerors, who developed the caste system to perpetuate their dominance and ensure acquiescence on the part of the lower classes. The taboos, restrictions, and rules imposed by the cast system created a society that was conservative beyond anything seen in the West. In the eternal trade-off between progress and growth on the one hand, and stability and order on the other, Hindu civilization chose a position biased in the extreme toward the latter. Jones (1988, pp. 103–4) cites the caste system as "the limiting case of rigidified institutions" and notes that "personal achievement is excluded in principle." It is, as always, hard to sort out the causality links exactly: was India a conservative society that bred a suitable religion, or was Hinduism responsible for India's backwardness? Hindu doctrine held that promotion to a higher caste was possible through reincarnation if an appropriately resigned and obedient life was led, a fiendishly clever and almost failure-proof incentive system to protect the status quo. The result was that despite their obvious skills in metallurgy, high-quality textiles, and hydraulic engineering, the Indian subcontinent does not figure prominently in the history of technological creativity.

Of course, India's backwardness cannot be blamed entirely on religion. India was repeatedly devastated by civil wars and powerful invaders. Yet most creative societies have bounced back within decades after even the most devastating blows. India simply fell behind, and eventually it fell prey to more advanced enemies, first the Mughals, whose use of artillery in the 1520s defeated much larger Hindu armies, and then the British. But it is not the political consequences that ultimately matter; India, where daily life was conducted for millennia at the very margins of subsistence, is a reductio ad absurdum of the notion of an equilibrium, so beloved by economists. Yet this equilibrium was regarded as an arrangement established by the Gods, a perfect world in which everything and everyone had its place, a society in which poverty was holy and action was vanity. In such a world, technological creativity had little chance.

Technology was not the only means by which people sought to affect their natural environment. Magic, astrology, and alchemy—for the most part regarded today as irrational activities—were practiced by people whose scientific credentials leave no doubt as to their "rationality." Isaac Newton was one of the most famous alchemists of all times. The Neapolitan Giambattista della Porta (1536–1605), one of the first to realize the possibilities of steam power, was a full-time magician and even named his book *Magiae Naturalis*. The spillover effects of astrology into astronomy are too obvious to require repetition. It will not do to dismiss magic as irrational, because rationality is conditional on the information available, and without modern science it was impossible for people to know what worked and what did not. The important difference between technology and magic is not that technology works and magic does not. The difference that matters here is that magic does not control nature, it begs favors from it. Rather than exploiting regularities and natural laws, it seeks exceptions to them by taking advantage of an imaginary capriciousness of the universe. Moreover, technology, if it worked, worked for everyone, whereas magic was confined to qualified practitioners. The sorcerer's apprentice had no access to his master's powers. To be sure, magic and alchemy had, at times, positive spillovers to technology, but their importance vanishes with the Enlightenment of the eighteenth century.

VALUES

As far as collective influences on behavior are concerned, perhaps the most pervasive influence on the propensity of a society to experience technological progress is the hierarchy of values. Economists are accustomed to thinking of human behavior in terms of utility, which is the valuation attached to goods and services consumed, and the conditions in which people live. Beyond this, however, there is a collective set of valuations that determines the relative prestige of various activities or attributes within a society. Although prestige is usually correlated with wealth, it is not identical to wealth. Through the ages, the various activities that conveyed prestige have included, beside the creation of wealth (economic activity), the military, the arts, worship, sports, administration, learning, and teaching. Different societies have ranked these various activities differently. It seems plausible that the higher labor, production, and the accumulation of wealth are in this ranking, the more susceptible a society is to technological progress. Among the great civilizations of antiquity, the Greeks had an appreciation for sports and learning, whereas the Romans placed the highest value on military and administative ability, and the Jews emphasized worship and later learning and textual exegesis of the scriptures. All of these societies appreciated wealth, but it was as good as or better to be brave or wise than to be rich.

Moreover, wealth could be held in different forms. In some societies, wealth was primarily measured by direct power over people rather than economic power over resources. In some slave and feudal societies, for example, wealth was measured primarily by control over other people.[18] Power over people and control over economic resources overlap, of course, a great deal in any society. But they are not identical. A slave society, in which wealth and social prestige were measured by the number of slaves, would be less inclined to adopt labor-saving machinery, which, although more efficient, would make slaves redundant. On the other hand, in a slave society in which wealth was counted in terms of all resources, and in which slaves were held solely because they represented a good way to accumulate wealth, technological change was as likely to occur as anywhere else. To use Hirsch's (1976) terminology, technological progress only increases material goods, while it leaves positional goods (such as social prestige and political power, which are measured strictly relative to others' consumption) unaltered by definition. The more wealth is measured by the consumption of positional goods relative to material goods, the less attractive technological change will look and the lower the prestige that will be associated with economic production.

A relatively low ranking of economic activity in this hierarchy is inimical to technological progress for a number of reasons. First, it channels the creative energies of the best educated and most successful persons into activities that do not increase the productive capacity of the economy. Whether they become priests, philosophers, or generals, they are not likely to be interested in the mundane problems of agriculture, ironworking, or tanning. What we call "production" was stigmatized by Graeco-Roman society as a lower-class activity, possibly associated with slaves, necessary but dirty. Insofar as invention consists of the twining together of hitherto unconnected strains of knowledge, some pre-existent set of economic objectives is essential. In other words, for an inventor to realize that an idea might be useful to a particular branch of economic production, he or she

18. A good example is the case of the Nunu people in Zaire, for whom the word for poor person is the same as the word for a person alone. See Harms (1987, p. 65).

has to have some prior conception of how production is carried out in the first place.

Second, because productive work was left to an uneducated and inarticulate class alienated from the elite, technological progress faced serious handicaps. In a society in which those who are educated do not work and those who work are not educated, the inarticulateness of the productive classes will thwart the diffusion and adoption of new technology in the unlikely event that it emerges. During most of history, children who received an education were kept away from practical matters. Whether they studied horsemanship, Latin, theology, geometry, or the Talmud, matters commercial and industrial were rarely part of the curriculum. An invention by a slave or a lowly farmer might occasionally benefit his direct master or landlord, or even himself. How likely was it, however, to spread outside the immediate neighborhood unless the educated elite took some interest in the details? The exclusion of large segments of society from the group whose ideas enjoyed some measure of acceptability was, however, not confined to slave societies. For example, the number of known female inventors is very small, although women were deeply involved in most spheres of production. Some cases of women displaying a knack for practical invention are known, but their social environment, which considered mechanical skills a male monopoly, did not permit their talents to develop, and female inventors remained in the shadow of their male partners.[19]

Moreover, in some cases a new insight evolved from a game or experiment carried out for its own sake. To convert such an insight into technological progress requires a pragmatic bend of mind. Whereas the act of insight occurs at the level of the individual (often subconsiously), the likelihood that the individual will ask how something can be made useful depends on the value that his or her environment attaches to things that are useful as opposed to, say, beautiful or virtuous. Technological progress depends on the extent to which homo creativus was also homo economicus. Relative to other societies, Europe approached the new knowledge it generated with a more pragmatic attitude than other societies. Although this pragmatism was made fully explicit only by the seventeenth century, es-

19. One example is Julia Hall, elder sister of Charles Hall, the inventor of the electrolytic process in aluminum production and founder of Alcoa. Julia Hall played an important role in the discovery process, but her role has been downplayed by contemporaries and historians alike (Trescott, 1979). Griffiths (1985) has argued that the male domination of technological progress dates from the Industrial Revolution, but clearly the phenomenon predates both modern industry and capitalism. It is plausible, however, that the sexual lopsidedness of invention became more pronounced in the nineteenth century.

pecially in the writings of Francis Bacon, it doubtless existed in implicit form long before. Marx's famous dictum that his purpose was not just to understand the world but to change it applies to thousands of tinkerers, mechanics, and engineers who built the windmills, clocks and fully rigged ships of medieval Europe. This pragmatism did not manifest itself in the economic realm alone. As we have seen, the Europeans approached the geographical expansion of the world with much the same attitude. In commerce, war, and politics, what was functional was often preferred to what was aesthetic or moral, and when it was not, natural selection saw to it that the useful won out in the long run. It need hardly be stressed that such pragmatism was never entirely absent in any society. Hierarchies of values are not the stuff of absolutes. But the difference in degree between European and most non-European societies was decisive.

The contempt in which physical labor, commerce, and other economic activity were held did not disappear rapidly; much of European social history can be interpreted as a struggle between wealth and other values for a higher step in the hierarchy. The French concepts of *bourgeois gentilhomme* and *nouveau riche* still convey some contempt for people who joined the upper classes through economic success. Even in the nineteenth century, the accumulation of wealth was viewed as an admission ticket to social respectability to be abandoned as soon as a secure membership in the upper classes had been achieved.

INSTITUTIONS AND PROPERTY RIGHTS

The institutional background of technological progress seems, on the surface, more straightforward. Rosenberg and Birdzell (1986) place institutional change at the center of events. They point out that for technological change to be effective and sustainable, the authorities must relinquish their direct control over the innovative process and decentralize it. This creates an important, though insufficient, condition for technological change to occur: the opportunity for the successful innovator to enrich himself. But, as they point out, decentralization was equally important because it meant that search and experimentation were carried out by many independent units, possibly over and over again. This duplication of effort was not the most cost effective way of engaging in technological progress. In fact, it involved much waste. Case after case of unnecessary duplication can be documented, not even counting the huge number of research efforts that were fruitless. But this system minimized the probability of a technological opportunity being missed, as "it reduced the risk that a desirable proposal would be rejected because of a viewpoint peculiar to a single decision maker" (ibid., p. 29). Nelson (1987, p. 9) adds that a pluralist system generates a wider variety of new departures and lets ex post selection separate the wheat from the chaff. Technological progress involves not just uncertainty, but ex ante differences in opinion. There may be more than one way to skin a cat, but only one way is the most efficient. Nelson argues that ex post selection and experimentation is costly and painful, but "given the nature of technological uncertainties, perhaps this is the best we can do" (ibid., p. 120).

The definition of property rights on new technology constitutes another condition for innovators to cash in on technological progress. There are two separate issues here. One is the general question of property rights. It seems widely agreed that if the ownership of assets is insecure, so that they can readily be confiscated or stolen by brigands or by the authorities, economic progress of any kind is not likely to occur. People will hold their assets in nonproductive, liquid forms, and the accumulation of productive capital will be reduced. Landes (1969, ch. I), Jones (1981), and North (1981) have pointed to differences in security as a key factor in European development. The other issue is the need to secure sufficient incentives to make innovators invest in new technology. Patents, monopolies, grants, pensions, prizes, and medals have provided would-be innovators with the rewards deemed necessary to keep up a high level of inventive activity. Patent systems, however, did not emerge until the fifteenth century, and were firmly entrenched only by the late eighteenth century. Even then, the patent system turned out to be a double-edged sword for inventors, as we shall see.

A patent system may have been a stimulus to invention. but it was clearly not a necessary factor. After all, many inventors were able to capture enough of the increased consumer surplus to make the effort worthwhile even if their inventions were soon imitated. Second, imitation may have been difficult for a variety of reasons, thus creating a "natural patent" that yielded a stream of rents guaranteeing the innovator some return, though it may have represented only a fraction of the social surplus created by the innovation. In a few cases it is quite clear that financial incentives were not a major consideration for the inventor, whose utility function may have included fame, the sheer satisfaction of solving a difficult problem, or even altruism.[20] Such cases were unusual, however; the profit motive was

20. A classic case in point is the invention of the mining safety lamp by Humphrey Davy in 1815. Davy, who was something of an eccentric, refused to patent this invention, claiming he made the discovery purely pro bono publico. Yet he jealously defended his primacy in developing the lamp, and was made a baronet for it. Another

probably as strongly present in invention as in any other economic activity, which is to say, it was almost always there, though it rarely operated entirely by itself. It is therefore plausible that societies that rewarded invention were likely to have more of it, but that there may have been more than one way of rewarding innovators whose activities led to major gains in productivity.

RESISTANCE TO INNOVATION

A different set of social factors affecting technological change may best be referred to as the political economy of technological change. Although technological progress is by definition a net improvement to the economy, it is almost always the case that there are some groups whose welfare is reduced because of it, or who at least believe so ex ante. Technological change shocks the labor market, alters the physical environment, makes existing human and physical capital obsolete, and unambiguously reduces the producer's surplus of the innovator's competitors. In a repeated game, the gainers might have tried to compensate the losers. By its very nature, however, technological change is a nonrepeated game, since an invention is only invented once. Once an invention is made, an inventor often needs protection from those who stand to benefit from the suppression of the invention. The dilemma is sharpened by the fact that the benefits are usually heavily diffused, while the costs are concentrated. Thus the losers will find it easier to organize, and are quite likely to try to squelch technological progress altogether. Resistance to technological change occurred in many periods and places but seems to have been largely neglected by most historians, though Morison (1966, p. 10) views it as "the single greatest matter of importance and interest in this whole process [of invention]."

Some examples: as early as 1397, tailors in Cologne were forbidden to use machines that pressed pinheads. In 1561, the city council of Nuremberg, undoubtedly influenced by the guild of red-metal turners, launched an attack on a local coppersmith by the name of Hans Spaichl who had invented an improved slide rest lathe. The council first rewarded Spaichl for his invention, then began to harass him and made him promise not to sell his lathe outside his own

example is the invention of chlorine bleaching by Claude Berthollet, who had no interest whatsoever in the commercial exploitation of his idea and voluntarily shared useful technical information with businessmen like Matthew Boulton (Musson and Robinson, 1969, p. 266ff). Gail Borden reputedly became interested in preserving food when, upon his return from the Great Exhibition in England in 1851, he witnessed great suffering among the children on the ship when the cows on board became so sick that they could not be milked.

craft, then offered to buy it from him if he suppressed it, and finally threatened to imprison anyone who sold the lathe (Klemm, 1964, p. 153). The ribbon loom was invented in Danzig in 1579, but its inventor was reportedly secretly drowned by orders of the city council. Twenty-five years later the ribbon loom was reinvented in the Netherlands—though resistance there, too, was stiff—and thus became known as the Dutch loom. A century and a half later, John Kay, the inventor of the flying shuttle, was harassed by weavers. He eventually settled in France, where he refused to show his shuttle to weavers out of fear.[21] But the prolonged opposition of vested interests against the flying shuttle in Britain was ineffectual. Resistance to new technology was traditionally strongest in the textile trade, but appeared in less expected places as well. In 1299, an edict was issued in Florence forbidding bankers to use Arabic numerals (Stern, 1937, p. 48). In the fifteenth century, the scribes guild of Paris succeeded in delaying the introduction of printing into Paris by 20 years. In the sixteenth century, the great printers revolt in France was triggered by labor-saving innovations in the presses.

These negative reactions to technological progress can only be understood as an attempt by those with an investment in certain techniques to prevent the decline in the value of their skills. Economies in which the institutional setup protects the inventor from such threats, or in which distributional coalitions, which protect the selfish interests of small groups at the expense of the large majority, are comparatively weak will have a far better chance for technological success. Inventors or manufacturers who perceive that innovating is a thankless and possibly dangerous activity will lose interest, and technological change will peter out.

POLITICS AND THE STATE

It is difficult to determine what kind of political structure is most conducive to technological progress. A strong, centralized government, secure enough to withstand riots and political pressure from coalitions representing losers, may have been able to resist the pressures exerted by the technological status quo. Yet it is equally plausible that a weak government that succumbed to demands to legislate technological progress away would be powerless to enforce such laws, and thus by default leave the decision to market forces. This dilemma was recognized by North (1984, p. 260) when he wrote that

21. The commonly repeated tale that Kay had to leave Britain because disgruntled weavers threatened his life seems to be apocryphal (Wadsworth and Mann, 1931, p. 456).

"if you want to realize the potential of modern technology you cannot do with the state, but you cannot do without it either." At first glance, political stability may seem necessary for technological progress. But Olson (1982) has argued that sweeping political changes prepare the ground for economic progress by eliminating reactionary institutions used by pressure groups with a vested interest in the status quo, whereas long-term political stability permits the crystallization of such groups. Monopolistic market structures among the innovating firms (which would be stimulated by a well-functioning patent system) would thus be advantageous in this regard. The more monopolistic the industry, the larger will be the share of social benefits captured by producers at the expense of consumers, meaning that lobbying would more likely be met by successful counterlobbying. In a competitive industry, on the other hand, the gains from innovation are passed on to the consumers in the form of lower prices, and the benefits are thus diffuse.

Even apart from protecting new technology from its victims, the state plays a central role in technological change. The ambiguity of this role is inevitable, because "the state" always consisted of layers of authority that were rarely coordinated and often in conflict with each other. A central government might have had different interests and attitudes from local authorities. Nongovernmental organizations such as guilds, trade unions, and chambers of commerce, often exercised considerable influence over the shape of the technology used. The impact of deliberate policy may not always have been the same as intended. For example, some governments, eager to stimulate technological progress, may have actually hurt it by imposing tariffs to protect local manufacturers. In the majority of cases, however, diffuse power seems on the whole to have been advantageous to technological progress.

It seems that as a general rule, then, the weaker the government, the better it is for innovation. With some notable exceptions, autocratic rulers have tended to be hostile or indifferent to technological change. The instinctive need for stability and the suspicion of nonconformism and shocks usually dominated the possible gains that could be attained from technological progress.[22] Thus both the Ming dynasty in China (1368–1644) and the Tokugawa regime in Japan

22. Finley (1973, pp. 75, 147) recounts two tales of Roman emperors who were offered important inventions. One inventor, possibly apocryphal, invented unbreakable glass and offered it to the evil Emperor Tiberius, who after ascertaining that no one else knew of the secret, had the man executed. More fortunate was the inventor who came to the wise Emperor Vespasian with a new device for transporting heavy columns. The Emperor rewarded the inventor, but then refused to use the device, asking, "How can I feed the population?"

(1600–1867) set the tone for inward-looking, conservative societies. Only when strong governments realized that technological backwardness itself constituted a threat to the regime, as in the cases of Peter the Great's Russia, post-1867 Japan, and, to a lesser extent, Napoleon's France, did they decide to intervene directly to encourage technological change. When rulers are weak, they are typically unable to halt technological progress, much as they may try. As we shall see, the weakening of central power in Europe following the collapse of the Roman Empire may help to explain the resumption of technological progress after 500 A.D.

Another reason politics matters is that technological change is notoriously subject to market failure, that is, the free market system left on its own is unlikely to produce a desirable level of innovation. This is the result of the public good property of new technological knowledge: once created, using it does not reduce the supply available to others, and it therefore has a zero marginal social cost. As far as new technology is concerned, therefore, society has found it necessary to supplement the usual market mechanism by additional institutions. In the past, governments have tried to correct the many failures in the market for technological knowledge by organizing exhibitions, awarding monopolies and other rewards for specific inventions, and subsidizing potential inventors by offering them pensions and easy positions in government service. In more recent times governments have taken it upon themselves to create the educational and research environment that they believe is needed to bring about technological change. Governments enforce the rules by which the game of innovation is played, and they often set them as well.

Politics also matters because the ruling elites, be they emperors, high priests, parliaments, or councils of regents, set an agenda of priorities. If these priorities divert too many resources, and in particular talent and creativity, into nonproductive or destructive uses, economic performance and innovation may be harmed. Thus, a string of devastating wars between 1550 and 1650 ruined the technological infrastructure of the most advanced parts of Europe, notably southern Germany, the southern Netherlands, and Bohemia. Moreover, political factors determined whether initiative and ingenuity would be directed toward productive purposes or not. There was nothing automatic about this. Economists have recently emphasized the dangers of the "rent-seeking" society, in which the tax lawyer and the political lobbyist have replaced the inventor and the engineer as the entrepreneur's main instruments towards higher profits. Political maneuvering is a zero-sum game at best, whereas technological change is a positive-sum game. It is the political sphere that determines which game attracts the best players (Baumol, 1988). One explanation of the decline of Rome has recently been recast in those terms (MacMullen, 1988). Corruption and venality led to a decline in the efficacy of the public goods provided by the government (primarily defense), and diverted energies and talent away from more productive purposes.

Finally, governments set the tone for society's attitudes toward nonconformists. Inventors are by and large unconventional people, who rebel in some way against the status quo. Morison (1966, p. 9) has gone so far as to suggest that invention is something of a hostile act, a dislocation of existing schemes, a way of disturbing comfortable bourgeois routines. Technological progress requires above all tolerance toward the unfamiliar and the eccentric (Goldstone, 1987). As Cipolla (1972, p. 52) put it, the qualities that make people tolerant also make them receptive to new ideas. The factors determining the degree of tolerance toward deviants in any society are not much understood. Persson (1988, p. 57) suggests that the size of the economy is the critical variable here. In small economies the potentially harmful acts of individuals are relatively more important. By that logic, Russia and Spain should have been much more tolerant than the United Provinces and Denmark, and China the most tolerant of all. This obvious historical contradiction illustrates the basic problem with his argument, which is that "size" is not well defined. Was an Amsterdam merchant in 1560 a part of the Amsterdam society, the county of Holland, or the Spanish Empire? Society, like politics, consisted of partially overlapping entities, and the mere fact of being, say, French or Polish may not be enough to determine whether one lived in a tolerant society or not. It was also important whether one lived in a town or the countryside, whether one lived in the diocese of a zealous bishop, and which social class one was born into. This makes Persson's notion of size as a determinant of tolerance impractical for our purposes.

The archenemy of tolerance and pluralism is conformism. The proclivity of individuals to conform to social norms and to force others to do so is deeply rooted in individualistic behavior. Recent work by economists has tried to reconcile it with rational (that is, utility-maximizing) behavior (Stephen Jones, 1984). Conformism can be explained in two ways. One is that imitating past practices provides an efficient way of learning. The other is that conforming to existing norms is important in being accepted by the existing society. All societies developed to some extent a disapproval of young members who do not conform to existing practices. Some individuals will nevertheless question accepted tradition, and at that stage the issue of society's attitude toward rebels and deviants becomes crucial. The more hostile this attitude, the more likely conformist attitudes will dominate and the new generations will be just like the old ones, producing technological stasis. Widely observed phenomena such as tradition and social inertia become understandable if conformism is assumed to be part of human behavior. Natural selection is likely to have favored conformism, since species in which the young did not conform to the established ways of gathering food would not survive. Yet such conformism is precisely what has to be overcome if technological progress is to occur. It is easy to use Stephen Jones's model to demonstrate this result, because in his model (formulated in the context of a workplace, but readily applicable to technological change) the utility of an individual depends on the ability to please and be accepted by other workers, a function of the severity of the disapproval of novelty. What is not quite clear is which factors determine the level of disapproval. Clearly, however, disapproval of nonconformists has historically been high. One student of the history of technology (Cyril S. Smith, cited by K. D. White, 1984, p. 27) sighs that "every invention is born into an uncongenial society, has few friends and many enemies, and only the hardiest and luckiest survive." Whenever religious and intellectual intolerance spread through Europe, as they did in the fourteenth century, their advent coincided with a temporary slowdown in technological development.

WAR

Of special interest to the connection between the state and technology is the question of spillover effects from military to civilian technology. War has always been permeated with technology, and the concentration of effort that the high stakes involved would suggest a positive correlation between military effort and technological success.[23] It has repeatedly been claimed that military needs stimulate and inspire technological change. Whether in fact innovations in military technology provided substantial benefits in the production of peacetime goods and services so that war can be thought of as an agent of technological progress is far from easy to determine. Some spillover effects are, of course, undeniable. Bessemer's converter grew out of the search for a more effective shell. Wilkinson's boring machine, without which Watt and Boulton would not have been able to manufacture their machines, was designed for cannon.[24] The early

23. The original statement of this view can be found in Sombart's famous *Krieg und Kapitalismus.* For more recent restatements in this tradition, see McNeill (1982) and Guilmartin (1988).

24. The inventors of the boring machine were a Swiss father-and-son team of gunfounders, both named Jean Maritz, working for the French government. The son perfected the machine in the 1750s (McNeill, 1982, p. 167).

blast furnaces of the fifteenth and sixteenth centuries were set up in part for the manufacturing of cannon. The invention of firearms led to significant spillover effects in metalworking, such as the development of rotary metal cutters, gauges, metallic screws, and so on (Foley, 1983). Interchangeable parts and mass production in the nineteenth century also had well-documented military origins and were originally known as "armory practice." Yet the military was hardly a milieu that tolerated technological creativity. In his recent book on military technology, Van Crefeld (1989, p. 220) points out that on the one hand the military hierarchy and frame of mind left little room for flexibility and tolerance for innovators, so that "the military represent an exceptionally unfavorable environment for . . . invention." On the other hand, the military is well placed to research, develop, and bring to fruition some ideas about which the civilian sector is hesitant. Yet the examples he provides come from the twentieth century. Before that, the needs of military technology seem to have intersected but little with those of civilian production. Innovations in ballistics, the technology of fortification, gunpowder, and military communications had little to contribute to economic welfare. Even in shipbuilding, the increasing specialization of merchant ships and men-of-war after 1500 meant that spillovers were small.

In any event, any suggestion that there is a positive correlation between the bellicosity of a society and its inventiveness must encounter grave skepticism. The wonderful war machines of the Romans, the catapults and the ballistas, did not lead to any clear advances in productive technology. Medieval metalworking, which was largely aimed at weapons and armor, did not change noticeably: welding, forging, and shaping did not benefit much from any learning-by-doing effects. The metallurgy learned during casting iron cannon could have been picked up from casting frying pans or oven grates. In spite of its technologically revolutionary nature at the time it was introduced, gunmaking did not lead to many civilian applications with the exception of hunting, and the machine-tool industry owed more to clocks and instruments than to guns. Civilian uses of explosives were limited; only after 1627 was gunpowder used to blast ore mines in Slovakia, and gunpowder helped build the first tunnel during the construction of the Languedoc Canal, completed in 1681 (Hollister-Short, 1985). However, not until dynamite was invented was an explosive available that could be employed on a large scale for peaceful purposes. In China, too, gunpowder was rarely used in mines (Golas, 1982). And as we have seen, Huygens's hopes to find a peaceful use for gunpowder in an internal combustion engine remained unfulfilled.

Indeed, it is surprising how little evidence there is that military activity created positive externalities for civilian production given the apparent opportunities. The huge armies of Louis XIV, with their demand for uniforms, blankets, arms, munitions, and supplies, should have been enough to trigger an Industrial Revolution in France. Yet Colbert's factories led to a dead end. Of the most bellicose nations of the century before the Industrial Revolution—Charles XII's Sweden, Louis XIV's France, Frederick the Great's Prussia—none saw many technological benefits of the expensive wars conducted by their sovereigns. The Industrial Revolution itself was hindered rather than aided by the wars of the time. Brunel's Portsmouth blockmaking machines were too specialized to be of use for much else than naval vessels, and most of the other positive technological externalities of the Napoleonic Wars have been disputed or shown to have been unimportant (Mokyr, 1985, p. 15). Even in the technology of the organization and disciplining of men, in which military commanders anticipated the needs of modern industry by centuries, the beneficial effects of military experience are hard to find. Prince Maurice of Nassau and Frederick the Great of Prussia perfected the military drill, but it was not until after 1800 that a disciplined factory labor force was created—without much insight gleaned from military experience.[25] During the wars of the Revolution and the Napoleonic era, the French iron industry had to satisfy a vast demand for iron, yet recent scholarship has confirmed that the adoption of new metallurgical techniques in France was slowed down in this period (Woronoff, 1984). Similarly, the American Civil War produced few noticeable spillovers to civilian technology. Mumford's description of war as an "agent of mechanization" notwithstanding, its technological benefits before 1914 were modest. Moreover, the net effect of war on technological change has to take into account the costs as well; there can be little doubt that the balance here is negative. I see no reason to revise the conclusions reached by Nef (1950, especially pp. 220–21) in his classic work, which argued that it was peace, not war, that was the innovating force in manufacturing, and that war and military preparation did not add conspicuously to the material prosperity of Europe.

The idea that war could have technological spin-offs and thus be in some sense beneficial to the economy is a curiously Eurocentric notion. For the rest of the world, war was an unmitigated scourge.

25. Two important French inventions made shortly before and during the French Revolution—hot-air ballooning and the semaphore telegraph—had important military potential. Yet Napoleon had no interest in ballooning, and Claude Chappe, the inventor of the semaphore, received no assistance from the military and was eventually rescued from bankruptcy by the French national lottery, which used his system to communicate winning numbers to the provinces.

Most of the Eurasian continent was repeatedly invaded and devastated by Mongols or related Asian tribes between 1200 and 1800 (Jones, 1988, pp. 108–15). The military methods of these tribes were so destructive of human and physical capital that they have been blamed for the permanent backwardness of the Middle East and Eastern Asia. India, Persia, and Mesopotamia were the victims of the hordes of Tamerlane. It took some regions decades or even centuries to recover from the most devastating wars. Only those parts of Eurasia that were spared the conquests of Mongols—Japan and western Europe—were able to generate sustained technological progress. As Jones points out, the negative impact of these wars could easily be overstated, and the damages were insufficient to explain entirely the divergence of Europe and Asia: some regions that were only lightly touched by the Mongols failed to generate much new technology either. Recovery might in some cases have taken decades, but if it took centuries we may suspect that other factors were at work as well.

Although military needs served at times as "focusing devices," to borrow Rosenberg's term, weapons were more often borrowers of civilian technology than sources of inspiration for it.[26] In general, military and civilian technology tended to be nurtured from the same sources of technological creativity. Societies that were creative in making clocks, plows, and spectacles were also good at making tributchets, guns, and men-of-war. The historical correlation between advances in the production of guns and that of butter does not prove that more or better guns "caused" more butter or the other way around; it could just as well be consistent with an overall increase in capacity that allowed society to have more of both. After all, that is precisely what technological progress is all about.

OPENNESS TO NEW INFORMATION

Beyond the nature of the state lies a more complex factor that, for lack of a better term, we may call openness to new information. When two previously unconnected civilizations establish contact, technical information is exchanged that may yield potential economic gains to both. Earlier I referred to these effects as "exposure effects." Not all societies were capable of taking full advantage of exposure effects. Human history is full of examples of societies holding others in utter contempt and despising people who looked different, spoke a differ-

26. The molding and casting of bronze cannon, for a long time the superior way of making artillery, was derived directly from techniques learned in bell founding in the early Middle Ages (Tylecote, 1976, p. 73).

ent language, or believed in a different God. The ancient Greeks and the medieval Chinese shared a loathing for "barbarians," from whom, they thought, they had nothing to learn. Medieval Europe probably felt just as hostile to Islam, but the hatred of the Saracen did not extent to the knowledge and the useful devices that Moslems could produce. Thus, Europe adopted many inventions from Islam, from the lateen sail to Arabic numerals.[27] In the seventeenth and eighteenth centuries, when the West had pulled ahead of Islam technologically, the Moslems did not return the favor. In spite of their geographical proximity, the Moslems allowed European innovations to filter in only slowly and selectively. The first book printed in Arabic script appeared in Istanbul in 1729, almost three centuries after the invention of moveable type. The Koran was not printed until the twentieth century, quite a contrast with the great Gutenberg Bibles. Military defeats alerted the Ottoman Empire to the need to keep up technologically with the West, but because the rest of the economy was so backward, military instructors had to be brought in from the West, which gave rise to resentment among local officials. As Lewis (1982) points out, the Islamic world's ignorance of the West was profound, and contrasts with the keen fascination Western culture had for Islamic civilization.[28] A saying attributed to the Prophet Muhammad has it that "whoever imitates a people becomes one of them," which was interpreted in the Ottoman Empire as a stricture against Western technology.

The number of inventions Europe is said to have borrowed from other cultures, in particular China, is large. Europeans appreciated useful knowledge regardless of the source; Asian cultures, with the exception of nineteenth-century Japan, did not. In the early Middle Ages, Europe borrowed techniques from other cultures the way an underdeveloped country today emulates the industrialized world. After 1500, as Pacey (1975, p. 189) notes, the exposure effect operated largely through showing Europeans technological possibilities that

27. Medieval rulers such as the Emperor Frederick II, King Alfonso X ("the Learned") of Spain, and King Roger II of Sicily brought Islamic engineers to their courts and had their works translated. A telling example of the differences between the Western and Middle Eastern attitudes is the fate of the works of the great tenth-century Persian physician and scientist, Al Razi (Rhazes), who compiled a huge textbook of medical science that, according to White (1968, p. 98), is "probably the biggest single book ever written by a medical man." After being translated into Latin in 1279 as *Liber Continens,* it became a fundamental book of reference for Western doctors for many generations. Yet no Arabic copy of the book survives, and after a few centuries it was practically forgotten by the Islamic world.

28. The first chair in Arabic studies at Cambridge University was established in 1633. Throughout European culture after 1650, Islamic motifs played a major role in literature, art, and music.

simply had not occurred to them. Iron suspension bridges, seed drills, and porcelain are examples of Oriental ideas that were taken up by Europeans and subsequently improved and perfected. Europeans had no sense of shame in borrowing foreign technologies, as illustrated by the many products and processes named after their alleged foreign origins. Thus we have Europeans producing chinaware, calicoes, satin, damasks, and Japan (a black varnish); using Arabic numerals; eating turkeys (or *d'Indes*); and rigging lateen sails. An attempt to close a country to foreign influences, as took place, for example, in Japan and the Islamic world, would never have succeeded in Europe even if it had occurred to anyone.

As Landes put it (1969, p. 28), "good innovators make good imitators." But in the final analysis good innovators and good imitators are both produced by a society in which material and practical values are held in high esteem. If something works, it does not matter where it came from. When Europeans were exposed to new information, their sense of wonderment was, so to speak, soon replaced by the thought of how to exploit the new knowledge. When they "discovered" the world after 1450, their chief purpose was to acquire wealth, either directly or by way of new information used to produce goods creating wealth. Gold, silver, spices, sugar, tea, and furs were directly imported into Europe; potatoes, tobacco, and maize were successfully transplanted. The test was always: is it useful? can it enrich me (or my king)? Typical of the Europeans' approach was the great Leibniz, who implored a Jesuit traveling to China "not to worry so much about getting things European to the Chinese, but rather about getting remarkable Chinese inventions to us; otherwise little profit will be derived from the China mission" (cited in Bray, 1984, p. 569). Newton wrote that neither pride nor honor should stand in the way of the principle that the important thing was "to learne, not teach" (Landes, 1969, p. 33). This approach to newly found lands differs radically from that of the Chinese, who toured parts of the world a few decades before Europe's explorations began in earnest. For the Chinese, the purpose of long voyages was to demonstrate the wealth and glory of China to the barbarians by means of lavish gifts, a laudable but ultimately prohibitively expensive policy. No wonder the Chinese terminated their explorations, while the Europeans carried on.

Equally important, the Europeans were willing to learn from each other. Inventions such as the spinning wheel, the windmill, and the weight-driven clock recognized no boundaries. The printing press was no more a German invention than the telescope was a Dutch one or the knitting frame an English one. European shipbuilders often travelled on board ships to see the types of ships in use elsewhere (Unger, 1980, p. 23). From the fourteenth century on, sons of northern European merchants traveled to Italy to study the *arte della mercadanta,* including commercial arithmetic and bookkeeping (Swetz, 1987, p. 12). In spite of the seemingly high barriers to long-distance communication, technological "news" traveled well and fast in Europe, except, as we have seen, in the case of agriculture. Technologically creative societies started off as borrowers and typically soon turned into the generators and then the exporters of technology. In the seventeenth century, England was regarded as a backward society that depended on foreigners for its engineering and textile industries; by the nineteenth century, the directions were reversed. Modern-day East Asia finds itself in the same position.

Societies also differed in their willingness to challenge the accumulated knowledge of earlier generations and in their tolerance toward "heretics" who did. Some societies, especially Islam and Judaism, eventually developed a notion that earlier sages had already discovered everything there was to discover, and that challenging their knowledge was sacrilegious. Lewis (1982, pp. 229–30) argues that Islamic tradition eventually came to believe that all useful knowledge had been acquired and all questions had been answered so that all one had to do was repeat and obey. From the later Middle Ages on "Muslim science consisted almost entirely of compilation and repetition." The inferiority complex vis-à-vis earlier generations imposed a constraint on the generation of new knowledge.[29] Lewis points out that in the Islamic tradition the term *bidaa* (innovation), acquired the same negative connotation as "heresy" did in the West. A particularly bad form of *bidaa* was imitating the infidel; the only exception allowed was military technology used in a holy war. Whether this conservatism can be blamed for the slowing down of technological progress is not entirely clear. It is possible that both scientific and technological progress ran into a barrier of conservatism that had causes outside religion, and this barrier was reinforced by reactionary religious elements. After all, in its early centuries the Islamic world had been curious and almost obsessive in its thirst for the learning of other civilizations, including technological knowledge. Moreover, new ideas could have emerged under the guise of exegesis.

A different way of looking at the problem was formulated by Ayres

29. A famous dictum from the Jewish *Chazal* (earlier sages) has it that "if those who were before us were like angels, we are but men; and if those who were before us were like men, we are but asses."

([1944] 1962), who distinguished between the dynamic forces of technology and the conservative forces of ceremony and ritual. These two forces oppose each other in dialectical fashion, and what is required for technological creativity is the right blend of accumulated knowledge of past generations and the ability to shed the stifling burden of past institutions. Ayres speculated that this blend existed in "frontier" societies such as medieval Europe or the nineteenth-century United States.

Beyond the basic assumptions in each society, the communicability of ideas and experiences was of importance. Europe always had a lingua franca, first Latin, then French—as did the Islamic and Chinese worlds. But beyond the language, trade, and communications that allowed information to flow (in the form of letters and published books), there were shared standards of verifiability and applicability, derived from a common epistemological heritage. A certain methodology accepted by all is necessary, so that results achieved by one scientist or engineer can be accepted by others without having to duplicate everything. The scientific world of Renaissance Europe was truly cosmopolitan, and as we have seen, many of the technological successes of the Industrial Revolution were cooperative international efforts.

DEMOGRAPHIC FACTORS

One factor that has received support from a small but highly articulate group in recent years is the effect of population change on technological progress. The most influential spokesperson for this approach has been Boserup (1981), who points out correctly that the distinction between invention and the adaptation and adoption of existing technique is blurred. Her main point is that population pressure leads to shifts to more labor-intensive techniques, which will offset and possibly nullify the effects of diminishing returns on income and output per capita. Some of the phenomena discussed in Part II, particularly in agriculture, may well be described by such a factor-substitution process. But it is also clear that most technological advances involved innovations that saved labor as well as other factors, and were no more due to the pressure of labor on land than the other way around. If Boserup is right and population growth leads to land- and capital-saving changes in technology, population decline should be equally conducive to technological change. Mechanization and the machine are by their very nature intended to save labor more than other factors. Whether such factor price-induced biases are confirmed by the available data remains an open question. Boserup's evidence is drawn almost exclusively from agriculture and is often inconclusive.[30] In any event, such theories cannot deal with many of the processes I have described. How could the invention of the telescope, the full-rigged ship, or the knitting frame be explained by population change?

Boserup hints at, but does not fully develop, a second possible link between population growth and technological change. She points out that population density determined the size of the urban sector, and that high urbanization led to externalities in the development of transportation technology. One could well use this idea to develop a full-blown model of scale economies in technological change, as Simon (1983) has done. Several elements of this model make sense. The size of the population determines the supply of potential inventors (though the ratio of actual to potential inventors may be the more important variable here). A larger population creates a larger market for the improved or cheaper good, but this source of scale economies seems soon exhausted for most products: it hardly matters whether one has half a million or five million potential customers for a new product. Greater population size will often imply a finer division of labor and regional specialization. Although these effects might be indirectly associated with technological progress properly, their primary effect is through Smithian growth and not through changes in knowledge. A finer division of labor is often asserted to have strong technological effects as well, but these effects are rarely backed up with evidence (for a recent example, see Persson, 1988). Economies of agglomeration may have been important. Inventors interact with and learn from each other, so that larger units with good internal communications would be more conducive to the development of useful ideas that work. Boserup argues that cities played a crucial role in technology outside agriculture and mining. Urbanization may not have been a necessary or a sufficient condition, but from about 1200 on, many of Europe's technological advances increasingly emanated from cities. In towns, greater occupational specialization allowed potential inventors to focus on specific soluble problems. The clockmakers, shipwrights, and specialized metalworkers were largely an urban class. Yet, as we have seen, the urban environment also fostered groups who stood to lose from technological change and tried to thwart it. Craft guilds, originally often the promotors of new techniques, became after 1500 increasingly a conservative force. Moreover, the importance of agriculture and mining, and the rural location of most good waterpower sites

30. The introduction of the heavy plow in Europe in the sixth and seventh centuries, when population was at its nadir, seems inconsistent with the theory, and Boserup has to rely on the difficulties in dating the widespread diffusion of the innovation to square it with her theory (Boserup, 1981, p. 96).

limited the actual importance of cities in this regard before the Industrial Revolution.

An original theory connecting population, the physical environment, and the rate of technological change has been proposed by Wilkinson (1973). In this argument, based largely on anthropological evidence, technological change occurs when the ecological equilibrium between population and resources is disturbed. When population growth occurs, maintains Wilkinson, "society will try to find ways of developing its technology to increase the yield from its environment" (ibid., p. 56). Additional power sources, for instance, were "necessary responses to increasingly difficult subsistence situations." The British Industrial Revolution, in his view, was the result of acute resource shortages resulting from the resurgence in population growth in the last third of the eighteenth century (p. 113). The historical evidence simply does not support his interpretations. Britain, a net exporter of agricultural goods until late in the eighteenth century, cannot be said to have been suffering from an acute land shortage. The timber famine, on which Wilkinson bases his argument, never existed, as we have seen. A large proportion of the technological innovations discussed earlier, from the mechanical clock to the marine chronometer, simply had nothing to do with ecological pressures and occurred in societies in which population was in fact more or less stagnant. Above all, Wilkinson fails to explain why in some cases population growth led to matching technological advances that prevented a Malthusian disaster, whereas in others overpopulation led to such disasters.

In what follows, three comparative examples—the growth of technological change in the medieval world compared to classical antiquity; the growth of European technology in the late Middle Ages, Renaissance, and baroque periods, compared with the technological stagnation in China; and the success of Britain as the cradle of the greatest example of technological change of all times, the Industrial Revolution—will illustrate some of the factors affecting technological change.

CHAPTER EIGHT

Classical and Medieval Technology

Why did classical society, with its indisputable intellectual superiority, achieve so little by way of technological innovation compared to the crude and illiterate peasants of medieval Europe? To start with, the slowness of technological progress of classical society should not be equated to, or even associated with, economic backwardness or poverty. We know very little about living standards in classical times, but there is no evidence that they were lower than in the Middle Ages or in the early modern period in Europe. Smithian growth, supported by the convenient location of the centers of gravity of classical civilization around the Mediterranean, and by their relative cultural and later political unity, could provide improved living standards. Moreover, technological progress in the West did not inevitably lead to permanently higher living standards, because population growth often took the place of increased consumption of goods and services. The best we can do is to note that if classical society had experienced more technological progress, it could have enjoyed a higher living standard than it did.

Nonetheless, in view of the striking contrast in the technological creativity of Graeco-Roman and medieval societies, there is something to be explained. It is clear that no single explanation will do, and that the picture is likely to remain blurry. But we can quickly dispose of a few hypotheses. To start with, the notion that somehow the Romans and Greeks attained the level of technology they needed and that no pressure existed to improve is unconvincing.[1] As Lee

1. See, for example, Bernal (1965, vol. 1, p. 222): "The decisive reason [for the failure of the Hellenistic period to produce the machinery of the Industrial Revolution] was lack of motive. The market for large-scale manufactured goods did not exist. The rich could afford handmade goods, the poor and the slaves could not afford to buy anything they could do without." Burford (1960, p. 18) explains the inefficient harnessing of horses in antiquity by maintaining that horses were largely a frivolous item of conspicuous luxury and not intended for work. Hence the harness used was per-

(1973, p. 77) notes, the Romans clearly could have made good use of a ship that would sail better to windward. For all their genius in physics and mathematics, the Greeks never improved their agricultural tools, and the Hellenistic astronomers never learned to apply their clever astronomical instruments to down-to-earth uses such as orientation at sea. The inefficient harnessing of horses, the tardy adoption of waterpower, and the backwardness of Roman metallurgy reveal the ancients' weakness on two important technological fronts: energy and materials.

Furthermore, the argument that the physical geography of the Mediterranean environment was not conducive to technological change will not get us very far. True, timber and running water, the building blocks of medieval technology, are comparatively scarce in southern Europe, though wood was certainly more abundant in the Mediterranean area in classical times than it is today. After Julius Caesar, most of western Europe was under Roman control and the empire used resources from Spanish silver to Cornish tin. Moreover, Mediterranean geography as such was not antithetical to technological progress. As we have seen, much of the technological change in medieval and Renaissance Europe originated in Italy and the Mediterranean region.

Another view holds slavery directly responsible for the slowness of technological change. Most slave societies we know of (ancient Egypt was a possible exception) were not particularly conducive to innovation. But then, neither were most nonslave societies. The argument raised by some (e.g., Lilley, 1965) that slaves were simply cheaper than machines seems hollow. It is incorrect to suppose that slave labor is substantially cheaper than "free" labor. True, at times classical civilization was wasteful in its use of slave labor, as in the use of slaves in the processing of lead ore, in which mortality was shockingly high. Yet slaves had to be fed, housed, clothed, and eventually replaced. Examination of the evidence for Roman slavery suggests that the history of slavery seems irrelevant in explaining technological development, and it seems difficult to find a connection between the availability of labor and attempts to conserve it (MacMullen, 1988, p. 18). The existence of slavery can hardly explain the failure to use horses appropriately, the primitive level of metallurgical technique, or the clumsy construction of ships in antiquity.

fectly adequate. Moreover, "people knew certain ways of doing necessary jobs [and] . . . could see no reason for other methods to come into existence." A more qualified statement is made by Usher (1954, p. 101), who notes that early society tended to be "oppressed by tradition" and that the rationalism of Greek society did little more than sublimate the respect for custom into a search for immutable truth. Such a society is not "interested" in technological change. Yet he then adds, without explanation, that the value of inventions was more justly appraised than appears on the surface.

Was slavery responsible for the low esteem in which practical technology was held? It probably exacerbated it somewhat, but it seems unlikely to have been the sole factor. Early Greek philosophy, Ionian and Pythagorean, carried a distinct pragmatic flavor, only to be superseded by a "divorce between science and philosophy on the one hand and the productive processes on the other" (Finley, 1965, p. 33). This change seems to coincide approximately with the expansion of slavery in ancient Greece (fifth century B.C.).[2] Some slaves, however, were educated. Seneca pointed out that most inventions of his time were naturally the work of slaves, as only slaves were interested in such things (cited in Klemm, 1964, p. 23). Slaves occupied positions of management and responsibility. Frontinus, the supervisor of the Roman water supply, was assisted by slave architects, and the sons of freedmen often rose rapidly into the ruling class (Finley, 1973, p. 77). Outside Italy, in any event, slaves were used largely for domestic services by urban elites, rather than for field work or manufacturing.

Yet dismissing slavery altogether as a factor seems premature. Technological change depends on the incentive structure of production. Almost by necessity, slavery implied a relationship of coercion between labor and management. The introduction of new techniques in hierarchical structures requires cooperation between the factors of production. New ideas may come from above or below, but unless management is willing to accept and reward new technical ideas from below, they will come to naught; unless the monitoring of labor is very tight, workers will be able to sabotage ideas implemented from above. Moreover, because the incentive structure for slaves is usually based on punishment rather than rewards, slave labor tends to be simple and routinized (Fenoaltea, 1984). The advantage of routinized and simple tasks is that such tasks reduce the cost of monitoring the workers. Such routinization tended to ossify technology. If slavery dominates the labor market, skilled labor will be in short supply, and it is plausible that a scarcity of technically trained workers makes deviations from the technical status quo more difficult. All the same, there is little evidence that in the ancient world slaves were in any sense less skilled or productive than free workers, or were necessarily at the bottom of the labor hierarchy. Moreover, during the later centuries of the Roman Empire, slavery declined and most of the labor force consisted not of slaves but of free peasants, urban laborers, and craftsmen, sailors, merchants, and so on. If the hierarchical structure of slavery inhibited technological change,

2. The apparent change in attitudes in the fifth century B.C. is mirrored in literature. Writing in about 450 B.C., Aeschylus had Prometheus boast of his beneficial technological discoveries. A few decades later, Sophocles and Euripides were far more subdued, demanding man to be humble and grateful toward nature (Kahn, 1970).

why was the free sector not more successful in developing new techniques?

With the decline of slavery during the later empire, other institutions emerged that may have had something to do with the decline of technological creativity. In an attempt to buy stability and tranquility at the expense of long-term progress, the reforms of Diocletian and Constantine (fourth century A.D.) tied many workers to the occupations of their fathers. Although never as watertight as the Indian caste system, an ascriptive hierarchy in which economic and social status were inherited rather than achieved would naturally inhibit technological progress. De Camp (1963, p. 280) deems these reforms "fatal to technological progress," and asks rhetorically, "what baker's son will fool around with an idea for a bicycle when he is compelled to make bread all his life willy-nilly?" Yet enforcement of these decrees must have slackened as the western Empire disintegrated, and the disappearance of central authority in the Middle Ages meant precisely that such draconian measures increasingly remained a dead letter.

It is often argued that slavery reduced the economic prestige of manual labor and deepened the chasm between the educated and the productive class. As Lee (1973) points out, classical antiquity was not inherently different in this from other civilizations, except perhaps in degree. The channels linking the upper classes and the working classes have always been narrow and easily clogged; in antiquity they barely existed. Intellectuals were discouraged by the ruling hierarchy of values from making their ideas useful. In a famous passage, Plutarch praised Archimedes for "regarding mechanical occupations and every art that ministers to needs as ignoble or vulgar, he directed his own ambition solely to those studies the beauty and subtlety of which are unadulterated by necessity." This statement tells us perhaps more about Plutarch than about Archimedes. Aristotle wrote in his *Politics* that "no man can practice virtue who is living the life of a mechanic or laborer." The engineering handbook *Mechanika* is attributed to him, but was probably written by someone else. Many classical scholars have noted the abstract and detached nature of Greek science, where experiments were unimportant because of the presumed separation between abstract reality and its manifestation in the real world.[3] Classical natural philosophy submitted that the natural world could be understood by rational principles, but the insights were their own reward. Science, when it stooped down to the

3. Pleket (1967, p. 2) points out that experiments, when performed at all, were only intended to confirm preconceived theories, and that experimental design was too oversimplified to warrant the conclusions drawn from it.

real world, either created toys and gimmicks for the rich and famous, or confined itself to classification and taxonomy. The technological barrenness of classical natural philosophy must be contrasted with the practical and applied bent that slowly surfaced in medieval science.[4]

The educated classes of the Roman Empire have come in for even more severe criticism. One recent author (Oleson, 1984, p. 405) has argued that "the typical Roman intellectual and aristocrat, like a child incapable of extensive analysis of his data . . . hedonistic in his dislike of pain, change, or risk, wandered through a dreamlike haze until the opportunity for progress—however limited—had been lost." This may be an overstatement, but technological progress does require an intellectual and psychological commitment. Few elites through history have satisfied these conditions.

One difference between the agrarian economies of feudal Europe and the economies that preceded them is that the medieval landlords left a considerable portion of the land to their peasants to till as they saw fit. Once the peasant had paid his dues in goods, labor services, and (later) money, he was free to allocate his own labor and that of his family as he saw fit, subject, of course, to the constraints that the village community imposed. This freedom, however limited, meant that the cultivator himself reaped some of the fruits of improved cultivation techniques. To be sure, attempts toward improved cultivation had to face the greed of lord and priest, as well as the inertia imposed upon each cultivator by the quasi-cooperative nature of medieval farming, in which many technical decisions were made jointly by the village community. These long odds notwithstanding, from about 900 A.D. on medieval peasants succeeded remarkably well in improving their techniques, aided by favorable climatic changes that occurred at about this time. It is significant to recall that one of the most important advances in medieval agriculture—the substitution of horses for oxen in haulage—was more pronounced in peasant agriculture than in demesne (landlord-operated) farming (Langdon, 1986, p. 255).

The medieval situation contrasts with that of the classical world as described by Hodges (1970, pp. 179–80), who blames management rather than labor for the technological stagnation of the classical world. In this view, day-to-day management of workshops and farms was divorced from ownership, and managers were often members of a "very mediocre and often dishonest class," who behaved like bureaucrats, settling into a routine and doing as little as they could to upset the quiet life. Columella, a Roman writer living in the first century

4. For a summary of this literature, see Oleson (1984, pp. 397–408).

A.D., complained in his *De Re Rustica* that when a rich man bought a farm, he selected from his entourage "the most enfeebled in years and vigor . . . disregarding the fact that this particular job demands not only knowledge, but the liveliness of a man in his prime . . . to cope with its hardships" (reprinted in K. D. White, 1977, pp. 20–21). K. D. White himself notes with some astonishment that the great landowners, who in his opinion alone possessed the resources to initiate technological improvements, did little to increase productivity or reduce labor costs even in the face of an increasing shortage of labor (White, 1975, p. 221). This explanation does not apply to public enterprise. Engineers and architects engaged in public projects were under heavy pressure to succeed and enhance the prestige and political standing of their employers. Why did this not happen in the private sector? The competitive process must have worked to some extent in the classical world, in which the market economy was of central importance. The real question is, did competition work well enough to ensure the survival of only the most efficient? And if so, did the competitive process extend to the market for technological ideas?

No simple monocausal theory can explain a thousand years of history. The difficulty is that the weakness of classical civilization was in part its inability to generate altogether new technologies (e.g., horse harnesses, high-quality iron, the mechanical crank, hydraulic oil presses), and in part its unwillingness or inability to find new and imaginative economic applications for the new ideas it did generate. The limited use of the water mill, and the difficulties the Romans had in rigging their sailing ships efficiently, illustrate the slowness with which technology progressed in Roman society. The importance of political centralization obviously increased over time as the Roman Empire grew more powerful. The empire biased the direction of technological change toward the public sector, thus encouraging trade and through it Smithian growth, as well as public sector technologies such as hydraulic engineering and roadbuilding, but at the expense of technological progress in the private sector.

It is often maintained that the empire stressed authority and conformism, at least during the later periods. By Romanizing the upper classes of the barbaric peoples they conquered, the Romans imposed certain common values. These values tended to stress distinctly noneconomic achievements: military, intellectual, administrative, and artistic success were appreciated more than production. Our limited knowledge of classical technology is in part symptomatic of the attitudes of intellectuals of the time. Classical writers rarely concerned themselves with the mundane details of technology; history, poetry, philosophy, and politics occupied them. A distinction should be made between the Greeks on the one hand, and the Hellenists and Romans on the other in this respect. Hellenistic and Roman engineers were more willing than the Greeks to apply the abstract knowledge generated by mathematicians. They used mechanics in the design of war machines and geometry in hydraulic engineering and the layout of towns. Hellenistic and Roman authors were also more willing to write about technical matters.[5] Yet when these subjects were discussed, it was usually incidental to other matters; by the second century A.D. discussion of technical issues had become rare. As I have stressed earlier, such a chasm between the educated and literate classes on the one hand, and the working portions of the population on the other was in itself damaging.

In a more pluralistic world this ranking of priorities might have been challenged, but as the empire became more powerful and its culture more influential, it became less pluralistic, and for centuries no effective center competed with Rome for power and influence. True, the values we are observing were upper-class values, and may not have filtered down to the masses actually involved in production. In some ways the Roman Empire resembled the Soviet Union today in its conflict between political centralization and technological change. The difference is that the Soviet bloc today is eager to borrow from the West the technological progress it cannot generate itself; Greece and Rome seem to have had little interest in alien technologies.

The influence of the Roman hierarchy of values on non-Romans may provide us with a clue to the comparative technological performance of imperial Rome and the medieval Occident. As I have noted, the Celts were in some respects more creative than the Greeks and Romans. As they fell under Roman influence, however, the elites of Celtic society tended to imitate Roman culture and adopt Roman values. Schools taught Latin and classical literature, and Celts found careers in the Roman bureaucracy and army. When the empire split into two, the Roman heritage in the West was increasingly diluted by Celtic and Germanic elements, which may well have given technological creativity a new lease on life. The purer form of classical civilization survived in Byzantium, whose record of technological progress is undistinguished.

The Romans and the Greeks were comparatively tolerant of de-

5. Vitruvius's *De Architectura* and Pliny's *Natural History* are rich sources on technology, as is the *Geographia* by the Hellenistic writer Strabo. Two major treatises on agriculture, by Columella and Varro (both entitled *De Re Rustica*) appeared in the first century B.C. These books deal with the practical needs of large estates rather than small independent peasants, and viticulture figures prominently in them. After Varro, the tradition stopped. Subsequent writers on farming, such as Palladius (fourth century A.D.), produced little more than exegesis.

viants and people who believed in other Gods. Such tolerance is in itself a desirable characteristic for the diffusion of new ideas. Graeco-Roman religion was, however, heavily animistic and anthropomorphic, and possibly discouraged substantial insults against the environment. Religions that regarded nature as a personal force implied that tinkering with its rules was dangerous and sinful. True, the desire to prevail over the forces of nature lies deep in the human soul, and, as all ancient mythologies show, was shared by pagans. But with the desire came deep-seated fears and guilt feelings that impeded technological progress. The myth of Prometheus, the quintessential manipulator of nature, illustrates this frame of mind well: although the use of fire, indispensable to survival, was accepted, the guilt feeling associated with controlling something that was properly the domain of the Gods was reflected in the awful punishment meted out to Prometheus and the banes of Pandora's box. Moreover, animism, as Lewis Mumford pointed out half a century ago, is the mortal enemy of technological change. If every stream, every tree, every patch of land is populated by spirits, the environment remains capricious, unpredictable, and uncontrollable. Any attempt to alter it may raise the awesome ire not only of Zeus but of small-time local deities as well. To be sure, the educated philosophers and playwrights with whose work and thinking we are familiar, did not take Graeco-Roman religion very seriously. But the uneducated classes, who were actually involved in production and whose beliefs thus matter a great deal, probably continued to believe in these Gods for centuries after the zenith of Greek philosophy.

Such feelings slowly started to dissipate before the advent of Christianity. Over the centuries the influence of pagan religion diminished, challenged by various deistic philosophies. The Stoic philosophers taught that the rational design of the world includes provision for human activities. People, in this view, were the natural caretakers of the world, and rational human efforts make the earth more beautiful and more serviceable for mankind (J. D. Hughes, 1975, p. 97). Classical writers such as Pliny and Galen reflect more anthropocentric attitudes. During the later empire, however, belief in the supernatural and mysticism became more influential, and Stoicism declined. Astrology and magic were central to the elaborate rituals of secretive and conspiratorial sects that acquired increasing influence on the religious scene for centuries. Many of these sects were offshoots of Oriental religions. Mysticism and superstition led to a decline in Roman science and philosophy, and it seems logical that they also affected technology. Yet the relationship between classical science and technology was always tenuous, and just as the flourishing of Hellenistic science had few favorable effects on production methods, its decline did not itself produce technological stagnation.

Medieval Christianity, as it eventually evolved, reinforced these tendencies. It is absurd to argue that Christianity in its original form was favorable to technological progress. In some views, it was little more than the most successful of these Oriental cults. During its first five or six centuries it had its share of mysticism, asceticism, rejection of material life and wealth, and outright condemnation of labor and all worldly activity.[6] Yet Western Christianity contained the seeds of future technological progress. Here Lynn White (1978, pp. 217–53) and others have stressed that the way people perceived their relationship with nature and with their creator, and which actions were deemed virtuous and which sinful, determined their technological activism. As White sees it, Western Christianity increasingly developed a view of a rational and calculating God, the designer of a huge intricate mechanism called the Creation, who "commands man to rule the world and to help to fulfill the divine will in it as a creative cooperator with him." Christian beliefs began to take it as axiomatic that nature has no reason for existence save to serve mankind. Benz (1966, p. 124), from whom White borrowed the original insight of a connection between Christianity and technology, argued that in the technologically creative but deeply religious societies of early medieval Europe, the belief gradually took root that technological efforts were justified by the destiny of man as the image of God and his fellow worker.

Some scholars feel that Benz and White may have overstated the case.[7] To be sure, there was always a view in the Latin church that took a more moderate position, advocating a harmonious relationship with the environment and holding mankind responsible for the stewardship of nature (Ovitt, 1986; 1987, pp. 85–87). St. Francis of

6. Pope Gregory the Great (about 600 A.D.), one of the pivotal figures in the history of the Church, burned libraries of classical writings to prevent such writings from distracting the devout from reflecting on pious subjects.

7. Benz (1966) points out that the notion of God as the Creator and architect of the universe, an essentially technological view, contrasts sharply with Buddhism. He adds (ibid., p. 122) that "significantly, Buddhism has not produced a technological culture but has a distinctly anti-technological attitude in many of its schools." It may well be true that the Buddhist notions that disengagement from reality is necessary for salvation and that the path to bliss leads through control of oneself rather than one's environment were not conducive to technological changes. But things were not that simple. Buddhism is divided into many sects, some of which were more receptive to technological changes than others. Similarly, Confucianism—which is hard to define as a religion at all—seemed consistent with technological progress during some periods of Chinese history and inconsistent with it in others.

Assisi rebelled against the exploitation of nature. The Latin church did not always speak with one voice, and the changes that Benz and White discerned were long in the making and encountered stiff resistance, both in Christian dogma and in popular attitudes. But even if White has overstated the case somewhat, a comparison with other religions reveals the importance of Western Christianity in the emergence of Western technology.

Proving causality here is, of course, impossible. It is plausible that the Christian attitudes toward nature were as much a compromise with existing techniques as their cause. But the very essence of technological change is such that some mental change had to occur first. As I have stressed above, invention of any kind is a game against nature, and before anyone undertakes to play such a game certain mental conditions have to be satisfied. Most importantly, the player must have some idea of the nature of the game played and the payoff involved. If nature is perceived as a hostile and jealous adversary, or if the only payoff that really matters is the salvation of the soul, then there seems little point in playing the game at all. But if the universe is subject to logical, mechanical forces that can be controlled and manipulated in ways that do not involve committing a sin, and if it becomes clear that the payoff is a standard of living above the very minimum of subsistence, the first condition necessary for the beginning of technological progress is fulfilled. There is some dispute as to when attitudes began to change. Lynn White (1978, p. 79) would trace the change in attitudes to the centuries before 1000 A.D., while Ovitt (1987, p. 44) is of the opinion that Western values before 1100 A.D. were static and unreceptive to change. Such controversies are hard to settle because the evidence on attitudes and mentality is vague. It seems hard to imagine that the many innovations before 1100 A.D. were not preceded by any change in outlook even if official Church doctrine lagged behind. Whatever its origins, the belief in a controllable, mechanistic universe in which human beings may exploit the laws of nature for economic purposes, was growing; by the end of the Middle Ages, it had triumphed. Thomas Aquinas recognized that man, created in God's image, held power over the natural world. In the later Middle Ages, God is increasingly depicted as an engineer or an architect in technological terms. Thus, in the fourteenth century Nicole Oresme, bishop of Lisieux, depicted God as having started the world the way a man starts a watch, winding it up and then letting it run on its own (Benz, 1966, p. 252). Pacey (1975) has emphasized that in the thirteenth and fourteenth centuries, religion led to the building of cathedrals, which had considerable technological spillover effects. Moreover, he points out, innovation is above all an imaginative act, and if medieval religion did anything, it stimulated the imagination. In this view, much of the advanced technology in the Middle Ages was thus the result of noneconomic causes (ibid., pp. 85–86). Pacey may well be right that the Baconian notion, according to which the final objective of all knowledge was not the understanding of the world but its manipulation, was the end result, rather than the beginning, of an intellectual tradition. I see no way, however, in which this hypothesis can be tested. It is beyond doubt that whatever the psychological origins of the spinning wheel, the fully rigged ship, and cast iron, their economic consequences were profound.

A good example of a change in mentality that favored technological progress in the West concerns the attitude of "if God had wanted man to fly, he would have given him wings," enshrined in the legend of Daedalus and cited by no less an authority than Landes (1969, p. 24) as a strong current in Christianity. The rebellion against this view came from within the Church. It is well known that the famous Franciscan friar Roger Bacon wrote prophetic essays in the thirteenth century envisaging airplanes, steamships, and automobiles. It is less well known that more than two centuries earlier another monk, Eilmer of Malmesbury, directly defied human winglessness by jumping off a tower in a self-made glider. The ill-fated monk broke his legs and was lame ever after. Yet it is significant that Eilmer did not feel he had committed any sin by trying to fly, but blamed his misfortune on his having failed to put a tail on the glider (Lynn White, 1978, pp. 59–73). Eilmer's flight was more than merely the reckless game of a bored eccentric. As White shows, the depiction of Christ's ascension to heaven, at first described as a majestic rise often carried by angels, is pictured in the early Middle Ages increasingly as "almost jet-propelled, zooming heavenward so fast that only his feet appear at the top of the picture, while the garments of his astounded disciples flutter in the air currents produced by his rocketing ascent" (ibid., p. 66).

It is not a coincidence that much of the early development of technology was associated with the regular clergy, i.e., monks. Much has been made of the importance of monastic orders in the emergence of Western technology. The connection between the strict daily schedule that required monks to know the time with some accuracy and the development of the weight-driven clock is well known (see Landes, 1983, pp. 61–70 for a restatement and survey of the literature). But the monastic orders, or at least some of them, went beyond their immediate needs and established a sound if narrow bridge between the educated and the productive classes. In the fourth and fifth centuries, a belief that productive labor was virtuous started to take root. In about 530 A.D., St. Benedict, the founder of the Benedictine order, wrote the Benedictine Rule, which earned him the unlikely accolade of being "probably the pivotal figure in the history of labor" (Lynn White, 1968, p. 63). White proposes to explain medieval technology by emphasizing the respect for work and production that the Benedictines were taught. Idleness is the enemy of the soul and to labor is to pray, taught St. Benedict. The transition in attitudes was far from abrupt: at first labor was less a positive virtue than a recommended form of penitence (LeGoff, 1980, pp. 107–21). Eventually, "the concept of penitential labor was supplanted by the idea of labor as a positive means to salvation" (ibid., p. 115). Because monks belonged to the educated classes—indeed *were* the educated classes for centuries—there is some basis for White's belief that "for the first time the practical and the theoretical were embodied in the same individual . . . the monk was the first intellectual to get dirt under his fingernails" (Lynn White, 1968, p. 65). By establishing the moral acceptability of physical labor and production, the Benedictine Rule was the first challenge to the classical notion that identified work with depravity. Benedictine monks had enormous influence on medieval life, playing a major role in education, in agriculture, in land reclamation, and in the techniques of arts and crafts.[8] "The Benedictines established an ethic that would, by the twelfth century, accept productivity and wealth as the natural by-products of labor," writes Ovitt (1986). With time, the example set by the regular clergy spilled over to other writers.[9] When the Benedictine order became rich and lax in its ascetism, the purist flame was carried on by the Cistercian monks. It took a very long time before labor was secularized, to use Ovitt's term, that is, before the virtuousness of economic production was assessed by Church doctrine in terms of its output rather than in terms of the effects labor had on the virtuousness of the laborer's soul. But that can hardly have mattered a great deal. The main point about producing goods cheaper and better is that they satisfy economic demand, which in the Middle Ages primarily meant satisfying basic physical needs. Better and more shelter, food, clothing, energy, and transport were their own reward.

Ironically, the bestowing of some measure of respectability on

8. A technologically inclined monk was Theophilus, a German Benedictine, who wrote a famous treatise on crafts and technology, entitled *De Diversis Artibus,* around 1122. Theophilus' work on technology underlines an important aspect of Christianity, namely, that the application of human industry was no longer sinful or reprehensible. His study, according to his own statement, was written "neither to seek praise nor from a desire for worldly rewards . . . but to afford help to many for the glory of God and for the exaltation of His name" (cited in Klemm, 1964, p. 65). Technology itself served a divine purpose of demonstrating God's wisdom and hence was a legitimate activity.

9. For instance, twelfth-century Anglo-French writers, such as Yves of Chartres and John of Salisbury, assured that "all labor ennobles man and pleases God" (cited by Fossier, 1982, p. 883).

physical work led eventually to its replacement. As White put it, "The goal of labor is to end labor." Only after intelligent and educated people roll up their sleeves and engage in physical labor will they get inspirations concerning how human muscles and sweat can be replaced by machines. Mechanical arts, although still far from prestigious, achieved a status of virtue hitherto unknown. The Victorine monks in Paris in the twelfth century included *mechanica* in their scheme of useful knowledge, challenging no less an authority than St. Augustine in the process. In the West, first organs, then clocks, became part of the religious ritual: machinery was no longer sinful. The Cistercian monks, who devoted every monastery to the Holy Virgin, did not fear that the creak of waterwheels would offend her: some of their abbeys had four or five mills. The "cultural climate," to use White's term, that the Latin church created in the West, helped transform the environment in which technological change operated. In this cultural climate artisans and scholars began to find a common ground and establish communications. As a result, people involved in technology began to write. Theophilus, the author of the first important surviving medieval text on technology, wrote in the 1120s. Villard de Honnecourt, who lived in the early thirteenth century, left a set of designs probably intended for teaching. In the later Middle Ages and the Renaissance, the literate and working classes were no longer totally disjoint, and the boundaries between the two began to break down. The importance of contact between the educated and the producing classes for technological progress seems too obvious to need additional emphasis.

The thirteenth and fourteenth centuries gave rise to craftsmen-scholars, such as Roger Bacon and Richard of Wallingford. With the advent of the printing press and the custom of writing in the vernacular instead of in Latin, the opportunities for people with technological knowledge to communicate with others grew rapidly. The great engineering and medical treatises of the Renaissance were the culmination of a tremendous change that occurred in Europe. The roots of this change are by now widely understood to be medieval, though once again the dating is a matter of dispute.

Cultural environment and *mentalité* cannot by themselves account for the spectacular expansion of medieval technology. We must come to grips, for instance, with the failure of the Byzantine part of Christianity to generate anything like what the West produced. Blaming the differences on the more contemplative and mystical bend of Eastern monasticism, as Lynn White does, may seem special pleading, for it fails to explain the fact that other Judeo-Christian religions eventually developed characteristics like xenophobia and conservatism that stifled technological creativity.

In its original form Islam was as receptive to technological progress as Western Christianity, but, as we have seen, it started to become less so around the twelfth century. The great Islamic philosopher and theologian al-Ghazali (Algazel) (1058–1111) concluded that science and technology were incompatible with Islam and attempted to reconcile mysticism with their religion, at a time when Western theologians were making Christianity more rational and compatible with classical science. Once more, correlation does not prove causality here. But it is telling, for example, that Moslem captains in the Arabian Sea whose *dhows* had trapezoid lugsails that could be maneuvered against the wind, preferred nonetheless to wait for the monsoons. According to an old Arab proverb, "No one but a madman or a Christian would sail to windward" (Landström, 1961, p. 213).

One difference between East and West, heavily emphasized in the writings of economic historians, was that the West was politically fragmented between more or less autonomous units that competed for survival, wealth, and power. Long before the states system emerged, Europe was a pluralistic and diverse society in which many political units jealously guarded their independence from each other. Rosenberg and Birdzell (1986, pp. 60–62) link European pluralism to feudalism, arguing that feudalism ensured the political decentralization needed for pluralism. The demise of feudalism and the rise of the state did not destroy pluralism in Europe (though it might well have endangered it for a while, as, for example, when Philip II tried to stamp out heretical movements in the Netherlands). In other places (such as Japan) feudalism did little for technological creativity, whether it led to pluralism or not.

The struggle for survival guaranteed that in the long run rulers could not afford to be hostile to changes that increased the economic power of their realm because of the real danger that an innovation or innovator would emigrate to benefit a rival. Technological improvements made abroad were pursued and imitated, foreign artisans were tempted and bribed to immigrate. Regimes that did not follow this course, such as Spain and the Ottoman Empire, fell behind and lost their economic and political power. Comparatively tolerant states, such as England and the Netherlands, became the cutting edge of economic progress and acquired political influence out of all proportion to the size of their populations. Competition between states stimulated innovative activity directly through government intervention. States following mercantilist policies in seventeenth- and eighteenth-century Europe offered prizes to inventors who solved technical problems and enticed foreign engineers and craftsmen to immigrate to their countries. Thus, competition among the different states was important not only for the creation of new techniques but also for the absorption and diffusion of ideas generated elsewhere. This, of course, was precisely where classical civilization was weakest.

The economist's intuition frequently leads to a comparison of the political fragmentation of Europe with the competitive market model. Because economic competition leads to efficiency, political competition is assumed to do the same. The analogy is misleading because economic competition is usually assumed to be a costless mechanism, whereas political competition leads to expensive military buildups and destructive warfare. Where political fragmentation may have been decisive is in allowing Europe to escape the trap of what I shall call Cardwell's Law. Cardwell (1972, p. 210) has pointed out that "no nation has been (technologically) very creative for more than an historically short period."[10] As stated, Cardwell's Law is no more than an empirical regularity, and a crude one at that. The definition of a "nation," let alone that of a "short" period, is ambigious. What is more, Cardwell gives us no inkling as to the possible economic and social reasons his law might be true. Yet if we accept it as an empirical regularity, Europe's advantage becomes clear. Europe always consisted of many nations. The technological center of gravity of Europe moved over the centuries, residing at various times in Italy, southern Germany, the Netherlands, France, England, and again in Germany. Political fragmentation did not inhibit the flow of information from technological leaders to followers in Europe, and so it came to pass that the technology used in Europe always eventually settled on the best-practice technique in use regardless of where it had been invented. Jones (1981, p. 124) put it eloquently: "Books might be burned and scientists tried by the church, machinery might be smashed by mobs, entrepreneurs banished and investors expropriated by governments, but Europe as a whole did not experience technological regression. The multi-cell system possessed a built-in ability to replace its local losses . . . and was more than the sum of its parts." Furthermore, political pluralism in Europe—first under feudalism and later under dynastic states—made nonconformist thought and behavior more acceptable than it was in large empires, even if it was not always easy. In medieval and Renaissance Europe, scientists, engineers, philosophers, and religious nonconformists challenged conventional wisdom, taking advantage of the states system to protect themselves against reactionary authorities.

The price that Europe paid for political fragmentation was, of course, war. As I have argued earlier, war's gross positive effects on

10. Carr (1961, p. 154) made a similar statement in a more general context. "If I were addicted to formulating laws of history, one such law would be to the effect that the group—call it a class, a nation . . . what you will—which plays the leading role in the advance of one period is unlikely to play a similar role in the next period."

technological progress were limited, and its net effects, taking into account both benefits and costs, were surely negative. Political competition did not inevitably lead to war any more than competition between firms inevitably leads to illegal actions between them. Perhaps the best of all possible worlds was one of peaceful threats without major conflagrations, as characterized Europe between Waterloo and Serajevo. Insofar as war accompanied political fragmentation, it was a price paid, not a benefit enjoyed. In any case, political unification did not necessarily mean peace, as Chinese history suggests.

One of the more perplexing dilemmas confronting the historian interested in the political economy of technological change is that the effect of political competition on technological progress is inherently ambiguous. Nations that worry about their political standing in the world are more likely to be subject to the "Sputnik effect," the discovery that a society has fallen behind in technological terms and is consequently threatened. From Peter the Great's Russia to Meiji Japan to the United States after the launching of the first Soviet satellite, nations have embarked on efforts to advance the techniques they employed primarily for political reasons. Some measure of competition between states is therefore healthy for technological progress. Unlike economic competition, however, political competition may degenerate into military expansion, war, and destruction, which negate any possible beneficial effects of political competition. There is thus clearly a subtle optimum point between the advantages and the hazards of competition between states.

Political fragmentation was thus not a sufficient condition for technological progress. In some cases (the city-states of ancient Greece, Islamic Spain, and the fragmented political units of medieval India come to mind) political decentralization led to more destruction than innovation. As in microeconomic theory, then, competition in and of itself does not guarantee efficiency. Nor was pluralism a sine qua non, since some progress obviously did occur within imperial Rome and imperial China. And yet in Europe, provided the units were able to preserve their independence and were not crushed by the economic burdens of defense, political fragmentation guaranteed that no single decision maker could turn off the lights, that the capriciousness or piety of no single ruler could prevent technological advances and the economic growth they brought.

CHAPTER NINE

China and Europe

The greatest enigma in the history of technology is the failure of China to sustain its technological supremacy. In the centuries before 1400, the Chinese developed an amazing technological momentum, and moved, as far as these matters can be measured, at a rate as fast as or faster than Europe. Many of their innovations eventually found their way to Europe, either by direct importation or by independent reinvention. Some of the Chinese achievements are summarized here.

1. Major improvements in the cultivation of rice revolutionized Chinese agriculture. Better control of the wet-field techniques allowed a tremendous expansion of rice cultivation in the south. The control of water through hydraulic engineering (dams, ditches, dikes, polders, walls) allowed the draining and irrigation of lands. Sophisticated sluice gates, pumps, and *norias* (water-raising devices that used a chain of buckets propelled by the stream itself to create a perfectly automatic pump) controlled the flow of water and prevented silting. It has been estimated that between the tenth and fifteenth centuries the number of water control projects in China increased by a factor of seven while population at most doubled (Perkins, 1969, p. 61).

2. The old Chinese scratch plow was replaced in the sixth century B.C. by an iron plow that turned over the sod to form furrows, and consisted of eleven different parts, some of which were adjustable to set the desired depth of the furrow. Later (in the eighth or ninth century) this plow was converted to be used in wet-field rice cultivation.

3. Seed drills, weeding rakes, and the deep-tooth harrow were introduced during the Sung (960–1126 A.D.) and Mongol (1127–1367 A.D.) dynasties. Chinese agriculture learned to use new fertilizers, such as urban refuse, mud, lime, hemp stalks, ash, and river silt. Insect and pest control used both chemical and biological agents with great success. A unique feature of Chinese agriculture was the large number of tracts and handbooks published dealing with agricultural technology. The *Annals* of the Sui dynasty (581–617 A.D.) mention the existence of eight treatises on veterinary medicine. Later, such

Figure 40. Fertilizing rice seedlings in medieval China, from the *Keng Chih* (Agriculture and Sericulture Illustrated), 1145.
Source: Thien Kung Khai Wu 1/15b.

masterworks as *The Essentials of Farming and Sericulture* and Wang Chen's massive *Treatise on Agriculture* (published in 1313) appeared. Chen's book contained 300 highly detailed illustrations, from which it was possible to reconstruct the implements depicted.

4. The Chinese led the Europeans by a millenium and a half or more in the use of blast furnaces, enabling them to use cast iron, and to refine wrought iron from pig iron. The casting of iron was known in China by 200 B.C.; it arrived in Europe at the earliest in the late fourteenth century. Although the exact dates of the begin-

ning of cast-iron production in China are unknown, there is no doubt that in the Middle Ages iron production in China far exceeded Europe's even on a per capita basis. This advantage rested on double-acting bellows that used pistons and were driven by waterpower; coal; refractory clays (to generate very high temperatures); and a superior knowledge of metallurgy.[1] One fortuitous factor was that Chinese

Figure 41. Chinese seed-drill used for sowing cereals, from the *Thien Kung Khai Wu* (Exploitation of the Works of Nature), 1637.
Source: Thien Kung Khai Wu 1/15b.

1. The Chinese succeeded, for instance, in making cast-iron bells, quite a difficult enterprise. In Europe, even after casting was known, bells were made primarily of bronze, since the Europeans did not recognize the special physical needs of bells (Rostoker et al., 1984).

Figure 42. Water-driven blowing machine in Chinese smelting furnace, from the *Nung Shu* (Treatise on Agriculture) by Wang Chen, 1313. Note the use of the crank in the transmission mechanism.
Source: Joseph Needham, *Science and Civilisation in China,* Vol. 4 part 2, Cambridge University Press.

ores tended to be high in phosphorous, which reduced the melting point of iron and made casting easier. China also led the West in steel production, in which they used cofusion and oxidization techniques.

5. In textiles, the spinning wheel appeared at about the same time in China and the West—the thirteenth century (possibly somewhat earlier in China)—but advanced much faster and further in China. The Chinese applied central power sources to the spinning of yarns for which this application was relatively simple, such as the throwing of silk and hemp, and the spinning of ramie, a Chinese fiber plant. In cotton, the application of central power sources was not solved until the British Industrial Revolution, but the Chinese did manage early on to develop a small multispindle spinning wheel not unlike Hargreaves's spinning jenny. Sophisticated weaving equipment came even earlier: draw looms that wove complicated patterns in silk were used in Han times (around 200 B.C.) and were later used for cotton as well. Cotton was ginned by mechanical gins. By the end of the

Middle Ages, it appears that China was about ready to undergo a process eerily similar to the great British Industrial Revolution.

6. The adoption of waterpower in China more or less paralleled that in Europe. Reynolds (1983) has demonstrated that before the third century A.D., the Chinese primarily used water levers, primitive devices that create reciprocating motion by means of a pivot with a chute at its end that receives water, thus tipping it over. As early as the eighth century A.D. the Chinese were using hydraulic trip hammers, and by 1280 they were fully committed to the vertical waterwheel.

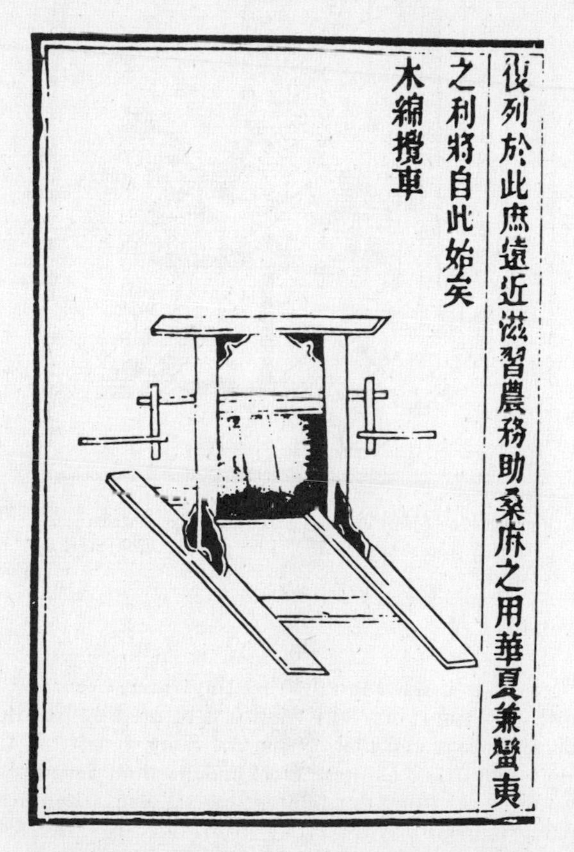

Figure 43. Chinese cotton gin, in use during the Yuan dynasty (thirteenth to fourteenth centuries). *Source: Nung Su*, 1313.

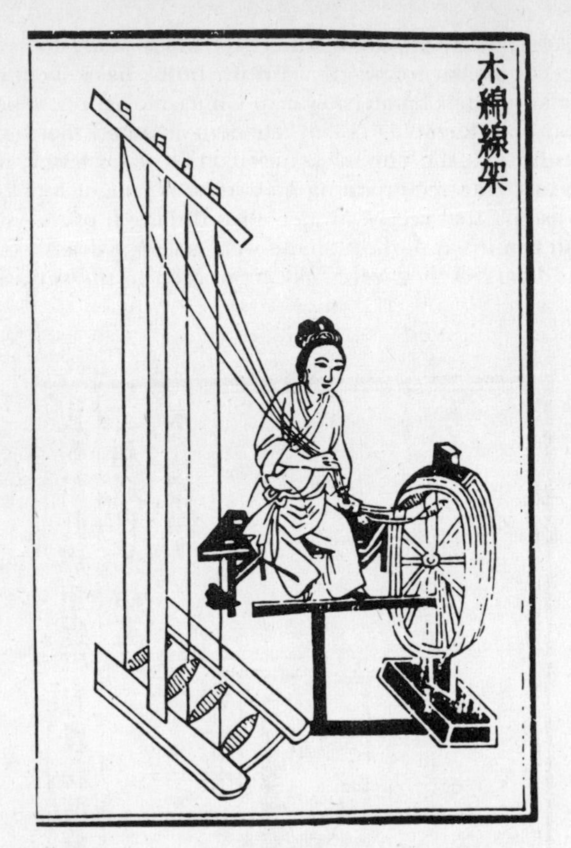

Figure 44. Spindle wheel for doubling (twisting together two separate yarns). The wheel is operated by a treadle.
Source: Nung Su, 1313.

7. For centuries it was thought that the Chinese learned to measure time from Westerners. This view was refuted by Needham and his associates, who showed that during the Sung dynasty, in the tenth and eleventh centuries, Chinese clockmakers built accurate and ingenious water clocks using escapement mechanisms (though their escapement mechanism was different from the verge and foliot that regulated the European weight-driven clock). The Chinese achievements in time measurement reached their pinnacle in the construction of Su Sung's famous clock in 1086 A.D. Su Sung's clock was probably the most sophisticated water clock ever built, measuring 40

ft. high, and displaying not only the time but also an impressive array of astronomical variables, such as the positions of the moon and the planets. Although it is not quite correct to see in the Chinese water clocks harbingers of the European mechanical clock, these instruments far exceeded in mechanical complexity, mastery of materials and mechanism, and accuracy of measurements anything that Europe had to offer circa 1100 A.D. (Landes, 1983, pp. 17–36).

8. Chinese achievements in maritime technology were also impressive. The Chinese invention of the compass (around 960 A.D.) is known to every schoolboy, but the compass was by no means their only success. In ship design and construction, the Chinese led Europe by many centuries. Their ocean-going junks were much larger and more seaworthy than the best European ships before 1400.[2] The Chinese ships were carvel built (planks laid out edge to edge), were equipped with multiple masts, and had no keel, sternpost, or stempost. The ships were built using a technique called bulkhead construction, which

Figure 45. Multiple spinning frame for ramie (hand-operated), used in medieval China. A larger version of this wheel was water-driven.
Source: Nung Su, 1313.

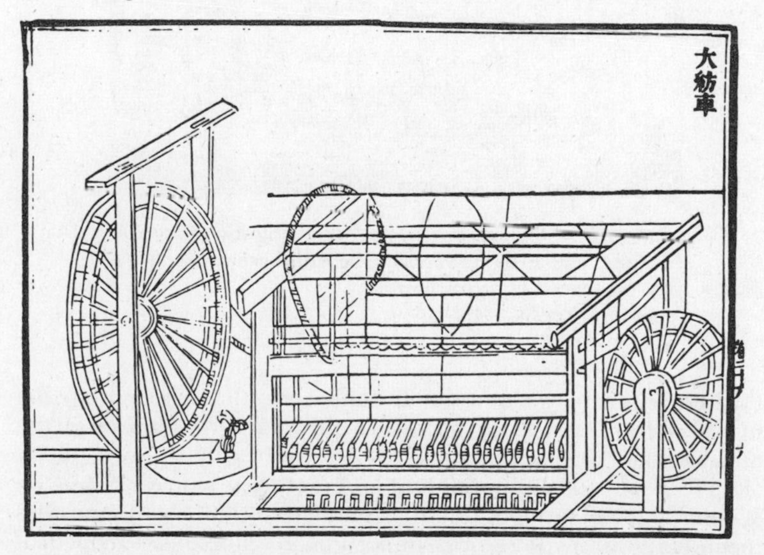

2. Marco Polo wrote in glowing terms of Chinese ships in the 1290s. The Moslem world traveler Ibn Battutah, echoed Polo's impressions half a century later. Elvin (1973, p. 137) notes that after 1000 A.D., foreign merchants preferred Chinese to foreign ships for travel.

Figure 46. Model of the "Great Cosmic Engine" made by Su Sung in 1086. The clock had a mechanical escapement, but was turned by water power.
Source: Science Museum, London.

used watertight buoyancy chambers to prevent the ship from sinking in case of leaks. Needham suspects that this idea was inspired by the nodal septa of the bamboo stick (Ronan and Needham, 1986, p. 66). Despite its obvious advantages, the technique was not adopted in the West until the nineteenth century. Needham (1970, p. 63) concludes that Chinese ships were of "a much more solid construction than that found in other civilizations."

Chinese ships were also maneuverable to a degree not equalled elsewhere before the development of the triple-masted caravel in the mid-fifteenth century. As early as the third century A.D. the junks were equipped with trapezoidal lugsails with fore-and-aft rigging, thus permitting the junks to sail to windward in much the same manner as the Western lateen sail. The Chinese also developed the sternpost rudder long before it appeared in Europe, an achievement all the more remarkable because Chinese ships had no sternpost properly speaking.[3]

9. The Chinese invented paper, an invention that took more than a millennium to reach the West. The invention was traditionally attributed to Tshai Lun, around 100 A.D., but modern scholarship has shown that paper was used several centuries earlier (Tsuen-Hsuin, 1985, pp. 40–41). Tshai Lun did pioneer the use of tree bark as a raw material. Paper was used for more than writing: high quality and durable paper found its way to the manufacture of clothing, shoes, and military armor. Paper money and wallpaper were used in medieval China, centuries before Europe, and as early as 590 A.D. the use of paper in Chinese lavatories was widespread (Needham,

Figure 47. Principal sail types. Sails of type D, D′, and E were the lug and sprit sails used on Chinese junks.
Source: Joseph Needham and Colin A. Roman, *The Shorter Science and Civilisation in China*, Vol. 3, Cambridge University Press.

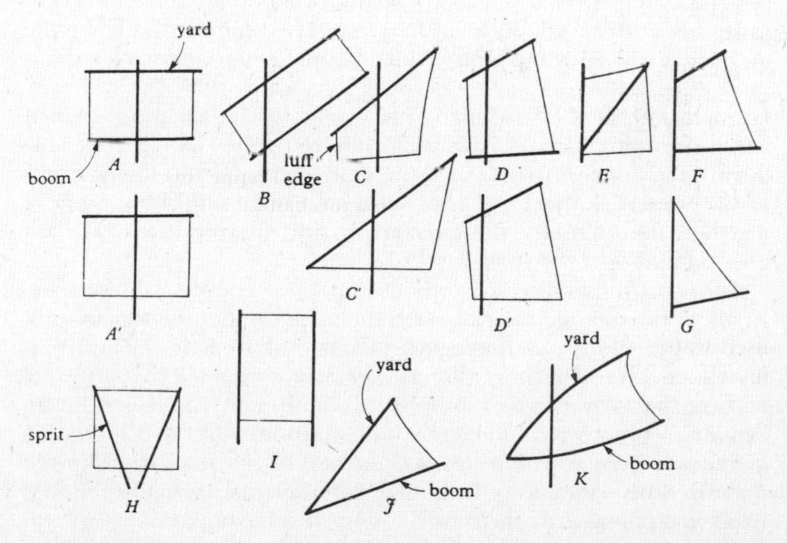

3. An archeological find in 1958 unveiled a pottery model from the Han dynasty (first or second century A.D.) that clearly shows an axial rudder (Needham, 1970, pp. 257–58).

1970, p. 373). Printing probably began in the late seventh century.[4] At first the Chinese used block printing, a technique known as xylography that used a wooden plank carved in reverse, but in 1045 A.D. Pi Shêng invented moveable type made of porcelain. Metal moveable type was used in Korea in about 1240. Whether or not Gutenberg knew of these inventions is still in dispute; the evidence and logic are against it. Yet even if Europeans invented it independently, the triumph of Chinese technology stands undiminished.

10. Many other examples of Chinese precocity appeared between 700 and 1400 A.D. Genuine porcelain appeared in China in the T'ang period (618–906). In the early fifteenth century the famous porcelain pagoda of Nanking was built, nine stories (over 260 ft.) high, with outer walls cased with bricks of the finest white chinaware. The chemical industry in China produced lacquers, explosives, pharmaceuticals, copper sulfates (used as insecticides), and metallic salts. The idea of attaching a wheel to a stretcher and dispensing with the second bearer does not seem to have occurred to anyone in Europe before the twelfth century, whereas in China wheelbarrows were used from 232 A.D. on, and probably earlier. The Chinese drilled deep holes (up to 3,000 ft.) in the ground for the extraction of brine in Szechuan Province. Many centuries before Eilmer of Malmesbury, the Chinese used kites for manned flights. In military matters, the crossbow and the trebuchet were standard equipment in China centuries before their adoption in the West. The Chinese developed the modern horse collar in about 250 B.C., an invention that took a thousand years to find its way to Europe (Needham, 1965, pp. 311–12). In medicine the Chinese made important breakthroughs, some of which (like acupuncture) have been fully recognized in the West only during the last few decades. In the daily comforts and amusements of life, the West owes to China mundane but useful ideas such as matches, the umbrella, the toothbrush, and playing cards. The list goes on and on.

And yet China failed to become what Europe eventually became. At about the time we associate with the beginning of the Renaissance in Europe, China's technological progress slowed down, and ultimately came to a full stop. China's economy continued to expand, to be sure, but growth was mostly of the Smithian type, based on an expansion of internal commerce, monetization, and the colonization of the southern provinces. Some techniques that had been known fell into disuse, then were forgotten. In other cases great beginnings were not pursued to their full potential. The implications of this

4. The first known printed book was discovered in 1907 in a cave. It is a Buddhist text printed in 868 A.D. (Huard et al., 1969, p. 287).

failure for world history are awesome to contemplate. The Chinese were, so to speak, within reach of world domination, and then shied away. "China came within a hair's breadth of industrialising in the fourteenth century," writes Jones (1981, p. 160). Yet in 1600 their technological backwardness was apparent to most visitors; by the nineteenth century the Chinese themselves found it intolerable.[5]

The slowdown in the rate of technological change should not be interpreted as economic stagnation. Until the nineteenth century, Manchu China was able to feed a growing population, apparently without a decline in living standards. Yet this growth had none of the technological dynamism of the Tang and Sung periods. It is difficult to time the waning of China's technological progress with any accuracy. A gradual deceleration took place that by the nineteenth century resulted in unmistakable backwardness compared to Europe. In 1769 a British visitor to Canton, William Hickey, observed that, when shown how greatly they might benefit from adopting European technology, the Chinese "without hesitation admit our superiority with the utmost sangfroid, adding in favour of their own habits, 'truly, this is China's custom' " (cited in Jones, 1989, p. 16). An often-cited example of the gap between Western and Chinese technology is the Opium War between Britain and China in 1842, when superior technology allowed Britain to impose scandalous terms on a huge and proud empire.[6] The awareness that the West was gaining on them must have been a constant torment to informed Chinese officials long before that. The dilemma they faced was, as Cipolla put it, one of Hamletian doubt: should they copy the West or ignore it?

For centuries China chose to ignore. From the rise of the Ming dynasty in 1368 until the end of the nineteenth century, the Chinese economy expanded primarily through population growth, deforestation, commercial expansion, and ever growing intensification of agriculture, in an environment of increasingly stagnant technology.

5. One of the most amazing features of the great Chinese decline was that despite continuous contact with European civilization, it was difficult for them to admit how much they were falling behind Europe. A Chinese text written at the end of the eighteenth century claimed that apart from some progress in land surveying and irrigation, the achievements of the West had been confined to toys and oddities (Cipolla, 1967, p. 89). Yet from the middle of the sixteenth century on, the Chinese knew that Dutch and Portuguese guns, clocks, and instruments outperformed theirs. Even Western ships were regarded by the Chinese and Japanese as superior (Cipolla, 1965, p. 122).

6. As Headrick (1981) describes in detail, a handful of steam-driven British gunboats entered the Yang-Tseh River, and their powerful guns and unexpected ability to sail upstream created havoc among the Chinese, thus deciding the outcome of the war in Britain's favor.

The Chinese experience is a powerful counterexample to the Boserup-Simon theory that population pressure leads to technological progress.

Among the cases of Chinese technology that were actually lost or forgotten, that of time measurement is the most interesting, because of the technological spillover effects that the mechanical clock is alleged to have brought to Europe. By the sixteenth century, the Chinese had no memory of Su Sung's masterpiece. Nor did they ever manage to develop anything close to the weight-driven mechanism of the European clocks. The Jesuits arriving in China in the 1580s reported that Chinese time measurement was primitive, and craftily used clocks as a bait to the Chinese authorities to gain entrance into China. The Chinese expressed joy and wonderment over the novel device, but regarded it as a toy rather than a useful instrument.

In ocean shipping, China's decline relative to the West was abrupt. Less than a century after the great voyages of Cheng Ho, the Chinese shipyards were closed and seagoing junks with more than two masts were forbidden. The technology of building large, seaworthy junks capable of long-distance journeys disappeared from China. In the iron industry there is less direct evidence of technological decline. In 1690 there is some evidence of a cold blast being applied in steel making, a sort of proto-Bessemer converter. Yet even an admirer of Chinese technology like Needham is forced to concede that "in modern times the world has seen China as a culture of bamboo and wood" (Needham, 1964, p. 19). Or consider the *sao chhê,* a silk-reeling machine that was used in China as early as 1090 A.D. (Needham, 1965, pp. 2, 105–8). Yet by the middle of the nineteenth century raw silk, which comprised about 35 percent of China's export, was entirely hand reeled, yielding a product of uneven quality that had to be rereeled in Europe (Brown, 1979, p. 553).[7] Or consider coal, which had been mined in China since medieval times and was reported with some amazement by Marco Polo. By the nineteenth century Chinese coal mining was primitive, took place in shallow mines, and was devoid of any machinery for ventilation, pumping, or elevation (Brown and Wright, 1981). Between the medieval era and the modern age something was lost in China.[8]

Equally striking is the inability of Chinese technology to press for-

7. It is possible, as Kuhn (1988, pp. 400–4) has maintained, that there was little gain in productivity by adopting mechanical reeling except for the increased uniformity of quality. Yet by the end of the nineteenth century, China adopted the Western silk-reeling machinery.

8. The decline of the Chinese iron industry has been ascribed in part to the Mongol conquest of the thirteenth century and the reduced military demand for iron by the Yuan dynasty (1264–1368), whose military technology was not iron based.

ward in areas in which they were very close to making a breakthrough. Moveable type, for instance, did not catch on in China, where wooden block printing continued to dominate. A likely explanation is that moveable type was less suitable to an ideographic script than to the simpler alphabet of the Western world.[9] But how can we explain the failure of Chinese spinners to develop a proper spinning jenny? As Chao (1977) has pointed out, multispindle spinning, adopted for ramie, never found an application to cotton, where small spinning wheels spun three or four spindles, but never more. The critical element in Hargreaves's spinning jenny, missing in Chinese cotton spinning technology, was the draw bar, a device that imitated the human hand in drawing out a large number of rovings at the same time. It seems hard to believe that such a relatively simple device never occurred to some ingenious Chinese, but if it did, there is no sign of it. Similarly, the Chinese developed a treadle loom in the Ming period (1368–1644), but after that weaving remained unchanged until the end of the nineteenth century. Something like the flying shuttle, a simple device that increased the productivity of weaving by a large factor, never seems to have occurred to them.

In many other areas, the Chinese were unable to press their advantage. Consider, for example, military technology. In the tenth century A.D. the Chinese used gunpowder in rockets and bombs. In spite of their knowledge of explosives and their superiority in siderurgy, they apparently had to learn to use cannon from the West (in the mid-fourteenth century) and they failed to develop Western military techniques any further.[10] When the Portuguese reached China in 1514, the Chinese were deeply impressed by Portuguese muskets ("Frankish Devices") and swiveling naval cannon and adopted them readily (Needham, 1981, p. 44). Yet the Chinese were unable to keep up with the continuous progress made in firearm technology in the West. In the seventeenth century, the Ming emperors had to ask the Jesuits in China to aid them in purchasing cannon from Macao to defend their country against the Manchu Mongolians. In the 1620s, Chinese officials repeatedly advised the adoption of Western cannon by the Chinese army. As late as 1850 the Chinese army still used

9. In Korea, where moveable type first appeared, fifteenth-century printers invented a phonetic alphabet known as *Hangul,* which could have made printing far easier. Yet vested interests insisted on clinging to the old Chinese characters, and consequently few books were printed in Korea until the nineteenth century (Volti, 1988, p. 141).

10. It is possible that someone in Szechuan province actually invented a cannon in the eleventh or twelfth century. Yet there is no evidence that firearms were used in China before they arrived from Europe. I am grateful to Prof. Lynda Schaffer of Tufts University for communicating this information to me.

weapons of sixteenth-century vintage, and only the pressing need of civil war during the Taiping Rebellion (1851–64) compelled them to buy modern firearms in the West (Hacker, 1977).

In waterpower technology progress never quite stopped altogether, but their accomplishment "does not compare to the European accomplishment, particularly in the period from the eleventh to the sixteenth century" (Reynolds, 1983, p. 116). In agriculture, the Chinese came into contact with new crops from the American continent through Portuguese traders and Chinese settlers in the Philippines. Their record in adopting these crops is mixed: some crops, such as sweet potatoes and peanuts, which thrive on marginal soils, were widely cultivated. But the major dryland staples, such as potatoes and corn, were adopted only slowly, in spite of the advantages of these crops (Perkins, 1969, p. 47; Bray, 1984, p. 458).[11] A specialist in ecological history, (Jones, 1981, p. 171) judges that Chinese progress in agriculture "in no way compares with Europe's record of technological achievements." Another historian of Chinese agriculture (Chao, 1986, p. 195) states flatly that "the invention rate [in Chinese agriculture] declined sharply after 1300 and finally came to a complete halt after 1700." Opportunities for improvements were missed. European piston pumps were brought to China, and would have been of great value in the irrigated farming regions of northern China, such as Hunan and Shansi provinces. They were rarely used however, presumably because of the high cost of copper. But even a simple device such as the Archimedes screw pump, which was brought to China by Jesuits and which the Chinese at first showed some ingenuity in adopting, was little used because of the high cost of metal (Elvin, 1973, p. 303). How it came to pass that a society that pioneered metallurgical technology in the Middle Ages had to forego the adoption of simple and useful gadgets because of the cost of materials is still a mystery. Even in the dissemination of technical knowledge there appears to have been retrogression: the great technical encyclopedia, the *Thien Kung Khai Wu* (Exploitation of the Works of Nature), written in 1637 by Sung Ying Hsing, ("the Chinese Diderot") provided an excellent summary of Chinese technology from weaving to hydraulics to jade working. The work was destroyed,

11. Corn had the advantage that it could be grown in relatively barren mountains, which had low opportunity costs. The same is true a fortiori for potatoes, which are labor intensive and provide a high yield per acre, making them an attractive crop for high-density populations. The alleged reason these crops failed to be adopted more widely in China is that the Chinese disliked their taste to the point where "eating potatoes is considered to be an act of desperation preferable only to starvation" (Perkins, 1969, p. 48) and "positively disliked [maize] . . . ," regarding it a "last resort" (Bray, 1984, p. 458).

probably because of the author's political views, and has survived only thanks to a Japanese reprint.[12] Wang Chen's great *Treatise on Agriculture* was published in 1313, but by 1530 there was only one surviving copy.[13]

It seems tempting to explain the waning of China's technology after 1400 as merely a relative change. The time at which China's technological change slowed down happened to coincide roughly with the century in which Europeans learned to cast iron, to print books, and to build ocean-going ships. Some historians (e.g., Hucker, 1975, p. 356) have tried to explain China's backwardness in modern times by arguing that the technological slowdown in China after 1400 was quite natural, and it was Europe's spectacularly fast advance that must be explained. But this relativistic view of Chinese history is not wholly satisfactory. First, China's lack of progress after 1400 is striking not only in light of Europe's success, but also compared with its own performance in the previous centuries. Second, such a comparative approach really only begs the question further. The European experience seems to suggest that nothing succeeds like success: upon the slow but relentless progress of the early Middle Ages, the engineers of the Renaissance built further achievements, which in their turn paved the road for the inventors of the Industrial Revolution and for the complete technological superiority that Europe had attained by 1914. Why does such a cumulative path-dependent model not work for China? After all, the technological problems the two civilizations had to solve—how to secure the necessary fertilizer for arable farming, how to produce textiles, how to utilize sources of kinetic and thermal energy and how to secure the supply of high-quality materials for tools and construction—seem similar. The solutions were indeed so similar that Needham has repeatedly argued for the transmission of innovations from the leader (China) to the follower (Europe).

In a related vein, Bray (1986) has argued that China's apparent failure is to some extent an optical illusion. Technological change in

12. The many beautiful and illustrative pictures of mechanical devices in the work clearly refute Cipolla's (1967, p. 33) generalization that the difference between the European and the Chinese environment was exemplified by Francesco di Giorgio and Leonardo Da Vinci drawing mills and gears, whereas their Chinese counterparts drew flowers and butterflies.

13. A comparison between two agricultural treatises published by the government half a millenium apart highlights the change. The *Nung Sang Chi Yao,* dating from 1273, is a practical handbook, providing useful information and guidance on how to grow the right crop in the right soil. The *Shou Shih Thung Kao,* published in 1742 and distributed throughout China, is largely a piece of imperial propaganda aimed at the glorification of the emperor and stressing the ceremonial over the practical (Bray, 1984, pp. 70–74).

the labor-intensive rice economies simply took a form that was different from the labor-saving inventions of Europe. Population growth led to an intensification of agriculture through multicropping and other labor-intensive techniques. Bray criticizes "Eurocentric" models of historical change as biased against the kind of changes that were occurring in the Orient. Up to a point, there is merit in her views. Wet-rice fields achieved yields per acre far in excess of anything in the West. Yet output *per person,* the ultimate arbiter of economic success, remained at best stable until 1800, after which it succumbed to population pressure. The difference between the European and the Chinese experiences can be illustrated by the fact that between 1750 and 1950 Europe's population grew by a factor of 3.5 while China's grew by a factor of 2.6. Yet Europe fed itself with ease and managed to create a vast surplus above subsistence, improving living standards beyond the wildest dreams of earlier ages. In the rice economies of Asia and their wheat-producing neighbors, poverty and malnutrition became increasingly frightful. Europe and Asia differed not because one was capitalistic and one was not, or because one had developed larger-scale cereal- and- livestock farms that lent themselves better to mechanization than small-scale rice paddies. The real difference was that the West, or at least a significant part of it, was technologically creative and managed to stay so for a longer period than any other society. Unlike the Chinese, the Europeans did not just save land and capital, using labor more and more intensively. European inventions were at times labor saving, at times land saving, at times neutral. Their main feature was that they produced more and better goods.

The immense difficulty of the question of why China fell behind is illustrated by the obvious weaknesses of some of the solutions that have been suggested. Many hypotheses have been proposed to explain the failure of specific innovations to emerge in China. Chao (1977, ch. III), for example, explains the failure of China to employ a jenny-type spinning machine in its cotton industry by arguing that such a device would require three persons to work simultaneously, making it unsuitable for household production. The argument is unpersuasive, but even if it were correct it would explain but one aspect of a much larger problem.[14] Similarly, the decline of ocean shipping has been blamed on the political victory of the antinaval clique at the Imperial Court after 1430. In agriculture, the lack of fertilizer has been seen as a cause for the stagnation of productivity

14. Apart from the fact that in Britain the spinning jenny was originally intended for and used in household production, Chao wholly ignores the large-size businesses in China that manufactured iron, ships, tea, and textiles, using hired labor. A labor-exchange system existed even in rice farming (Bray, 1986, p. 120).

relative to Europe. Such an hypothesis merely begs the questions, however, because it fails to explain why the Chinese were unable to increase livestock production through fodder crops, as was done in Europe, or why they were so slow in adopting corn or potatoes. In her monumental work on Chinese agriculture, Bray points out that wet-rice cultivation did not lend itself to mechanization because of its small optimal production scale and the difficulty of developing machinery that would replace labor in the wet fields without reducing yields (Bray, 1984, p. 613). But how does this theory explain the lack of progress of agriculture in the dry-field cereal economy of north China? Jones (1981, p. 221) emphasizes the importance of internal migration, which operated as a safety valve. The opportunities of the southern forest lands drew entrepreneurs away from the technological frontiers of Sung times, "setting Ming and Manchu on a course of static expansion." But in the West colonization and internal migration do not seem to have interfered with technological innovation. In the twelfth and thirteenth centuries, when western Europeans colonized the areas east of the Elbe River, progress seems to have been as rapid as ever.

An ambitious theory to account for the Chinese decline has been suggested by Elvin (1973), who tried to explain the stagnation of the Chinese economy in terms of a "high-level" equilibrium trap. Elvin's model assumes that the opportunities for technological change in agriculture were limited, and that population growth shifted demand from nonagricultural goods to agricultural goods. Moreover, he suggests, population pressure reduced the supplies of indispensable materials, such as wood and metals, which reduced the opportunities for technological change. It is interesting to note that this approach is diametrically opposed to those theories in which technological change is considered a "response" to the "challenge" of necessity and scarcity. Instead, in this view scarcity impedes technological change. Elvin's theory, however, is hard to square with other facts, among them the sharp decline in population due to the epidemics that devastated China between 1580 and 1650 and reduced population by 35 to 40 percent by Elvin's own estimates (ibid., p. 311), though other sources estimate the decline to have been smaller. Furthermore, he assumes that population growth led to a falling surplus and per capita income, and therefore to declining demand for manufactured goods to explain why "profitable invention became more and more difficult" (ibid., p. 314). According to Perkins (1969) and Gernet (1982), the significant decline in income or agricultural output asserted by Elvin did not occur. Moreover, the argument involves circular logic, as successful invention would have raised real incomes and thus made itself profitable. Finally, and most importantly, Elvin confuses *total* with *per capita* income: the determinants of market demand are both income per capita and the number of capita in the market. At least on the latter account, successful invention in China must have been more profitable than Elvin imagines, because population went from 75 million in 1400 to 320 million in 1800. Finally, Elvin's argument that demand grew mostly in the agricultural sector whereas the technological opportunities were in manufacturing ignores the linkages between nonagricultural technology and food supply through improved transportation and industrially produced implements. It also fails to explain why Chinese farmers were so slow to adopt steel-tipped plows or piston-driven water pumps in this period, or why they were reluctant to adopt a labor-intensive, high-yield crop, such as potatoes.

The problem seems so huge that it is tempting to resort to some exogenous but relatively simple theory to explain a massive societal behavior change. Physiological or dietary factors thus appear an attractive alternative to social explanations. China's ever-increasing dependency on rice as its main source of food could be related to protein deficiencies, particularly in view of the low consumption of meat and the total absence of dairy products. The shift from wheat to rice as the center of gravity of Chinese society moved southward may have been associated with a decline in average nutritional status. A telling comment of travelers was that the southern Chinese were substantially shorter than the northern Chinese, whose diets depended less on rice. Whether or not the increasing dependency on rice can be correlated with increased protein deficiency remains unclear, but the possibility seems worthy of further investigation.[15] Malnutrition need not have been confined to children, and it is possible that the overall level of food production in parts of China did not keep up with the rapid growth of population, leading not to mass starvation so much as to the lethargy and lack of energy characteristic of undernourished populations. An interesting related point is made (but not pursued) by Jones (1981, pp. 6–7), who notes that when the demographic center of gravity of China shifted south, much of the farm work was carried out in warm standing water and human feces were used as fertilizer. This led to an incidence of para-

15. Polished rice contains relatively small quantities of protein (seven percent), though its protein tends to be of a high quality. Unmilled ("brown") rice is richer, but because it is more difficult to cook and chew, it is less popular, especially for infants. Without knowing more about the quantities consumed by infants, any inferences concerning infant protein deficiency syndrome (IPDS) must remain speculative. The cereals in the Chinese diet were supplemented by fish and soya products, which provided protein supplements. In the dense agricultural regions of the south, however, the dependence on rice diets may have caused significant nutritional deficiencies.

sitic disease unparalleled in Europe. Debilitating endoparasitic infection—especially schistosomiasis, which is closely associated with wet-paddy cultivation—may have devastated the energetic and adaptable labor force required for sustained technological change. Explanations of macrohistorical events based on human physiology may seem farfetched and speculative. But the connections between changes in human biology and economic history have only recently begun to be explored, and future research will have to examine this issue in detail.

A popular explanation for China's backwardness has been that the Chinese frame of mind was somehow not suited to scientific and technological progress. In a famous essay entitled "Why China Has no Science," published in 1922, the Chinese philosopher Fêng Yu Lan argued that Chinese philosophy was inherently inward looking. It was the soul, not the environment, that the Chinese wished to conquer. "The content of [Chinese] wisdom was not intellectual knowledge, and its function was not to increase external goods," wrote Fêng Yu Lan (cited in Needham, 1969, p. 115). Confucian philosophy viewed the purpose of scholarship and public administration as the maintenance of harmonious relationships within society and an equilibrium between human beings and their natural environment. But the argument in this crude form is untenable. Before 1400 China not only had a science, but its science and technology were in many ways superior to those of the West. The Chinese never hesitated to put their inventions to use, creating the same kind of free lunches the West learned to generate in the Middle Ages. Even after 1400, there seems to be little evidence of any reluctance by the Chinese to tamper with their environment. In the eighteenth century, massive deforestation and colonization of southern woodlands took place, leading to soil erosion and environmental traumas typically associated with industry and rapid expansion. The worst error one can make in this area is to attempt to explain China as if it were an unchanging entity. The question most in need of an answer is not why China differed from Europe, but why China in 1800 differed from China in 1300.

Between the one extreme position (that China's technology was backward because of an aversion to manipulate and exploit nature) and the other (that there was no difference between China and the West), it is possible to occupy a middle ground. The difference between Chinese and European civilizations was one of degree, a degree that rose after 1400, when Europe's attitudes to the material world grew increasingly exploitative. Both civilizations believed in the right to use nature to satisfy human material needs and improve the human lot whenever possible. But the history of technology is not only the tale of taking advantage of opportunities, it is also one of creating them. Here the aggressiveness of the West and its belief in unbridled and unconstrained progress differed substantially from the more moderate Eastern view. Needham (1975, p. 38) cites with approval Lynn White's essay "The Historical Roots of our Ecological Crisis" (repr. in White, 1968), in which he places the responsibility for the ecological crisis squarely on Western religion. The belief in a personal God looking with approval upon the relentless exploitation of material resources was lacking in China. Instead, Chinese (primarily Taoist) natural philosophy sought to find an equilibrium between humanity and the physical environment. Neither the submission of humans to overwhelming forces of nature, nor their unquestioned dominance over nature in the anthropocentric view held by Western civilization was prevalent in China. Rather, a steady state in which a cooperative, harmonious relationship between nature and humanity prevails was held to be ideal. "The key word is always harmony," writes Needham (1975, pp. 35–37), "to use nature it was necessary to go along with her . . . there was throughout Chinese history a recognition that man is a part of an organism far greater than himself."

Even in this modified form, less extreme than the Fêng view, the importance of the influence of natural philosophy on technology has been disputed.[16] A possible scenario is that there was a gradual change in the Chinese attitude toward nature after 1400 or so. Some scholars believe that the rise of a "sterile conventionalized version of neo-Confucianism" may have led to a replacement of the vigor of the Tang and Sung periods by an introspective culture and political lethargy that was reflected in many branches of science and technology (Ronan and Needham, 1986, p. 147; Gille, 1978, pp. 466–67). An important example of this school was Wang Yang Ming, a sixteenth-century philosopher who taught that nature was but a derivative of the human mind and that outside the mind there are no principles or laws. Such a theory, had it been widely accepted, could help explain the slowdown of Chinese science, although Wang also taught that knowledge is nothing unless it is put into practice (Harrison,

16. The key word in Chinese is *wu wei,* which encapsulates Taoist wisdom. In Bertrand Russell's words, it means "production without possession, action without self-assertion, development without domination." Max Weber contrasted Confucian rationalism, which implied rational adjustment to the world with "Puritan" rationalism," which implied rational mastery of the world. Some modern writers, including Elvin (1984) and Jones (1981), tend to be rightfully skeptical of this simplified view of Chinese economic history, pointing to the contrast between Chinese farming, which made heavy use of irrigation, river control, terraced slopes, and similar manipulations of nature, and European rain-based farming.

1972, p. 336). In any event, metaphysical idealism never became dominant in Chinese thinking, and it seems of little help in explaining the slowdown in technology in post-medieval China.

An interesting difference between Chinese and Western thinking concerns logic. Unlike the West, China failed to develop a system of formal logic. Despite their achievements in algebra, the Chinese do not appear to have been interested in rigid logical structures such as "something is either A or not A." Instead, they were attracted to what we would today call "fuzzy logic," a rather recent branch of mathematics in which concepts such as "perhaps" or "somewhat" are allowable. Needham (1975, pp. 31–33) maintains steadfastly that this peculiarity in Chinese thinking had no effect on scientific developments and believes with Francis Bacon that "logic is useless in scientific progress." Clearly, however, that statement is one by which Bacon would not have liked to be remembered. His objection to mathematics as a tool of scientific inquiry was ignored by Galileo, Descartes, Newton, Leibniz, and almost every other leading scientist of his time.[17] In a similar vein, Hartwell (1971) has argued that Chinese logic was based on historical analogy rather than on the hypothetico–deductive method of the West. He admits that inference by analogy and induction can lead to successful engineering discoveries, but transformation of the modern world using inductive methods would have taken "several millennia" rather than "three or four centuries" (ibid., pp. 722–23).

The question is not whether the lopsided development of Chinese mathematics impeded modern science, but whether it had strong repercussions on technology. Although we have seen that mathematics began to penetrate the work of Western engineers in the seventeenth century, it seems exaggerated to attribute all of the divergence of Western and Eastern technologies to it. The question raised by Fêng is thus doubly ill-posed for our purpose: China *had* a science but it eventually became "mired in an obsolete traditionalism" (Gille, 1978, p. 467). It was also a technological leader for many centuries, but lost its leadership after 1400 to the West. However, the nexus between technology and science is much more subtle than earlier writers seem to assume. Needham and other writers on China may have overestimated the impact that Western science had on Western technology before the mid-nineteenth century. The inference that the West had a modern science and *ergo* had superior technology is unwarranted. We simply do not know what kind of logic was em-

17. It is interesting to observe that Needham changed his mind on this point. In a paper first published in 1964 (1969, p. 117), he wrote that "in Western Europe alone there developed the fundamental principle of the application of mathematized hypotheses to Nature, the use of mathematics in putting questions."

ployed by the unknown men and women who helped advance technology in the centuries before the Industrial Revolution. The correlation between scientific and technological development does not imply causation. The same kind of factors that brought about the growth of science in the West were also responsible for its technological progress. Fêng's question should have been why China lost its lead in science, not why China had no science at all. Yet the answer to that question, important by itself, is not necessarily the same as the answer to the question why China lost its lead in technology, which is my main concern here.

It seems thus all too readily assumed that China's technological backwardness relative to the West is attributable to the undeniable progress in European relative to Chinese science after 1500. Tang (1979) believes that Chinese agricultural technology ran into diminishing returns in the absence of scientific breakthroughs. But science explains little of the increase in European agricultural output in the century between 1750 and 1850, when population increased by close to 100 percent without any visible indications of diminishing returns. Tang (ibid., p. 9) argues that "the sort of tinkering and experiment and discovery by trial-and-error in which the Chinese got their early technological jump could not have led to a cumulative systematic knowledge base capable of advancing itself and sustaining an endless flow of ever more advanced applications." Yet such trial and error tinkering was precisely how Europe established its technological lead over China between 1500 and 1800. The differences between Europe and China in this regard are not overwhelming. Needham (1959, p. 154n) remarks that those who point out that China never produced men like Galileo or Descartes forget that China did produce men like Agricola and Tartaglia.

In terms of social explanations, no fully satisfying interpretation has been proposed. Regrettably, the planned eighth volume of Needham's *Science and Civilization in China,* in which he promised to deal with the social factors determining technological change, has not appeared. In an essay originally published in 1946, Needham (1959, pp. 166–68; 1970, p. 82) suggested that the main cause of China's failure to develop a European-style technology was the failure of merchants to rise to power in China. In Europe, Needham maintained, technology was closely related to merchants, who financed research in order to develop new forms of production and trade. In Chinese society, dominated by its imperial bureaucracy, little or no private profit could be gained from mechanics, ballistics, hydrostatics, pumps, or other forms of applied knowledge. Economic historians such as Rostow (1975, pp. 19–21) have found this view impossible to accept. Modern research suggests that Needham's premise is incorrect. Metzger (1979) has pointed out that in the Ming-Qing period (1368–1911), the social status of commoner groups, among which merchants were predominant, improved. Moreover, whatever the role of mercantile capitalism in the development of Europe's technology, it is not clear that the political position of European merchants was on the whole much more powerful than that of their counterparts in China, where trade and commerce were as well or better developed. Finally, the whole notion that research was "financed" seems oddly anachronistic.

In more recent writings, Needham seems more inclined to embrace views closer to those of Jones's and others (cf. Needham, 1969, pp. 120–22). China was and remained an empire, under tight bureaucratic control. European-style wars between internal political units became rare in China after 960 A.D. The absence of political competition did not mean that technological progress could not take place, but it did mean that one decision maker could deal it a mortal blow. Interested and enlightened emperors encouraged technological progress, but the reactionary rulers of the later Ming period clearly preferred a stable and controllable environment. Innovators and purveyors of foreign ideas were regarded as troublemakers and were suppressed. Such rulers existed in Europe as well, but because no one controlled the entire continent, they did no more than switch the center of economic gravity from one area to another.

Perhaps the best example of what a regressive government could do comes not from technological change, but from Chinese geographical exploration, which was completely halted after 1430 due to a decision by the imperial court. No single European government could have stopped exploration: when the Portuguese lost the initiative in their overseas empire after 1580, the Dutch and the English were all too happy to assume their duties. More was involved, however. The exploration and exploitation of the new territories by Europeans was a joint venture between private and public enterprise. Why did China not produce an organization like the East India Company—say, a West Europe Company—after 1430? The answer is not only that the government prohibited the construction of large ships, but more fundamentally that the Chinese lacked demand for foreign goods. Europeans had always wanted things that only the Orient could produce; the Oriental empires had little interest in Europeans and their products. All this changed in the middle of the nineteenth century when both Japan and China discovered the military capability of Western weapons. Even then, Japan adopted European technology rapidly lock, stock, and barrel, while China tried for decades to import European arms while preserving its old social and economic institutions (Hacker, 1977).

Can a similar argument be applied to technological change? Could it be that the Chinese simply lost interest? In other words, was the difference between Europe and China after 1400 one of preferences, of attitudes toward technological change and its consequences? Needham (1969, pp. 117–19) has argued that although a "stagnant" China never truly existed, there was a certain "spontaneous homeostasis" about Chinese society, which he contrasts with Europe's "built-in quality of instability." Chinese society in this view had an endemic preference for self-regulation, a set of feedback mechanisms that ensured the ergodic motion of Chinese technology. Although there is a certain attractiveness to a view that reduces the difference between East and West to the difference between an equilibrium and a disequilibrium model, as stated it is more a description than an explanation. The statement could be interpreted, however, to mean that on average the Chinese somehow valued stability more than the Europeans did. East and West may have differed in their aversion to change and in their attitudes toward the rate at which change occurs.

Why should this be so? As I argued earlier, technological progress is a positive-sum game, with winners and losers. Although by definition total gains exceeded total losses, the adjustment costs and possible political unrest may have constituted a price that some societies were not willing to pay. The evaluation of the social costs of technological progress is difficult; they could differ immensely from place to place. What may have appeared as a very cheap lunch in the West may have been regarded as unacceptably costly in China. A decline in the rate of technological change in China could thus be attributed to a change in social preferences in the direction suggested by Fei (1953, p. 74), who emphasized the desire of Chinese society to avoid the social conflicts often entailed by technological changes. Another possibility is that a shift in the distribution of power and influence within society in favor of more conservative groups took place. Here, perhaps, is one crucial difference between Europe and China. Luddism and outright resistance to technological progress did occur in China, though the number of documented cases is small (Elvin, 1973, p. 315).[18] Potential innovators may have sensed the dangers and shied

18. In 1870, a Chinese silk weaver by the name of Chen Chi-yuan built a steam-driven silk-spinning mill that he had copied from the French in Annam (Indochina). A protest from workers who felt that their livelihood was threatened led him to modify the machine into a smaller and cheaper unit that could be more readily acquired. Not many cases were necessary for the threat of violence to deter potential innovators. Xenophobia may have had as much to do with resistance to innovation as the protection of vested interests.

away from novel ideas, often associated with Western influence.[19] The guilds in China remained powerful, and have been blamed for blocking the adoption of improved technologies in mining, transport, soybean-oil pressing, and silk reeling (Olson, 1982, p. 150). One author concludes that "market forces could not overcome vested-interest opposition and ensure success even in the transfer of a demonstrably superior technology" (Brown, 1979, p. 568). In China's past, technological progress had typically been absorbed by the political status quo, without disturbing the existing order. Radical technological changes threatening the balance of power were carefully avoided.

The difference between China and Europe was that in Europe the power of any social group to sabotage an innovation it deemed detrimental to its interests was far smaller. First, in Europe technological change was essentially a matter of private initiative; the role of rulers was usually secondary and passive. Few significant contributions to nonmilitary technology were initiated by the state in Europe before (or during) the Industrial Revolution. There was a market for ideas, and the government entered these markets as just another customer or, more rarely, another supplier. Second, whenever a European government chose to take an actively hostile attitude toward innovation and the nonconformism that bred it, it had to face the consequences in terms of its relative status in the economic (and thus, eventually, political) hierarchy. Moreover, the possibilities of migration within Europe allowed creative and original thinkers to find a haven if their place of birth was insufficiently tolerant, so that in the long run, reactionary societies lost out in the competition for wealth and power.

Before 1400, the state played a far more important role in generating and diffusing innovations in China than it did in Europe. The government, for instance, deliberately attempted to monopolize the measurement of time and the calendar. As Landes (1983, p. 33) put it, "In China the calendar was a perquisite of sovereignty, like the right to mint coins. . . . The Emperor's time was China's time." Su Sung's great masterpiece was built by and for government officials at the emperor's instructions. In the great agricultural expansion of the Middle Ages, government played a central role in the coordination of hydraulic projects and the dissemination of technical infor-

19. In 1887, an American newspaper reported that "over a thousand telegraph poles of one line in China have been pulled down by the people, who say the telegraph is a diabolical European artifice" (cited in the *International Herald Tribune,* Aug. 29–30, 1987).

mation. Officials wrote and published books on farming and promoted the adoption of faster-ripening and more drought-resistant strains of rice, especially the Champa varieties introduced from Southeast Asia early in the eleventh century. Wang Chen and Hsü Kuang Chhi, the authors of massive treatises on farming, were government bureaucrats. As early as the Han period (221 B.C. to 220 A.D.) the government provided peasants with the capital they needed for agricultural improvements, including tools and draft animals, and actively promoted the use of better plows.[20] A millennium later, the Sung government offered financial incentives to farmers to invest in improvements (Bray, 1984, pp. 597–99). The government also played a major role in the development of transport technology and in the diffusion of medical knowledge. The Chinese imperial government established huge state-owned iron foundries that promoted the use of iron implements. Even in textiles, the Yuan and early Ming governments played a very active role in the diffusion of the use of cotton (Chao, 1977, pp. 19–21).[21]

At some point, government support was withdrawn. Europeans trying to develop Chinese mining in the mid-nineteenth century found that is was an impossible task without state support, but that such support did not exist. Chinese officials simply were not interested in technological advances (Brown and Wright, 1981, p. 80). During the Qing (Manchu) dynasty (1644–1911), the Chinese government ceased almost entirely to provide any kind of public services (Jones, 1989). It did not provide the usual elements of the infrastructure necessary for economic development, such as standardized weights, commercial law, roads, and police. In many areas, the private sector managed to substitute for the public sector in providing these services, but in technological progress this could not be done. Jones (ibid., p. 30) concludes that "China's political structure did not establish a satisfactory legal basis for economic activity." It would be more appropriate to restate this proposition in terms of change. Routine economic activities seem to have been carried out reasonably well, as long as existing technology and institutions were not changed.

Why the Chinese bureaucracy in earlier centuries came to play

20. Mencius, the Chinese philosopher who helped spread the teaching of Confucius in the third century B.C., mentions that "the Minister of Agriculture taught the people to sow and reap, cultivating the five kinds of grains" (cited in Fei, 1953, p. 65n).

21. In scientific development, too, the state played a major role in China, publishing books, keeping records, financing scientific expeditions, initiating medical research, and constructing scientific equipment. Some famous inventions are attributed to bureaucrats. Chang Hêng, a mathematician and inventor who lived in the second century A.D., and is credited with the first seismograph, was president of the Imperial Chancellery. Tshai-Lun, who invented the use of mulberry tree bark in papermaking, was a eunuch in the imperial service, in charge of instruments and weapons.

such an active role in technological progress is not easily explained. Its roots are often associated with the dependence of agriculture on public works in water control that led, in Wittfogel's famous term, to a "hydraulic despotism" that needed a bureaucracy capable of managing large projects (Wittfogel, 1957, pp. 22–59). Wittfogel insisted that such projects were aimed exclusively at social and political control. It is by now widely realized, however, that water-control projects were managed largely by local authorities and landowners.[22] The central government was involved in water-control projects largely through the diffusion of technological information and the legal aspects of water rights. Still, the notion that the ruler of a society had a responsibility toward its subjects, that there was a reciprocal exchange in the relationship between the population and the authorities, was an ancient Chinese concept firmly rooted in the teachings of Mencius whose influence grew during the Sung dynasty. The Chinese state produced its own requirements of iron, ships, and large-scale construction. Its granary system system stabilized food supplies. The state monopolized trade in some goods (such as salt), and foreign trade—insofar as it existed at all—was under state supervision. The pre-1800 Chinese state, in short, intervened directly in the economy, in part in an attempt to improve the economic well-being of the people. In Europe, the state's attempts to encroach upon the private domain of production usually ended in failure, and the idea of a social contract was not fully understood until the seventeenth century. In most cases the kings, bishops, city councillors, and whoever else represented the state in Europe were little more than another consumer, buying, selling, hiring, and borrowing at prices dictated by a larger market.

Perhaps the state assumed such a central role in China because the landed gentry and intelligentsia displayed little interest in technology, leading to a vacuum that had to be filled. The role of the Chinese elite in inhibiting the advance of technology is central to a number of interpretations. Fei (1953, pp. 72–74) advanced an argument similar to the one made earlier in connection with classical civilization. If the educated and powerful classes are not interested in production and lack technical knowledge, they will not make any effort to introduce technological improvements, and stagnation will ensue. Fei's central argument was that in traditional Chinese society the intelligentsia was a class without technical knowledge, interested mostly in the wisdom of the past, literature, and art. By regarding the world through human relations, he maintained, the Chinese intelligentsia were a conservative force, because in human relations the

22. For a summary of the Wittfogel debate, see Harris (1977, ch. 13).

end is always mutual adjustment whereas technological change leads to social disruption. Fei's explanation is somewhat vague: there is no specification of the dates he has in mind when he writes of "traditional China," nor does he state whether any changes in the outlook of the intelligentsia occurred during the Ming and Qing dynasties. Surely the lack of interest in technical matters is a common enough phenomenon among all intelligentsias. Nevertheless, the central idea that technological progress needs a bridge, however narrow, between the educated class and the working class seems a logical proposition. In China, this bridge was provided by the government.

Empires are thus not necessarily antithetical to technological progress. But the Chinese example provides us with some insight into why a negative correlation between the two has been observed. First, China has, in Needham's term, always been a "one-party state" and for 2,000 years it was ruled by the "Confucian party." In the Qing (Manchu) era, the bureaucracy did not encourage intellectual or political deviants, although the violent religious intolerance of Europe was alien to the Chinese. In contrast to Europe, there were no small duchies or city-states to which bright men with new ideas could flee. Moreover, China in the Ming and Qing eras was run by a professional bureaucracy that was at least in theory selected through competitive examinations (in practice the process was often rife with corruption and nepotism). Desirable as such a meritocracy may seem to the student of European history accustomed to rulers marked mostly by greed, violence, and incompetence, the mandarinate exacted a heavy price in terms of economic progress. By attracting, at least in principle, the best and the brightest from the commercial class, the system focused the nation's intellectual resources toward bureaucratic activity, which was by its very nature conservative. Although Needham's (1969, p. 202) conclusion that "the one idea of every merchant's son was to become a scholar, to enter the imperial examination, and to rise high in the bureaucracy" is somewhat exaggerated, it does highlight a crucial difference between Europe and China.[23] In Europe, engineers, inventors, merchants, and scholars rarely belonged to the ruling class. Talented men who were not born into the right families could not, as a rule, occupy positions of power, and thus channelled their energies elsewhere.

Most important, however, is the fact that technological change that

23. On the basis of official biographies, Wittfogel (1957, pp. 351–52) concludes that only a small minority of officials were commoners. But the proportion was rising: whereas in the T'ang period (618–907 A.D.) fewer than 10 percent of all officials were commoners, their proportion rises to 23 percent under the Ming dynasty, a trend that continued under the Manchus.

is generated in large part by public officials and a central government has the nasty weakness of depending on the government's approval. As long as the regime supports progress, progress can proceed. But the government can flip the switch off, so to speak, and private enterprise is then unlikely to step in. Innovation that is run largely by bureaucrats is thus not impossible, but it depends on their goodwill. Because most entrenched bureaucracies tend to develop a strong aversion to changing the status quo, state-run technological progress is not likely to be sustained over long periods. The Chinese miracle is indeed that it lasted so long. It ended when the state lost interest in promoting technological change.

Why the Chinese state changed its attitudes toward technological change is hard to determine. The Ming and Qing emperors were more absolute and autocratic than their predecessors. Before them, coups d'état and regicides occurred frequently, thus introducing an element of "competition" into the Chinese political market. Rigid etiquette and complete obedience and conformism became the hallmark of the Chinese government under the Ming emperors. At the same time the Chinese civil service became a major force in preserving the status quo. It learned to resist changes it did not want, and not even the most powerful emperors could implement progressive policies. The two great enlightened Manchu despots, K'ang Chi (1662–1722) and Ch'ien Lung (1736–1795), whose rules are invariably described as peaceful and prosperous, were interested in pacification, order, and administration. In their search for stability, their interests and those of the bureaucracy converged. The absolutist rule of an all-powerful monarch whose preference was for stability above all discouraged the kind of dynamism that was throbbing throughout Europe at the time. One specialist (Feuerwerker, 1984, p. 322) concluded that under Ming and Manchu rule the state "contributed little if anything toward modern economic growth."[24] This view may be somewhat exaggerated, as the imperial court was very active in bringing about the recovery of the late seventeenth century (Shang Hung-k'uei, 1981). Economic expansion did not grind to a halt in the eighteenth century. Agricultural growth occurred primarily through deforestation and clearing of the southern provinces, for

24. It is possible that part of the reason that the Chinese government lost its leadership role in technological change was that its resources shrank relative to the economy. The revenues of the Chinese central government declined after 1400 and amounted to only a few percentage points of national income, according to Perkins (1967, p. 487). Feuerwerker's (1984, p. 300) figures are higher, but they too imply a sharp fall from 13 percent of national income around 1080 A.D. to 4–8 percent by about 1750. Laffer-curve theories that associate economic dynamism with low taxes will have a hard time explaining this one away.

which the government provided active support. Commercial expansion was a natural consequence of the stability provided by the enlightened Qing despots. But there is little evidence for the kind of technological dynamism of Sung China or eighteenth-century Europe. By the fifteenth century, the role of the imperial government in both invention and innovation was far less remarkable than it had been in medieval times, and no other entity in China was in a position to replace the state in promoting technological progress. There were no substitutes for the state in China. In Europe, precisely because technological change was private in nature and took place in a decentralized, politically competitive setting, it could be sustained in the long run, it could make great leaps, and it could continue unabated despite serious setbacks and obstacles.

CHAPTER TEN

The Industrial Revolution: Britain and Europe

As we have seen, after 1750 the Industrial Revolution was initially concentrated primarily in Britain. Explaining Britain's headstart on its Continental neighbors has been a popular pastime among economic historians for many decades now, though a consensus has failed to emerge. In some global sense, the question may seem relatively unimportant compared to the much larger question of why such a deep gap opened between Europe and most of the rest of the world in the same period. Nevertheless, the variance within the West, the "successful" part of the world, remains puzzling as well. The difficulty is that during the British Industrial Revolution there were changes in many aspects of the economy that were not technological in nature, and it is impossible to disentangle demographic change, urbanization, enclosures, wars, social and commercial policy, and so on from technological changes and then compare the outcome with what happened on the Continent. In what follows, I shall confine myself to the questions of why Britain managed, for about a century, to generate and diffuse superior production techniques at a faster rate than the Continent, and serve as a model that all European nations wished to emulate and how and why it eventually lost its leadership in technology.

Technological success depended on both the presence of positive elements and on the absence of negative ones. Among the positive factors, the generation of technological ideas and the ability to implement them seem a natural enough point from which to start. The generation of ideas, as we have seen, was often an international effort. The British, to be sure, were prominent in providing technologically revolutionary ideas: there can hardly be any question that most of the truly crucial inventions in the period were made by

Britons. Yet Britain's relative role in invention was smaller than its corresponding role in implementation. Many important inventions that can be attributed to Continental inventors found their successful implementation in Britain. Such names as Berthollet, Leblanc, de Vaucanson, Robert, Appert, de Girard, Jacquard, Argand, LeBon, Heilmann, and Fourneyron deserve a place in the Inventors' Hall of Fame next to Newcomen, Arkwright, Watt, Cort, and their colleagues. In the eighteenth century, Britain did not have much of a reputation for being particularly original or inventive. Daniel Defoe remarked in 1728 that the English perfected other people's ideas. A Swiss calico printer, Jean Ryhiner, wrote in a treatise on his trade in 1766 that "[the English] cannot boast of many inventions but only of having perfected the inventions of others . . . for a thing to be perfect it must be invented in France and worked out in England" (cited by Wadsworth and Mann, 1931, p. 413). Much later, in 1829, John Farey, an engineer, told a Parliamentary committee that "the prevailing talent of English and Scotch people is to apply new ideas to use and to bring such applications to perfection, but they do not imagine as much as foreigners" (cited in Musson, 1975b, p. 81). Invention was not equivalent to technological change. More was needed.

One crucial difference between Britain and the Continent that helped Britain to establish its head start was its endowment of skilled labor at the onset of the Industrial Revolution. In his *Industry and Trade,* Alfred Marshall (1919, pp. 62–63) observed that the Englishman had access to a great variety of highly skilled artisans, with a growing stock of tools capable of work more exact than the work of the human hand. "Thus every experiment cost him less, and was executed more quickly . . . than it could have been anywhere else. When at last success had been fully achieved, the new contrivance could be manufactured more cheaply and . . . applied in production on a scale far greater than in any other country." In other words, by the middle of the eighteenth century, Britain had at its disposal a large number of technicians and craftsmen who could carry out the mundane but indispensable construction details of the "new contrivances."[1] These skills rested on an informal and antiquated system of apprenticeship and on-the-job training; they had little to do with schooling. If England led the rest of the world in the Industrial Revolution, it was despite, not because of her formal education sys-

1. Contemporary opinion supported Marshall's contention. In 1685, a French visitor called the British above all *adroit* (handy). In 1704, another observer wrote that the English are "wanting in industry excepting mechanicks wherein they are, of all nations, the greatest improvers" (cited by Hollister-Short, 1976, p. 159).

tem. In this, Scotland was different, as its schools were far superior to England's. In England the dissemination of technical knowledge took place through informal lectures, scientific societies, and technical literature.

Britain had been fortunate. In the late seventeenth century it had taken the lead in clock- and watchmaking. France, its closest competitor, had been "crippled by the exodus of some of its best practitioners fleeing a wave of anti-Protestant bigotry" (Landes, 1983, p. 219). By contrast, Britain welcomed men of technical ability whatever their religious persuasions. The mechanical skills of clockmakers became one of the cornerstones of the new industrial technology.[2] Another industry that produced skilled artisans was the shipping sector, with its demand for accurate and well-made instruments. British shipping grew enormously in the eighteenth century, and naval instruments were a chief output of men such as Harrison, Smeaton, and Ramsden. A third industry that helped to prepare the skills and dexterity necessary for the Industrial Revolution was mining. Pumps and transport equipment were crucial to mining, and both the steam engine and the iron rail were built first for use in the mines. By the end of the seventeenth century, British mining and metallurgical technology was still "between a hundred and a hundred and fifty years behind the best-practice techniques of the Continent" (Hollister-Short, 1976, p. 160). By 1760, it was at the forefront of Europe in these areas, giving it a technological advantage that has been fully recognized by historians only recently.[3] The net result of Britain's tradition in high-tech industry was that it could rely on engineers such as Wilkinson, Newcomen, and Smeaton to help build and improve machines conceived by others.[4] Other countries, of course, had some engineers of distinction. But most of these, such as the Swede Christopher Polhem, the Austrian Joseph Karl Hell,

2. The best-known inventors trained as clockmakers were Benjamin Huntsman, the originator of the crucible steel technique, and John Kay (not to be confused with the inventor of the flying shuttle of the same name), who helped Arkwright in developing the water frame.

3. Cardwell (1972, p. 74) points out that a number of basic technologies converge on mining (chemistry, civil engineering, metallurgy) and that mining sets the hard, "man-sized" problems, controlling powerful forces of nature and transforming materials on a large scale.

4. The relationship between inventor and engineer was often complex. Wilkinson's boring machines allowed Watt to build his steam engines with the required accuracy. But in 1782 the roles were reversed, when Watt constructed a small steam hammer for use in Wilkinson's ironworks. In 1768 the Carron works in Scotland could boast an advanced system of bellows using cylindrical blowers, originally conceived by John Roebuck but built by the ubiquitous engineer, John Smeaton.

and the Frenchman Jacques de Vaucanson, were relatively isolated.[5] In Britain, the number of engineers and mechanics was sufficiently large to allow interaction with each other, through lecturing, spying, copying, and improving. The famed Manchester Literary and Philosophical Society and the Birmingham Lunar Society were but two of the many mechanisms via which technical ideas and information were exchanged in Britain. Interaction among engineers, scientists, and businessmen created a total that was larger than the sum of its individual components. Technological change and the creation of new information are processes that do not obey the laws of arithmetic.

Britain did not have a significant scientific advantage that would explain its technological leadership. In view of the modest role that scientific progress played in technological progress, this observation is hardly surprising. One historian of science (Kuhn, 1977, p. 143) has even gone so far as to describe Britain as "generally backward" in the century around the Industrial Revolution, and concludes that science must have been unimportant in the technological changes of the time. Such a view, as we have seen, is somewhat oversimplified. Perhaps a better way to approach the role of science in explaining British leadership is to recognize that Britain did not necessarily have "more" science than other countries, but it did have a different kind of science. As Kuhn notes, the traditional view that British science was predominantly experimental and mechanical, whereas French science was largely mathematical and deductive seems to have withstood the test of time (ibid., p. 137). In the early stages of the Industrial Revolution, there was an advantage in having a science that was applied and down-to-earth, and a scientific community that maintained close ties with engineers and manufacturers. In Britain, as Jacob (1988) has recently stressed, scientists were not in opposition to the political status quo, nor its servants. Unlike the Continent, where scientists and *philosophes* either worked against the political establishment or were employed by it, British scientists and engineers worked together with commercially minded persons, who were more interested in money than in matters political or military. In this respect, Britain's advantage seems self-evident. Yet such an advantage was inevitably ephemeral, as the natural principles on which British

5. Britain's advantage in the supply of skilled technicians is demonstrated by the determined efforts of Continental countries, from Sweden to Portugal, to entice British workers to emigrate. See, for example, Ashton (1924, pp. 200–205). The French and Belgian textile industries went through their early stages with the help of Britons such as John Holker, William Douglas, and William Cockerill. It may have been that Britain had a large supply of technically adept human capital because it went through the Industrial Revolution first. But the causation is more likely to have run the other way.

technology rested were being uncovered during the nineteenth century. Indeed, Cyril Stanley Smith (1981, p. 36) has suggested that the successes of Britain in metallurgical technology (which had little to do with scientific insights) forced Continental countries to try to discover the science behind the practice in order to beat Britain eventually at her own game.

The formation of human capital in Britain before and during the Industrial Revolution depended on a rather unique social environment. By 1750, Britain already had a "middle class" of sorts, that is, people who were literate and well fed, and came from commercial or artisanal backgrounds. This class supplied most of the founders of large industrial undertakings in Britain (Crouzet, 1985), and there is no doubt that most of the creative technical minds also came largely from this class. The supply of creativity was channeled into industrial activities, in large part because of the lack of alternatives. Government and military services careers were closed to nonconformists, as they were, for all practical purposes, to anyone not born into a wealthy British family. Members of Parliament and army officers had to buy into their offices and maintain expensive lifestyles. The professional civil service was small, and the imperial bureaucracy was still embryonic in 1800. Moreover, by being exclusive it forced talented men born below it to search for the only key that could open the doors of politics, public schools, and landed estates: money. As far as the social elite is concerned, the landowning elite, which controlled political power before 1850, contributed little to the Industrial Revolution in terms of technology or entrepreneurship. It did not, however, resist it.

One development that may help explain the timing and location of the Industrial Revolution concerns the attitudes of the educated and literate elite toward technological change. MacLeod (1988, ch. 11) has recently argued that there was no linear progression from the Baconian notion of technological progress as a means of increasing wealth to the Industrial Revolution. In the late seventeenth century, in her view, attitudes toward inventions regressed, becoming more abstract and detached, and a concern for unemployment coupled with a sense of British inferiority emerges from the writings of economists and philosophers. Technological progress plays a much more modest role in the writings of Hume and Smith than it did in those of Bacon and Boyle. By 1776, however, the tide was turning again, and Smith's lack of enthusiasm for inventions was exceptional for his time. MacLeod interprets her evidence to mean that writers were influenced by the success of technology and the need to protect it against its detractors and enemies. The relationship was, however, more complex. Values and attitudes toward invention could and did differ greatly across time and space, as we have seen repeatedly. The causal chain ran as much from attitudes to achievements as it did in the opposite direction. Margaret Jacob (1988) has explicitly argued for this mechanism, namely that the Baconian ideas of progress were increasingly accepted by the British educated elite and constituted an essential precondition for the success of the Industrial Revolution. Moreover, she shows how other countries lagged behind in accepting these notions, thus providing a plausible explanation of British pre-eminence.

Yet advantages in human capital are fragile. With some exceptions, Britain's early inventors tended to be "tinkerers" without much formal technical schooling, whose genius lay primarily in their mechanical ingenuity. As it happened, most of the devices invented between 1750 and 1830 tended to be a type in which mechanically talented amateurs could excel. In many cases British inventors appear simply to have been lucky, although, as Pasteur once remarked, Fortune favors the prepared mind. The cotton, iron, and machine tool industries during the Industrial Revolution lent themselves to technological advances that did not require much scientific understanding of the physical processes involved. When, after 1850, deeper scientific analysis was needed, German and French inventors gradually took the lead, and the breakthroughs in chemistry and material science tended to be more concentrated on the Continent. Bessemer, Perkin, and Gilchrist-Thomas notwithstanding, the "amateur" stage in the history of technology was coming to an end by 1850. But Britain rode the wave high while it lasted.[6]

Another factor in Britain's head start in technology at the beginning of the Industrial Revolution was that Britain alone among the large European economies constituted a comparatively unified market in which goods and people moved easily. Compared to the European Continent, Britain had excellent internal transportation. Coastal shipping, canals, and roads provided it with a network unequalled by any Continental nation, with the possible exception of the Netherlands. Transport in Britain was itself becoming a specialists' occupation, run by professionals who increased efficiency, speed, and reliability (Szostak, 1986). Moreover, Britain was politically unified and cohesive. No tolls were charged on rivers and no tariffs were levied when crossing man-made lines (unlike France, for example, where before the Revolution internal tariffs were levied on

6. In 1868 a British dye manufacturer expressed his amazement to a Parliamentary commission that given the lack of scientific knowledge in Britain, the country held the position it did. "It is remarkable how well we do, considering how little we know," exclaimed another witness (cited by Julia Wrigley, 1986, p. 162).

goods moving within the country). As the technology of building roads and canals improved in the eighteenth century, Britain became an integrated market system.

Why did market integration matter to technological progress? Market size affected both the generation and the diffusion of new knowledge. In 1769, Matthew Boulton wrote to his partner, James Watt, "It is not worth my while to manufacture [your engine] for three counties only; but I find it very well worth my while to make it for all the world" (cited by Scherer 1984, p. 13). Some minimum level of demand was necessary to cover the fixed costs of development and construction. In very small and segmented markets insufficient demand may have impeded the diffusion of certain innovations that involved fixed costs. "The whole world" was perhaps an exaggeration; in the eighteenth century the British market was large enough to cover the costs of invention. Adam Smith, too, thought that market size and integration were crucial. Smith regarded technological progress as a consequence of specialization. Specialization depended on the division of labor and thus on the extent of the market. If workers concentrate every day on a particular task, they are much more likely to find "easier and readier methods of performing their own particular work." As evidence, Smith argued that "a great part of the machines made use of in those manufactures in which labour is most subdivided were originally the inventions of common workmen."[7] The Master's authority notwithstanding, it is far from obvious that a fine division of labor and a high degree of intrafirm specialization are conducive to learning by doing or to the generation of new techniques.[8]

Market integration has a more profound effect on the diffusion of new techniques. In a world of high production costs, inefficient and conservative producers are insulated from their more innovative competitors. A world of high transport costs is described by an economic model of monopolistic competition. One of the characteristics of such a model is that innovator and laggard can coexist side by side. In the region served by the innovator, lower production costs

7. To buttress his case, Smith tells of a boy who, while operating one of the first steam engines, tied a string to the handle of a valve, thus allowing it to shut and open by itself. As Cannan points out in his notes to *The Wealth of Nations* (Smith [1776], 1976, book I, pp. 13–14), the story is apocryphal. Landes (1986, p. 592) argues that the division of labor implies simplification and repetition and that these might suggest the imitation of manual skills by machines. Before standardization and interchangeable parts, however, the simplification of work brought about by the division of labor as such was not significant.

8. Brenner (1987, pp. 109–10) who is also quite dubious of the technological benefits of specialization, points out that Smith himself hardly practiced what he preached, writing with ease on half a dozen unrelated topics.

due to technological change meant a combination of higher profits for producers and lower prices for consumers. Nothing could force the laggards to follow suit, however, and the "survival of the cheapest" model so beloved by economists is short-circuited. High transport costs also made local oligopolies possible. A small number of firms in a market could facilitate conspiracies to stop new techniques, but as the market expanded and the number of firms with access to a given market increased, such "antitechnological cartels" became more difficult to organize and enforce without support from the authorities. In Britain, to a far greater extent than on the Continent, good transportation allowed competition to work, and the new technologies superseded the old sooner and faster than elsewhere. When diffusion lags occurred in Britain, they usually resulted from imperfections in the new techniques, rather than from entrepreneurial incompetence. Coke smelting and power weaving took decades to perfect, for example, but once their superiority was clear, the triumph of these new techniques in Britain was as assured and swift as the triumphs of the mule and the puddling-and-rolling technique, which were perfected comparatively quickly.

All the same, unified markets and a high degree of commercialization were not sufficient conditions for technological progress. Both Manchu China and Imperial Rome serve as counterexamples. They were commercially integrated economies in which Smithian growth was a substitute for technological progress. The Netherlands in the eighteenth and nineteenth centuries—the open economy par excellence, with a glorious tradition in shipping, highly developed commercial institutions, and a tradition of free trade—turned out to be singularly uncreative during the Industrial Revolution. It seems that Smithian growth served in many cases as a substitute for technologically driven growth, whereas in other cases it served as a stimulus and handmaiden of technological change. At this stage of knowledge it is not possible to specify the relationship between them more exactly. Nor is the nature of the relationship between competition and innovation clear. The Schumpeterian hypothesis argues that at some point less competition may actually be good for innovation. Although Britain's performance during the Industrial Revolution lends no support for this view, it remains to be seen whether the hypothesis can explain the post-1850 patterns.

Much has been made of the British political system as a cause of the Industrial Revolution. Perhaps the most distinguishing feature of Britain was that its government was one of, by, and for property owners. Economists have maintained that well-defined property rights were necessary to static efficiency. But what about technological change? The direct links between the British government and the rate of technological progress were few. Some scholars (McNeill, 1982) have maintained that government demand for military purposes led to innovation, but such effects were small. More important was the effect of patent laws on inventive activity. North (1981), in particular, argues that patents, which allowed the inventor to capture a larger part of the social benefits of his invention, were as important as a larger market. Here Britain led the Continent by a large margin. British patent law dates from 1624, whereas France did not have a similar law until 1791, and most other European countries established patent laws only in the early nineteenth century. The United States had a rather ineffective patent system from 1790, and the formal Patent Office was established only in 1836.

How decisive was the protection of the inventor's property rights by patents? Economic theory and contemporary empirical research suggest that the effect of a patent system on the rate of technological progress is ambiguous and differs from industry to industry (Kaufer, 1989). The ex ante positive incentive effects on inventors have to be weighed against the ex post negative effects on the diffusion of new knowledge, which will slow down as a result of an inventor's monopoly. Moreover, technological progress may be hampered by the closing of avenues in the development of new ideas if a particular ingredient has been patented by someone else.[9] A monopoly position awarded as compensation for inventions might discourage further activity if it led to increased leisure consumption, or if the profits of additional invention had to be weighed against the possible loss of monopoly profits currently enjoyed (Kamien and Schwartz, 1982, pp. 29–30). Yet patent rights could also lead to further inventions if they were used to finance additional research, a substantial advantage when capital markets were leery of innovators.

9. Two examples of patents that blocked other inventions can be found in the career of James Watt. One was a patent held by Watt that covered noncondensing engines that "wrought by the force of steam only." This claim effectively blocked the development of a high-pressure engine, even though Watt himself firmly opposed such engines and did not plan to develop them. The other was a patent held by another engineer that prevented Watt from converting the reciprocal motion of his steam engine into circular motion. The obvious way to do this was to use a crankshaft connected to the piston rod, but the competitor's patent covered the application of the crank to all steam engines. Watt's mechanical genius soon circumvented the obstacle by constructing the famous "sun-and-planet" epicyclic gear. Another case in which a patent created a temporary obstacle to technological progress was the invention of the telephone. Bell's telephone was primitive (messages could be sent over short distances only) and was much inferior to a device invented by Edison a few years later. Yet Edison could not get around Bell's patent, and the two engaged in a long struggle (though the improvements suggested by Edison eventually found their way to the telephone developed in the 1880s).

Certainly some great inventors had no doubts about the importance of patent protection. James Watt wrote that he feared that "an engineer's life without patent was not worthwhile," and Bessemer believed that "the security offered by patent law to persons who expend large sums of money . . . in pursuing novel invention, results in many new and important improvements in our manufactures." These statements may have been self-serving, coming from men who lived off patented inventions, and it is easy to find quotations from other major inventors indicating their disappointment with the patent system (Gilfillan, 1935, p. 93). But the very fact that so many inventors chose to patent their inventions in spite of the costs indicates that patents mattered. In countries in which there was no patent system, its absence was sometimes regretted. Goethe wrote that "we [Germans] regard discovery and invention as a splendid personally gained possession . . . but the clever Englishman transforms it by a patent into real possession . . . [and] is free to use that which he has discovered until it leads to new discovery and fresh activity. One may well ask why are they in every respect in advance of us?" (cited in Klemm, 1964, p. 173). Patents may also have encouraged entrepreneurs and investors to team up with patentees and supply venture capital; there is some evidence that Matthew Boulton, James Watt's partner, invested in the project only after he had the certainty of patent protection (Scherer, 1984, p. 24).

The operation of the patent system as an incentive to inventions was far from perfect.[10] Hargreaves's patent rights were denied by the courts on the technicality that he had sold a few jennies before applying for his patent. Arkwright, after a long and expensive case, lost all his patents in 1785, the same year as Argand. Similarly, Tennant lost his patent on bleaching liquid, though not on bleaching powder, on a technicality. A century later, J. B. Dunlop, the inventor of the pneumatic tire, was denied a patent on the grounds that the pneumatic principle had appeared before in an obscure patent taken out in 1845, though Dunlop clearly had come up with the idea independently. Despite these setbacks, Arkwright, Tennant, and Dunlop prospered. What this means for the importance of the patent system is ambiguous, however, because both Tennant and Dunlop relied on patents on secondary inventions.

Litigation over patent infringement could sap the creativity of great technical minds, and ruin inventors financially. Among the inventors

10. Charles Babbage wrote in 1829 that the British patent law was "a system of vicious and fraudulent legislation, which . . . deprives the possessor of his natural rights to the fruits of his genius" (cited by Robinson, 1972, p. 116).

who were destroyed by patent litigation were the flying shuttle's inventor John Kay, the engineer Jonathan Hornblower, and Charles Goodyear, the inventor of the rubber vulcanization process. Eli Whitney's patent wars over the cotton gin led to his arrest by his opponents and almost bankrupted him. In later days he claimed that the gin actually cost him more in litigation than he earned on it (Hughes, 1986, pp. 133–34). The Foudrinier brothers, who introduced mechanical papermaking into Britain, went bankrupt in 1810 and spent much of the rest of their lives in prolonged and expensive patent litigation. Robinson (1972, p. 137) maintains that British judges were in general unsympathetic to inventors, and that the laws were so ambiguous that the outcome of any patent litigation was unpredictable. Moreover, inventors were caught between two difficult options. On the one hand, the more detailed a patent application, the more likely it was to be approved: the law required that the application be specific enough to allow a third party to reconstruct the device from the application alone. Supplying that many technical details, however, would rule out the use of secrecy as an alternative barrier to entry should the patent be denied. Despite these problems, Robinson's negative verdict on the patent system in Britain is inconsistent with the fact that between 1770 and 1850 only 257 patent cases came before the courts, out of 11,962 patents granted (Dutton, 1984, p. 71).

In some cases society circumvented the patent system to reward innovators whose private benefits were deemed to be much smaller than the social benefits of their contribution: special grants from Parliament were awarded to Thomas Lombe, the inventor of the silk-throwing machine, whose patent was denied renewal; to Samuel Crompton, who never took out a patent on his mule; and to Edmund Cartwright, who lost the patent on his power-loom to creditors. The Swede John Ericsson, who made a crucial contribution to the invention of the screw propeller but who could not prove priority, was awarded £4,000 by the admiralty. Henry Cort also lost his patents through a series of financial mishaps and was awarded a small pension, but Richard Trevithick was denied a similar request. In the United States the South Carolina legislature awarded Eli Whitney $50,000 for his invention of the cotton gin. Prerevolutionary France routinely awarded pensions to the inventors of devices deemed socially beneficial and offered prizes for specific projects, such as the attempt to produce synthetic saltpeter between 1775 and 1794 (Multhauf, 1971). Between 1740 and 1780 the government paid out 6.8 million livres to inventors in subsidies and interest-free loans. The policy of encouraging invention may have been inconsistent and sometimes corrupt, but it did provide an incentive similar to a patent system.[11]

Despite the imperfections of the patent system, any other form of protection worked even less well. Secrecy was always a possibility, but by the nature of things could only be applied to a limited range of industries. Richard Roberts thought that "no trade secret can be kept secret very long; a quart of ale will do wonders in that way" (cited by Dutton, 1984, pp. 108–111). All the same, it was tried. Benjamin Huntsman was so compulsive about the secrecy of his crucible steel process that for a while he only worked at night. Henry Bessemer decided to keep one of his earlier inventions, bronze powder, secret rather than patent it because he believed that if the details of his system became known he would not be able to maintain the price (MacLeod, 1988, p. 95). If secrecy was an option, the decision to patent presented a particularly hard dilemma because patent application demanded that the inventor divulge all the technical details.

Of particular interest is the phenomenon of collective invention suggested by Allen (1983). When technological progress results from experimentation that is a by-product of investment, when it is difficult to patent, and when it proceeds gradually, it can happen that firms decide to share information freely and build on each other's results. Such cooperative efforts are hardly the rule, but Allen shows that the British iron-smelting industry conformed to this pattern between 1850 and 1875. Considerable progress was made in the design of the optimal blast furnace by the free sharing of information. Even in the iron industry, of course, what was patentable was patented and what could be protected by secrecy was protected. Collective invention occurred under rather special circumstances; normally firms will not release information that will increase their competitors' profits. Only when the number of firms engaged in Research and Development is small and free riders can somehow be excluded can firms work out cooperative arrangements and binding reciprocal contracts that permit the sharing of technological information. The information swapped tends to be, as could be expected, the type of information that does not endanger the firm's relative advantage over its competitors. Cooperation between innovators serves, however, as an important reminder that patents and property rights on new information were not strictly necessary for technological advance (Nelson,

11. In England, too, the government offered prizes for solving technological problems. The prize offered by Parliament for the sea chronometer in 1714 is the most famous instance of such an incentive. It should be mentioned that after the invention is completed there is little incentive for the government to pay out the prize money. Harrison, who won the prize, had to haggle with Parliament for a decade before he was paid, though his success was not contested.

1987, p. 79). Landes (1986, p. 614) claims that a large part, perhaps most, of the productivity increases in factory manufacture was the result of small, unpatentable improvements and concludes that patents were not the major incentive to invention. At this stage of our knowledge, such a claim remains premature.[12]

The importance of patenting differed from industry to industry. Overall, patents were more likely to be taken out in mechanical than in chemical inventions, and more in competitive than in concentrated industries. The initial wealth and place of residence of the inventor also played a role. Not all patents were inventions, and not all inventions were patented. MacLeod (1988, p. 145) assesses that that nine out of ten patents arose in industries that saw little innovation. The use of patent data as an indicator of technological activity, which has recently become fashionable again among economic historians (Sokoloff, 1988; Sullivan, 1989) should therefore be carried out with extreme caution. Inventions are an example of "fuzzy objects," and are notoriously difficult to count and measure. It remains an open question whether a bad approximation such as patent statistics is better than no approximation at all, and whether an econometric analysis of patents adds anything to our understanding of technological progress.

The verdict on the importance of the patent system in explaining technological creativity is thus decidedly mixed. Given the private and social costs of patenting and patent litigation, and given the many alternatives to patenting, its impact on the technological creativity of societies is far from clear.[13] The patent system may have been a relatively minor factor because of the "free lunch" property of technological progress, that is, because the social benefits of invention often dwarf the development costs. Consequently, it is not necessary for the inventor to capture all, or even most, of the social surplus created by his or her creation; a small fraction may suffice to make the effort worthwhile. A patent system grants the inventor a part of the surplus for a finite time, thus paradoxically encouraging primarily marginal inventions. Of course, if there are enough such marginal inventions, its effect on economic growth may still

12. Landes's assertion is open to two kinds of criticism. First, no study actually decomposes the increase in productivity between patentable and unpatentable inventions. Even if a definition of exactly what "unpatentable" means could be agreed upon, such a study would be fraught with difficulties. Second, Landes fails to realize that many small and unpatentable inventions may not have been meant to become that way, and were originally intended to follow the example set by successful patent holders.

13. Comparative studies of technological change in countries without patents have been inconclusive (Schiff, 1971), but seem to discount their importance.

have been significant. More importantly, the patent system encourages those inventions whose expected social surplus is low because of a very low ex ante probability of success, that is, it encourages ideas that represent radical departures from accepted practice, which I have termed *macroinventions.* It is thus important in generating the occasional spectacular breakthrough, the cases in which a crackpot hits the jackpot. Scherer's conjecture (1980, p. 448) is that such cases are rare. The social costs against which these breakthroughs must be weighed are the wasted hours of work of talented and original minds in search of the pot of gold at the end of the technological rainbow.

Was the patent system a factor in encouraging technological change during the Industrial Revolution? Dutton (1984) has argued that an imperfect patent system, such as the British system, represented the best of all possible worlds. Without patents, inventors would be deprived of their financial incentives. But if patent enforcement had been too perfect, the diffusion of inventions might have been slowed down. The patent system appeared to the inventors as providing more protection than it actually did. Such a gap between ex ante and ex post effectiveness may, indeed, have been beneficial. An economist's intuition would perhaps be that here, too, people cannot be fooled in the very long run. By that logic, however, Atlantic City would have gone out of business long ago. Moreover, invention is not exactly like gambling because by definition no two inventions are the same, and hence the information that a potential inventor can derive from the previous experiences of others is limited. It could thus well be that the patent system fooled would-be inventors into exerting more effort than they would have had they known how stacked the deck was against them. If that was indeed the case, it attained its goal.

Another possible reason technological progress was so much faster in Britain than on the Continent between 1760 and 1830 was that the Industrial Revolution happened to coincide with one of the stormiest episodes in the history of Europe. If the French Revolution and the subsequent turmoil significantly slowed down technological progress on the Continent, it may be that the small head start that Britain had achieved by the late 1780s became a veritable gap by 1815 merely because of this coincidence. The evidence for such an interpretation is ambiguous. It became more difficult for British technology to cross the Channel during the war years, and after 1806 trade relations between Britain and both its overseas territories and the Continent became hazardous. As Cardwell (1971, p. 150) points out, Britain had the more advanced technology, whereas France was the leading scientific nation. Scientific ideas and literature continued to flow across the Channel, but technology transfer was the victim of the disruption of trade relations between 1793 and 1815. Furthermore, on the Continent in general and in France in particular, political and military matters absorbed the creative energies of talented persons and slowed down progress. Some inventors' careers were disrupted by political events.[14]

Yet the French Revolution and Napoleon installed more forward-looking governments in Europe. In France, applied research was subsidized, prizes were awarded for useful inventions, and schools such as the École Polytechnique (established in 1794) and the École des Arts et Métiers (established at the initiative of Bonaparte in 1804) mobilized technological talent and applied it to current needs, usually determined by the government. Similar institutions were established in Prague (1806), Vienna (1815), Zurich (1855), and Delft in the Netherlands (1863). Mining schools like the one in Leoben, Austria (founded in 1840), represented the same idea. The culmination of this movement was the establishment of the famous German technical universities, the first of which was founded in 1825 in Karlsruhe. Moreover, the enormous disruptions in international commercial relations also bestowed certain advantages on the Continent. For two decades its industries found themselves protected from cheaper British manufactured goods, giving rise to "hothouse" industries, especially in cotton. Despite this temporary relief from British competition, the infant industries on the Continent did not generate anything like the British Industrial Revolution. Some towns, such as Ghent in Belgium and Mulhouse in the French Alsace, did manage to build the beginning of a cotton industry before 1815, but they never attained the momentum of Lancashire. Still, under pressure of the blockades and stimulated by the French imperial government, important original technological breakthroughs did occur on the Continent, among them the wet flax spinning process, the Jacquard loom, and sugar-beet refining.

Quite apart from political events, there is the matter of the social environment in which inventors and innovators operated. In France, for instance, very few inventors did well financially or otherwise. The

14. The most famous example is the great chemist Antoine Lavoisier, who was executed in 1794. Lavoisier had headed the gunpowder commission, which had tried for years (unsuccessfully) to solve the problem of the supply of French saltpeter, desperately needed for the manufacture of explosives. Another example is Nicolas de Barneville, who was active in introducing British spinning equipment into France. De Barneville repeatedly was called upon to serve in military positions and was "one of those unfortunate individuals whose lives have been marred by war and revolution . . . clearly a victim of the troubled times" (McCloy, 1952, pp. 92–94). A third case was that of Nicolas Cugnot, the first to build a steam-powered wagon in 1770. Cugnot was awarded a pension of 600 livres by the French government. The revolutionary government terminated the pension, driving the old inventor into poverty.

sad fates of Leblanc and Argand have already been mentioned. Robert (the inventor of the continuous paper process), Thimonnier (the inventor of the sewing machine), and Jouffroy (who built the first functional steamboat in 1783) also died penniless. Claude Chappe, the inventor of the semaphore, committed suicide in 1805 because of financial difficulties. Others, such as de Girard and Brunel, eventually emigrated. An exception was Joseph Marie Jacquard, the inventor of the loom that bears his name, who was rewarded by a pension and royalties and became something of a national celebrity. On a more modest scale, Nicolas Appert, the inventor of food preservation, was awarded 12,000 francs by the French Society for the Encouragement of Industry, set up by Napoleon. Of course, not all British inventors fared as well as Arkwright or Watt, but there were enough success stories in Britain to preserve a constant interest in inventive activity. The Continent seems to have suffered more from a scarcity of innovative entrepreneurs than from a scarcity of inventors. Manufacturers such as Wedgwood, Crawshay, Boulton, or Strutt, who did not create much new technology but knew a good thing when they saw it and moved rapidly, seem to have been in short supply on the Continent. It is arguable that though Britain may have had an absolute advantage in both inventors and entrepreneurs, it had a comparative advantage in entrepreneurs and skilled workers, and thus imported inventions and inventors and exported entrepreneurs and technicians to the industrializing enclaves of the Continent.[15] Before 1860, some of the greatest industrialists on the Continent had names like John Cockerill, Isaac Holden, Samuel Lister, and William Mulvany. The movement of technically skilled and enterprising Britons to the Continent demonstrates not only that a disequilibrium existed in the first half of the nineteenth century, but also that equilibrating forces were at work, spreading technological change from leader to follower. As long as that disequilibrium was maintained, Britain reaped a quasirent derived from a temporary advantage. Hence the prohibition on the enlistment of skilled workers to work abroad (abolished in 1825) and the export prohibition on machinery (abolished in 1843), though neither of these measures was particularly effective. Britain's temporary head start during the Industrial Revolution was not different from other periods in which

15. In addition to the inventions that Britain imported from the Continent, it imported some inventors, including Marc Brunel, Friedrich Koenig, and the Swiss engineer J. G. Bodmer. There are few cases of Continental businessmen trying their luck in Britain. The best known is Frederic Winsor (né Friedrich Winzer), a German adventurer who pioneered gaslighting in London. British engineers began emigrating to the Continent in large numbers after 1840, spreading the new technologies all over Europe (Buchanan, 1986).

a European region held a temporary technological advantage over others. Efforts of the leading region's government notwithstanding, technical knowledge spread rapidly across European borders.

How do we explain concentrated clusters of technological successes such as those that comprised the Industrial Revolution? If innovations occurred at random independently of each other, we should expect their time pattern to be distributed more or less uniformly over time. The difference between the Industrial Revolution and previous clusters of technological change was in the extent to which innovations influenced each other. First, there was an imitation effect: James Watt and Richard Arkwright became famous and wealthy men whom many tried to emulate. Invention and improvement became, in some circles at least, respectable. Second, there was a complementarity effect: the successful solution of one problem almost invariably suggested the next step, and so chains of inspiration were created. Many of the most useful inventions were indeed not more than radical modifications of earlier ideas: Cort's puddling-and-rolling process, Watt's and Trevithick's engines, Crompton's mule, all fall under this definition. Neither of these "one-thing-leads-to-another" theories constitutes an *explanation* of the Industrial Revolution. Both merely explain its time pattern, which is that when agents strongly affect each other, it is likely that success will appear in clusters. Clustering can occur when a critical mass is generated by the continuing interaction and cross-fertilization of inventors, scientists, and entrepreneurs, as Musson and Robinson (1969) have shown. Clustering phenomena are not, of course, confined to technology: Dutch painting in the seventeenth century and Austrian music in the eighteenth- and nineteenth centuries come immediately to mind. Although the emergence of talent may be uniformly distributed over time, its focusing and employment surely are not There were few British artists of much importance between 1770 and 1830. Apart from a cluster of romantic novelists and poets, centerstage was held by engineers, scientists, and political economists.

One avenue that has barely been explored by scholars interested in the question of "why England first" concerns the political economy of technological change. A widespread concern during the Industrial Revolution was that machines would throw people out of work, a misunderstanding that has persisted over the centuries.[16] A more legitimate concern was the fear of loss suffered by established firms in industries that were being mechanized. From hand-loom weavers to wagon drivers to blacksmiths, the Industrial Revolution

16. In a confused numerical example in his chapter on "Machinery" inserted in the third edition of his *Principles,* even Ricardo concluded that machinery could lead to technological unemployment.

forced firms to conform or go out of business because of competitive pressure. Resistance to innovation was more likely to come from existing firms than from labor (though in the case of craftsmen and hand-loom weavers that distinction is perhaps not very sharp). Technological progress reduces the wealth of those possessing capital (real or human) specific to the old technology that cannot readily be converted to the new. Resistance was therefore strongest in long-established, skill-intensive industries such as printing and wool finishing.

Resistance to innovation was exacerbated by the fact that the gains from the innovations in the Industrial Revolution were captured by consumers (for whom joint political action is very hard), while the costs tended to be borne by a comparatively small number of people, many of whom may already have been organized, or knew each other and lived in the same regions. The losers could try to use extralegal methods (rioting, machine breaking, personal violence against innovators) or the political system to halt technological progress. Either way, the diffusion of technological progress became at times a social struggle, and politicians and judges became arbiters of decisions that should have been left to market forces. In this regard, technological change was similar to free trade. The benefits being diffuse and the costs concentrated, the survival of free trade has always been in jeopardy, and its lifespan usually short. And although we understand the forces at work, it is difficult to predict outcomes or even fully understand why a particular outcome came about. What is clear is that between 1750 and 1850 the British political system unflinchingly supported the winners over the losers, on both matters of technological progress and, increasingly, free trade. On the eve of the Industrial Revolution the British ruling class had most of its assets in real estate and agriculture; it had no interest in resisting the factory and the machine.

Once again, the difference between Britain and the Continent was one of degree and nuance. Before and during the Industrial Revolution, there were numerous examples of anti machinery agitation in Britain. In 1551 Parliament prohibited the use of mechanical gig mills, used in the raising of the nap (a finishing process) of woolen cloth. The hosiers guild's opposition to William Lee's knitting frame (1589) was so intense that the inventor had to leave Britain. In 1638, the Crown declared a ban on the use of ribbon looms in Britain. John Kay's flying shuttle (1733) was met by fierce hostility from weavers who feared for their livelihood. In 1768, 500 sawyers assaulted a mechanical sawmill in London. Severe riots occurred in Lancashire in 1779, and in 1792 a Manchester firm that pioneered Cartwright's power loom was burnt down. Its destruction was said to have "inhibited the development of powerloom weaving for several years in this area" (Stevenson, 1979, p. 118). In the southwest of England, especially in Wiltshire and Somerset, resistance to advances in the spinning and weaving of wool was strong and may have contributed to the shift of the center of gravity of the woolen industry to the northern counties. Between 1811 and 1816, the Midlands and the industrial counties were the site of the "Luddite" riots, in which much damage was inflicted on machines. In 1826, hand-loom weavers in a few Lancashire towns rioted for three days, and in 1830 the "Captain Swing" riots, aimed at threshing machines in agriculture, took place in the south of England.

By and large, these attempts were unsuccessful. Gig mills, ribbon looms, knitting frames, and flying shuttles were all adopted by British industry (though perhaps somewhat more slowly than they might have been). The laws prohibiting machinery remained ineffective. In the eighteenth century, the government took an increasingly stern view of groups who tried to halt technological progress. In 1769 Parliament passed a harsh law in which the willful destruction of machinery was made a felony punishable by death. In 1779 the Lancashire riots were suppressed by the army, and the sentiment of the authorities was well expressed by a resolution passed by the Preston justices of the peace: "The sole cause of great riots was the new machines employed in cotton manufacture; the country notwithstanding has greatly benefited by their erection [and] destroying them in this country would only be the means of transferring them to another . . . to the detriment of the trade of Britain" (cited by Mantoux, [1905] 1961, p. 403). During the Luddite outbreaks in 1811–13, the British government deployed 12,000 men against rioters, a force greater in size than Wellington's original peninsular army in 1808. Riots were harshly suppressed and usually ended in hangings and deportations.

Attempts by legal means to stop the Industrial Revolution were no more successful. In 1780, cotton spinners petitioned Parliament to forbid cotton-spinning machinery, but the committee appointed to investigate the matter denied the petition. In 1794 wool combers petitioned against a wool-combing machine invented by Edmund Cartwright, but once more the employers carried the day. Other petitions were treated similarly, including one that sought to ban the gig mill on the basis of the 1551 law (Mantoux, [1905] 1961, pp. 403–8). Between 1803 and 1809 a battle between industrialists and workers raged in the woolen industry over the repeal of ancient statutes and regulations that were regarded by the industrialists as inimical to the new technologies. In 1809 the laws were repealed (Randall, 1986). In 1814, the 250-year-old Statute of Artificers was repealed, despite the demands of journeymen trying to protect their old technologies. The case of the workers was politically hopeless. The argument that stopping the new machinery would only lead to its flight abroad was persuasive, but there was more to it than that. The politically dominant classes in Britain were the propertied classes and the new technology did not threaten to reduce the value of their assets. Furthermore, not all workers in the traditional sector were initially made worse off. Although some workers, such as handloom weavers and frame knitters, lost out as a result of mechanization, many workers displaced by machinery eventually found employment in the factories. Some cottage industries that produced goods that were complementary to or inputs for the factories actually thrived for many years thanks to the new technology, until their turn came to compete with the new machines. The list of disturbances and riots, although long, is thus misleading. Most innovations were adopted without significant trouble, and not all trouble necessarily reflected resentment toward the new machines.[17]

Were things significantly different on the Continent? As in Britain, resistance came from guilds of skilled artisans and from unskilled workers fearful of unemployment. Before the French Revolution, craft guilds, which still existed in most regions, held some new techniques back. In part, this was done by outright banning of inventions when established interests felt threatened. The ribbon loom, known in England as the Dutch or engine loom, encountered resistance all over the Continent, in striking contrast to its progress in Lancashire after 1616 (Wadsworth and Mann, 1931, p. 104). In Brandenburg, guilds succeeded in keeping out frame-knitting well into the eighteenth century. In France and elsewhere, printing and cotton textiles were among the industries in which new techniques were successfully resisted by pressure groups.[18] Increasingly, the old urban guilds became a fetter on technological progress, less by outright resistance than through a vast body of regulations and restric-

17. Many workers' riots during the Industrial Revolution were aimed more at low wages and high prices than at new technology. The Luddite riots of 1811–16 were in part of that nature; the knitting frames destroyed in Nottingham were an old and established technology.

18. In 1772 Wilhelm Haas, a type founder from Basel, invented a printing press that was heavier and more stable than existing presses and could thus be built largely from heavy metal parts. The guild of Basel master printers had him legally restrained from building the press (Audin, 1979, p. 658). In France, a good example is the opposition faced by the paper industry when it attempted to introduce the *Hollander,* a machine that cut and beat the rags into a pulp. The *compagnonnages* (journeymen associations) of the paper industry successfully resisted the innovation using sabotage and arson.

tions on inputs and outputs.[19] Under these regulations, for example, it would have been difficult for a barber like Richard Arkwright to set up shop as a cotton spinner. Yet the powers of the guilds were already declining in the eighteenth century. After 1760, guilds came under pressure in France and Germany, and were abolished in 1784 in the southern Netherlands. The French Revolution abolished them in France in 1791 and subsequently in areas that fell under French domination. By 1815 guilds had either been fatally weakened or abolished altogether on the Continent. The political upheaval and disruptions incurred as the price for ridding the economies of obsolete institutions between 1790 and 1815 may in part explain the lag in the adoption of some techniques into Europe. However, the Revolution's long-term effect was to clear up the debris of the ancien régime on the Continent, thus assuring Europe's ability eventually to follow Britain in revolutionizing its production system.

Not that there was a lack of resistance. Fears of technological unemployment surface in the 1789 *cahiers* (grievances sent to the Estates General). Between 1788 and 1791, workers rioted repeatedly in protest against machinery that they felt threatened their livelihood. In the summer months of 1789, the city of Rouen was the scene of antimachinery vandalism that spread to other towns, including Paris and St. Etienne. The targets of the rioters' ire were spinning machines imported from Britain and locally made devices, such as pitchfork-making machines. Although the riots eventually blew over, the damage may have been long-lasting. One historian (McCloy, 1952, p. 184) has stated that "the rioters . . . set back the clock of time, at least industrially, some two decades; for it was not until after the wars . . . that France was able to introduce such machines again." Yet after the Revolution, under the forceful protechnology government of Napoleon, opposition to new techniques appeared as moribund on the Continent as it did in England.[20] The weavers of Lyons resisted Jacquard's loom in the first decade of the nineteenth

19. The Dutch shipbuilders guilds were initially technologically progressive and supportive of innovation, but became conservative and opposed to progress some time around the middle of the seventeenth century, thus slowing down the rate of progress in Dutch shipbuilding. See Unger (1978).

20. While visiting the textile town of Sedan in 1803, Napoleon was met by wool shearers, one of the best organized and most intransigent crafts, who called to him, "Long live the First Consul, down with the machines." Napoleon answered characteristically: "Your fears are groundless. Since the shearing machines lower the price of fabrics, consumption will increase with the lower price at a rate greater than that of the decrease of hands" (Payen and Pilsi, 1979, pp. 616–17). The emperor's judgment on technology was not infallible, however. He referred to gaslighting as "a folly," and its introduction into France was delayed until after 1815.

century, but to no avail (Ballot, [1923] 1978, p. 379). A decade later some resistance was encountered to the introduction of wool-shearing equipment in France. Nevertheless, some scholars believe that Luddism "surely cannot have impeded the introduction of machinery into French manufacturing" (Manuel, 1938).

It is misleading, however, to identify what is commonly called Luddism (machine breaking) with a rational resistance to a new technology by a group that stood to lose from its introduction. For one thing, machine breaking was often resorted to simply because machines were a convenient and vulnerable target in a labor dispute, and not necessarily because of a specific grievance against the new technology they embodied. More importantly, resistance to new technology often took a more subtle form that is not always easy to detect. Recent work in social history has emphasized the high level of organization of French workers. After 1815, the guilds were replaced by "mutual aid societies," which often served in secret the interests of the *défense professionelle.* Small independent masters often supported illicit unions in fighting back against innovative entrepreneurs who introduced cost-cutting machinery (Sewell, 1980, p. 182). In 1895 these societies had almost 400,000 members (Shorter and Tilly, 1974, p. 176.[21] The need to placate skilled craftsmen may well have steered France into choosing a technology somewhat different from Britain's. The industrial France that emerged in the nineteenth century thus continued to be based on skilled small-scale handicrafts producing for relatively local markets. Some economic historians have attributed this difference between Britain and France to differences in population growth, but there must be limits to the burdens that demography can bear. More so than in Britain, therefore, technological progress in France had to accommodate the artisans and to find compromises between their traditional skills and the needs of modern factories. In Belgium, Switzerland, Bohemia, and the Rhineland, resistance to new cotton-spinning technology crumbled. In the Netherlands, on the other hand, workers repeatedly smashed machines in the textile industries in the south. Though such cases were not numerous, they may well have deterred entrepreneurs from installing such equipment.

Technological progress was a multinational collaborative effort, in which Britain's advantage was qualitative and ephemeral. Yet there are important lessons to be drawn from comparing different expe-

21. A good example of a group of skilled craftsmen who tried to slow down mechanization and through it the devaluation of their traditional skills were the glassworkers in southern France. In the late nineteenth century these craftsmen formed a union to try to keep their jobs and preserve apprenticeship practices (Scott, 1974, pp. 91–107).

riences in Europe between 1750 and 1914. The comparison can shed some light, for instance, on why Cardwell's Law that no economy remains technologically creative for extended periods of time seems to hold. Britain, the cradle of the technologies that created the Industrial Revolution, lost its preeminence in the late nineteenth and early twentieth century. Although it is still debated by economic historians whether the slowdown occurred before or after 1900, there is no question that by 1914 the cutting edge of technology had moved elsewhere.

* * *

The political economy of technological change is only dimly understood. To some observers, technological progress seems to resemble the human life cycle: the vigor of youth is followed by the caution of maturity and finally the feebleness of old age or the "climacteric." Such an anthropomorphic view makes no economic sense without more specific detail. Societies are aggregates; they do not age like individuals. If we are to understand why the fires of innovation die down, we must propose a model in which technological progress creates the conditions for its own demise. To begin, recall that part of the social cost of innovation is caused by the process of "creative destruction." Schumpeter, who popularized the term, associated it with capitalism, but in fact it is part and parcel of technological progress under any economic regime. The continuous obsolescence of specific, nonmalleable assets, both physical and human, is the price a society pays for sustained progress. Insofar as the groups that benefit from the new technology coincide with the groups that pay the price (which is lower than the benefit), resistance to technological progress will be weak. The greater the divergence between these two groups, the greater the incentive for the losing groups to try to slow down progress. It seems likely that the obsolescence of human capital is a crucial variable here, because its malleability declines sharply over the life cycle. But it could apply to physical capital as well. As long as free entry into an industry is guaranteed, however, the ability of any group to keep out new techniques is limited. A "life cycle" of a technologically advanced society would thus consist of three stages. First is the youthful stage, in which the new technology manages to break through, supplanting the previous technology by means of its greater efficiency. This stage corresponds with the period between 1760 and 1830 in Britain. The next stage is one of maturity, in which the new technology is in control, but creative destruction by new techniques takes an ever-increasing toll, leading to a growing incentive by those currently in control to protect themselves. In the third stage, the by-now old technology develops social or political mechanisms with which to protect itself against innovation. If it is successful, technological creativity comes to an end. If it fails, the cycle begins anew.

Does such a schematic life-cycle theory of technological creativity have any value? Its predictive powers are very limited: without more information we cannot tell how long creativity will last, how it will end, or whether it will recur. Yet like most theoretical models, it suggests to the empirical economist and economic historian where to look for specific evidence. Societies that become conservative after a creative period need to obstruct innovation by setting up mechanisms to prevent existing firms from adopting new ideas, by preventing would-be innovators from entering an ossified industry, and by stopping the inflow of new ideas from abroad. Japan practically closed itself to Westerners in 1638. Islam and China, as we have seen, tried to block Western influence using a combination of ideology and control. But in the West itself such defenses often took more subtle forms. Protective tariffs have been used for centuries to allow obsolete technologies to survive (under the pretext of softening the pains of transition). Guilds, labor unions, professional and manufacturers associations, and licensing requirements have been used to police conformism within the industry and to try to prevent the entry of homines novi with new ideas. As the fixed capital requirements in manufacturing rose, credit rationing was used to create another barrier that tried to exclude new ideas and techniques.[22]

In late Victorian Britain the new ruling class that had come to power defended its position by closing Britain's elite to the same kind of entrepreneurs from whom they had descended. They tried to reorder the hierarchy of values so that production and technology would once again occupy the lowly positions they had occupied centuries before the Industrial Revolution. Such attempts were largely unsuccessful. Competition with other industrial powers and the internal competition of British firms against each other guaranteed that Britain could not be passed by the Second Industrial Revolution. What the British tried to do, however, was to live through the Second Industrial Revolution with the tools of the First, much like generals apply the tactics of the last war to the fighting of the cur-

22. During the dawn of the American automobile industry, R. E. Olds and Charles E. Duryea were refused small loans by suspicious bankers. When Will Durant predicted that some day 500,000 automobiles would be produced in the United States, a powerful banker, George Perkins, advised him "to keep those notions to himself if he ever wanted to borrow money" (Stern, 1937, p. 44). Henry Ford's treasurer, James Couzens, was thrown out of so many offices in Detroit trying to raise money that once he just sat down on the curb and wept (Hughes, 1986, p. 288).

rent one. The systematic application of the newly developing natural sciences was delayed by the survival of a tradition of amateurishness and tinkering, and by the virtual absence of technical education. On-the-job training through apprenticeship remained the chief mechanism through which skills were transmitted. Invention was largely the business of inspired outsiders, and British scientists (with some notable exceptions, such as Charles Parsons) were as little involved in exploring new opportunities as they were in the daily drudgery of production. As we have seen, Britain did not lose its central position in invention itself, but its position was reversed compared to a century earlier: during the heyday of the First Industrial Revolution, it was a net importer of technology; after 1850, it became an exporter. Technological creativity, defined as the application of new ideas to production, began to slow down in Britain. The ideas themselves were still forthcoming, but the economic environment slowly became less receptive to them.

An unexpected tool the ruling elite used for this purpose was the educational system. English public schools opened their doors to members of the new elite, but painstakingly avoided providing them with the kind of practical education that would enable them to threaten the technological status quo. Those who profited from the educational opportunities tended to choose careers in the professions. The educational system, never the most progressive of Britain's institutions, resisted the introduction of applied sciences into its curricula.[23] The old British tradition of informal training remained the principal means through which technological information was transferred. In contrast, most other European countries established technical schools that played a central role in the catching-up phenomenon. The graduates of the German *Technische Hochschule* amazed British businessmen and raised increasing fear about competition (Ashby, 1958, p. 795; Julia Wrigley, 1986, pp. 172–73). Similar schools existed, as we have seen earlier, in the Netherlands (Delft), Switzerland (Zurich), and France (for example, the *Institut Industriel du Nord* in the manufacturing town of Lille).

It can hardly be maintained that demand factors alone were decisive here: there was great interest in Britain for scientific and technical development that, after the mid-nineteenth century, they increasingly were forced to import. Thus, the British Association for the Advancement of Science commissioned von Liebig's famous 1840 treatise, and the English translation was enthusiastically received.

23. In Germany, too, the universities tried to resist reforms, and succeeded in barring engineering from their curricula. The German governments, however, circumvented the universities, building an independent system of technical colleges.

Liebig's student, August von Hofmann, was brought to London to teach chemistry at the Royal College of Chemistry and consulted to private companies as well as the British government. But after Hofmann returned to Germany in 1865 and particularly following the Paris Exhibition of 1867, it became increasingly clear that Britain was lagging behind. One Briton wrote in 1867 that at the Paris Exhibition "a singular accordance of opinion prevailed that our country had shown little inventiveness and made little progress in industry. . . . The one cause upon which there was unanimity of conviction is that France, Prussia, Austria, Belgium, and Switzerland possess good systems of industrial education . . . and that England possesses none" (cited by Ashby, 1958, p. 789). A number of royal commissions and select committees were appointed to look into the problem, and all recommended increasing the quantity and quality of scientific and technical education. Scientific education advanced significantly, but technical colleges had to wait for the Technical Instruction Act of 1889. Throughout this period, the overriding argument deployed by the supporters of technical education was that Britain would not be able to compete with other nations without it. The competitive states system was, as we have seen repeatedly, the most effective check upon the forces of technological reaction.

Of course, the weakness of British technical education should not be naively attributed to some kind of fiendish conspiracy on the part of the status quo. It is clear, however, that the existing political and economic order had little to gain from the increased state intervention that would be needed if technical education was to be brought to a par with the Continent. Just as that order successfully combatted state intervention in the form of protective tariffs, they resisted any increase in expenditures that would pay for education. But conservatism permeated not only the universities and government, but penetrated the workplace as well. What Landes has called the "mystique of practical experience," the idea that technical skills could not be taught as a part of formal education but had to be acquired on the job, worked against the establishment and expansion of the technical schools. The business world tried to establish an "old boy network" in an attempt either to bribe or to exclude those who would challenge the technological status quo. The "gentleman mentality" was resurrected, not as a quirk of British character, but as a defense mechanism against those who would do to the British elite as their grandfathers had done to others a century earlier. Once again, profits became a source of guilt rather than pride.[24] One tactic was to co-

24. For instance, William Perkin, who made a fortune from inventing aniline dye, reflected in his later days that "it was said that by my example I had done harm to

opt the would-be rebels into the old elite, by admitting them to public schools and universities, on condition they did not rock the boat. Those who did not go along were treated like social pariahs. Even men like William Perkin or Edward Nicholson (also a dye chemist), who made their fortunes from their scientific knowledge, subsequently retired to dedicate their lives fully to pure research.

After the middle of the nineteenth century labor, too, took an increasingly negative attitude toward machinery, though their motives were directed as much toward higher wages and better working conditions as toward keeping machines out to protect their special skills. Alfred Hobbs, the American lockmaker who introduced interchangeable parts to lockmaking, stated in 1857 that the "great obstacle in the way of the gunmakers of Birmingham in introducing machinery was the opposition of the workpeople to such innovations." Joseph Whitworth, in his 1854 report on the differences between American and British manufacturing, emphasized that British workers were far more hostile to new technology than American workers, because they were more skilled, better organized, and less mobile (Rosenberg, ed., 1969). The installation of a sewing machine was prohibited in the center of the shoemaking industry, Northampton, after three strikes against it in the late 1850s, and new machinery was successfully kept out in some centers of carpetmaking, printing, glassmaking, and metalworking where resistance was stiff (Samuel, 1977, pp. 9–10, 33). In most cases, however, resistance took the form of hard bargaining over changes in wages and working conditions resulting from any changes in the production environment.

Lazonick (1979, 1986) has shown in detail how strong labor unions in nineteenth century Britain succeeded in modifying the process of innovation and later in creating an atmosphere inimical to technological change. The bargaining power of skilled workers made it costly for manufacturers to introduce new machinery in the cotton industry, the ageing flagship of British industry. In cotton spinning in particular, workers forced the capitalists to retain the system of "minders" that had sprung up in the first quarter of the nineteenth century. In this system, the capitalist delegated some of the supervisory and recruiting functions to a male worker in charge of a mule. This system did not prevent the adoption of the self-actor, though it may have slowed it down and prevented the capitalists from using it to strengthen their bargaining power vis-à-vis the

science and diverted the minds of young men from pure to applied science, and it is possible that for a short time some were attracted to the study of chemistry by other than truly scientific motives." (cited by Beer, 1959, p. 45).

laborers. By the 1880s, however, the limits of the mule had been reached and ring spinning passed Britain by. Lazonick (1987, p. 303) concludes that "vested interests—in particular the stake that British workers had in job control and the historic underdevelopment of British management—stood in the way of . . . promoting the diffusion of advanced production methods." All this did not necessarily lead immediately to inefficiency in the narrow economic sense of a poor allocation of resources. We should recall that, as Cardwell (1972, p. 193) points out, the failure at the time (of late Victorian Britain) was not an economic failure, but a technological and scientific one.[25]

Economies reaching the third stage of their technological life cycle can choose different defense mechanisms, depending on their political system and social customs. A powerful ruler can be enlisted either to protect the status quo or to support the forces of progress. Sometimes a compromise is reached. As a result, the outcome is indeterminate and the technologically creative stages in the life cycle may last for longer or shorter periods. Still, this type of political mechanism can shed some light on the forces underlying Cardwell's Law. These mechanisms cannot be made more explicit without a more complete specification of the complicated series of games played between those who stand to gain from technological change and those who stand to lose. Yet what is clear is that the game is structured in a particular way. Every time a new technology emerges, it has to struggle with the status quo. It has some probability p of winning. If it does, it becomes the dominant technology and it will in turn be challenged in the next period. If it loses, the forces of reaction will establish a new antitechnological set of institutions that will make it far more difficult to be challenged, so that the next round the game is played the new technology has only a chance of p' to win, where $p' < p$. It is conceivable that $p' = 0$, so that a society locks itself forever into an existing technology, in which case a victory of the status quo would be a true absorbing state. There are some reasons to believe, however, that this is unlikely to be the end of the story. For one thing, p and p' are both likely to depend on the ratio be-

25. Modern economic historians have tried to absolve British entrepreneurship from having "failed" to take advantage of technological opportunities. The problem is, however, one of *creating* opportunities, not just exploiting them. Lindert and Trace (1971, p. 266) concur with the view that British entrepreneurs can be indicted or absolved depending on what one considers to have been under their control. "If scientific discoveries are viewed beyond their control, British entrepreneurs cannot be blamed for leaving the lion's share of British and world dyestuff markets to German and Swiss firms. . . . It would, however, seem valid to blame British entrepreneurship for not making the original discoveries." In any case, British society as a whole clearly lost its knack for taking advantage of the innovations associated with the Second Industrial Revolution.

tween gains and losses associated with a new technology. A more drastic change will increase both the gains and the losses but not necessarily pari passu, so that it may become possible for a sufficiently advanced technology to break through despite reactionary institutions. Moreover, in a global setting p' will depend on the size of the gap between the reactionary and the progressive economies. At some point the gap will become intolerable, and the reactionary forces will be defeated, as they were in Japan in 1868 and in China more gradually after 1898.

An alternative way of understanding why periods of technological creativity were finite and usually short centers on the connection between market structure and innovation. Economists have been discussing the Schumpeterian hypothesis for many years, and the literature is summarized elsewhere (Kamien and Schwartz, 1982; Scherer, 1980, ch. 15; Baldwin and Scott, 1987). In brief, Schumpeter argued that large firms with considerable market power, rather than perfectly competitive firms, were the "most powerful engine of technological progress" (1950, p. 106). Free entry, in his view, was incompatible with economic progress based on technological change. Much of this debate is of questionable relevance to anything that happened before 1914. The version of the Schumpeterian hypothesis associated with Galbraith, who tried to relate firm size to the propensity to innovate, is even less relevant here.[26] Before 1850, large firms were highly unusual and production took place largely within family firms, sometimes assisted by domestic servants and hired hands. It seems unlikely, therefore, that firm size will have much explanatory power for technological history before the modern age. At any rate, Schumpeter referred not to size but to the degree of competition. Here the historical record offers much more variation. Small firms do not guarantee competitiveness. If firms are catering to a small enough market (that is, if transport costs are sufficiently high), even a single artisan could be a monopolist. Moreover, competitive industries can devise cushioning mechanisms that mitigate the sharp edges of competition and eventually make the industry behave as a monopolist in some respects. The guild system, although not set up for that purpose, clearly carried out that task in Europe for many centuries. In the nineteenth century, competitive pressures were mitigated by cartels, professional associations, government regula-

26. In any event, the argument centering on firm size has always been the most fragile part of the Schumpeter hypothesis. There seems to be little theoretical or empirical support for this view. Indeed, the evidence suggests that small firms tend to be superior in the research and development process. See Kamien and Schwartz (1982, pp. 67–69). Only in unusual cases are the costs and risks of an innovation so large as to require the resources of a large firm to carry out the work.

tion, and unwritten gentlemen's agreements on what constituted proper business behavior. Perfect competition was rare. Instead, a bounded set of rules defined a game that favored innovation at some times more than others.

Much of the modern empirical literature on the market structure best suited to technological change is inconclusive and sometimes contradictory. The historical evidence is not much more helpful here. After 1750, Anglo-Saxon economies tended to be more competitive than others; cartels and formal barriers to entry were far more common on the Continent. In Germany and Austria, cartels were encouraged and forced by the state in the late nineteenth century; in the United States after 1890 they were illegal and in Britain they were extra-legal, neither enforceable nor illegal. In practice, however, most British firms remained small, and British industry fragmented and decentralized. In and of themselves these differences do not imply that the Continent was less competitive than Britain or the United States. Moreover, the record on technological change between 1750 and 1914 does not provide much support for any simple Schumpeterian hypothesis. Britain, the more competitive economy, gained an initial advantage in innovation, but then lost it after 1870 to the European Continent and the United States. Either we are mismeasuring competitiveness, and (appearances notwithstanding) Britain was becoming less competitive in the late nineteenth century, or the new technology associated with the Second Industrial Revolution was inherently different in this respect from the earlier technology. Inventions such as chemical dyes, the internal combustion engine, electrical appliances, and steel required considerable development and improvement after their conception. Here the less competitive German oligopolies, led by a small number of powerful investment banks, and the large firms that emerged in the United States by the end of the nineteenth century may have provided a better environment for technological change (Mowery, 1986).

In spite of the inconclusiveness of the literature on the connection between market structure and innovation, it may provide an additional insight into the causal roots of Cardwell's Law. The relationship between technological change and market structure is reciprocal. Market structure is said to affect the technological creativity of an industry or economy, but a feedback mechanism leads back from technological change to market structure. Here, too, the effects are indeterminate. Some forms of technological change tended to enhance competition—improvements in transportation and communications diminished local monopoly power, the availability of electrical power led to a decline in the optimal size of a firm, and the introduction of new products competed with established products. Other kinds of changes (for instance patents) reinforced economies of scale and created barriers to entry. Yet the remarkable conclusion is that whatever the relationship, the model implies that technological change is likely to be ephemeral. A highly simplified exposition of the model suffices to illustrate how the relationship between firm size and technological change might explain Cardwell's Law. Suppose there are only two forms of market structure, ***i***, which is conducive to technological change, and ***j***, which is not. This ranking of market structures according to their conduciveness to technological change depends itself on the technology in use, and technological change may cause the existing structure to switch from ***i*** to ***j*** as well as reverse the ranking. Suppose we start with market structure ***i*** and technological change takes place. There are now four possible outcomes: (1) the next period remains dominated by market structure ***i***, and ***i*** is still the more conducive to technological change; (2) the next period leads to a switch from ***i*** to ***j***, and the ranking of ***i*** and ***j*** is preserved; (3) the next period still finds the economy in ***i***, but the ranking is reversed; and (4) the next period leads to a switch from ***i*** to ***j***, but now ***j*** is the more favorable market structure for technological change. Outcomes (1) and (4) imply that the process is repeated; outcomes (2) and (3) imply that technological progress ends. If all four outcomes are thought of as random events with positive probabilities, the process will inexorably come to an end when either outcome (2) or outcome (3) occurs.[27] The use of such Markov processes in characterizing Schumpeterian dynamics was pioneered by Futia (1980), who has shown how industry structure and the chances for technological progress are determined by ease of entry and technological opportunities. This research could be extended to produce further understanding of Schumpeter's insight that technological change is temporary in nature.

In examining technological progress, the question arises whether we should consider creativity or stagnation the normal condition. The answer to this question determines the explicandum: is it the periods of progress or the periods of stagnation that need be explained? This question cannot be answered based on historical evidence alone. Clearly, over the entire human historical record, periods of technological change have been exceptional. The present and the future may be radically different, however, as a result of the Industrial Revolution. On the other hand, Cardwell's Law seems too pervasive a phenomenon to take anything for granted.

27. A similar argument is made by Kamien and Schwartz (1982), who believe that "self-sustainability" is the most important outstanding issue in this field. They point out that "if a market structure that is supportive of technical advance leads to one inimical to it, technical advance is not self-sustaining . . . the process of 'creative destruction' described by Schumpeter may turn on itself and destroy the foundation on which it is based" (ibid., pp. 218, 220).

PART FOUR

DYNAMICS AND TECHNOLOGICAL PROGRESS

CHAPTER ELEVEN

Evolution and the Dynamics of Technological Change

What do we know about the dynamics of technological progress? Economists have long recognized that the traditional tools of economic theory pivoting on the concept of equilibrium are not suitable to the analysis of technological change. It is not clear, however, what can serve as an alternative. One concept that has been employed in the economics of growth is the *steady state*, a form of growth that is itself constant and predictable, and thus can be regarded as a dynamic equivalent to the concept of equilibrium. It would seem natural to think of the steady state as an evolutionary process. Terms such as "technological drift" and "remorselessly creeping change, partly by trial and error" (Jones, 1981, p. 63) conjure up evolutionary progress. Yet the term is more frequently tossed about than actually employed in an operational framework. The word "evolution" takes on two different meanings: one as a synonym of "gradual" or "continuous" and one as a specific dynamic model governed by mutation and selection. The two definitions differ and can be contradictory. Modern theories of evolutionary change allow explicitly for chaotic bifurcations and catastrophes leading to unpredictable new steady states (Laszlo, 1987).

A number of historians (Cipolla, 1972, p. 46; Hindle, 1981, p. 128; Basalla, 1988) and an economist (de Bresson, 1987) have suggested making evolution in its second meaning the model for technological change. In what follows, I shall examine the potentialities of such an analogy.[1] The analogy is useful for understanding

1. The analogy between technological change and biological evolution has a long history, starting with Marx's chapter on "Machinery and Modern Industry" in Vol. I of

the dynamic aspects of technological progress. In particular, it can be used to answer the question whether or not technological progress took place in small incremental steps or large leaps. The analogy seems more apt today than ever because a lively debate between gradualists and saltationists has been raging for the past decade or so among evolutionary biologists, paralleling to some degree the debate among economic historians.

The idea of applying concepts from the theory of evolution to economics is of course not new, although evolutionary theories have fallen largely outside the mainstream of modern economic theory. Boulding (1981) made an attempt in this direction, but his analogy between economics and biology is somewhat vague.[2] In their pathbreaking work, Nelson and Winter (1982) tried to rewrite the theory of the firm using evolutionary models. They shun the traditional economist's concepts of profit maximizing and equilibrium, and instead assume that firms follow certain standard procedures or routines. The analogue to the species in the Nelson-Winter model is thus the firm. Competition still is central to the process, but the way it operates is through Darwinian selection mechanisms: firms that have more adaptive routines will prosper and grow at the expense of less successful ones. In general, applications of this nature have had little room for technological progress as an evolutionary process itself, that is, the firms are the units upon which selection occurs rather than the units that do the selecting.[3] An explicit comparison between evolutionary biology and technological change is proposed

Capital. A similar idea was suggested, but not pursued, by Gilfillan (1935, pp. 14–17). For a survey of other proponents of the biological analogy, see Basalla (1988, pp. 14–25). This chapter was largely completed before I had seen Basalla's book, and a number of the points below were developed independently.

2. In Boulding's words, individual phenotypes are "members of both biological and social species, including economic goods" (Boulding, 1981, p. 24) and mutations are any process by which the parameters of the system of ecological interaction change. The species, in Boulding's analogy, is the commodity, with production corresponding to birth and consumption to death (p. 33).

3. In a formidable new work, Foster (1987) develops a concept of *homo creativus* as an alternative to *homo economicus.* In his work, evolutionary dynamics are made explicit, but his book is not directly concerned with Darwinian mechanisms of natural selection in models of long-term historical change. Another economist directly interested in evolutionary models is Guha (1981), who believes that economic growth is not analogous to evolution but part of it. Guha argues that technological and other exogenous changes lead to changes in society, which adapts to changes in the environment in the same way species do. Society is thus the analogue of a species struggling for survival. Guha's view is somewhat similar to Ruse's (1986) notion of Darwinian epistemology, which maintains that science develops because natural selection tends to encourage intelligence and creativity. This somewhat eccentric approach cannot explain the long periods of scientific and economic stagnation, and will not be pursued here.

by de Bresson (1987), who makes some of the same points below. Yet perhaps because he is less interested in long-term historical change, de Bresson ends up concluding that "there are well-founded reasons for not adopting an evolutionary framework for an economics of technological change. . . . A careful examination would lead us to drop the biological analogy" (ibid., p. 759).

Whether or not analogies between evolution and cultural or scientific progress are useful has been the subject of a large and lively literature. The use of analogies from other disciplines has a venerable tradition in economics. The inspiration of classical Newtonian physics is at the base of equilibrium theory and comparative statics. The parallels are inevitably incomplete, and do not provide the researcher with a sharp analytical tool, however. Not every element in one area has an exact equivalent in the other: where is the economic counterpart of gravity?[4] Maynard Smith (1972, pp. 36–43) and Ruse (1986, pp. 32–35) discuss the advantages and pitfalls of such analogies. They distinguish between what Ruse calls "analogy-as-heuristic," in which the analogy between *A* (which is well-understood) and *B* (which we are investigating) suggests certain ways to approach the study of *B* and certain hypotheses to test, without implying that because they hold for *A* they necessarily hold for *B*. The stronger claim for analogies of this type is "analogy-as-justification," in which certain propositions are claimed for *B* because they are true for *A*. Biological evolution and technological change are similar enough to make a heuristic analogy interesting, and to suggest new ways of thinking about the economic history of technology. They are, however, sufficiently dissimilar so that no immediate inferences can be drawn from Darwinian theory to issues such as the Industrial Revolution. Such dissimilarities do not ipso facto invalidate the use of the analogy.

The approach I adopt here is that techniques—in the narrow sense of the word, namely, the knowledge of how to produce a good or service in a specific way—are analogues of species, and that changes in them have an evolutionary character. The idea or conceptualization of how to produce a commodity may be thought of as the genotype, whereas the actual technique utilized by the firm in producing the commodity may be thought of as the phenotype of the member of a species. The phenotype of every organism is determined in part by its genotype, but environment plays a role as well. Similarly, the idea constrains the forms a technique can take, but adaptability and

4. Some writers go beyond analogy and maintain that the basic underlying dynamic structures of change in natural and social phenomena, from physics to life forms to social change, obey similar general laws of motion. For a statement in that spirit complementary to the position taken here, see Laszlo (1987).

adjustment to circumstances help determine its exact shape. Invention, the emergence of a new technique, is thus equivalent to speciation, the emergence of a new species. The analogy is incomplete, and arguing from partial analogy can be misleading. Nevertheless, something can be learned from the exercise.

My basic premise is that technology is epistemological in nature. It is not something that somehow "exists" outside people's brains. Like science, culture, and art, technology is something we *know,* and technological change should be regarded properly as a set of changes in our knowledge. In recent years, a new school of evolutionary epistemology has gained considerable influence in which knowledge and culture are regarded as propelled by mechanisms similar to those that cause changes in species.[5] How does the theory of evolution apply to systems of knowledge? The fundamental idea is simple. Like mutations, new ideas, it is argued, occur blindly (Campbell, [1960] 1987). Some cultural, scientific, or technological ideas catch on because in some way they suit the needs of society, in much the same way as some mutations are retained by natural selection for perpetuation. In its simplest form, the selection process works because the best adapted phenotypes are also the ones that multiply the fastest.

In the context of technological progress, being retained for perpetuation means that at given factor prices, a technique can produce a good at the lowest quality-adjusted cost. Through this selection process, the best ideas survive, in some cases completely supplanting inferior ideas. The competitive game here is played not at the level of the firm but at the level of the techniques themselves. Winning this game means that a technique or a cultural trait is what evolutionary biologists call "adaptive."[6] Outside the realm of production, it has been difficult to give the idea of adaptiveness much operational content. The literature on cultural change has had to struggle with the ambiguities inherent in trying to identify cultural traits as "adaptive" or not. Consider the issue of clothing. The adaptive value of wearing a white dress in the desert or a fur coat in the Arctic is obvious. The wearing of a "provocative" dress by a woman is also adaptive, because it stimulates sexual activity (though the exact definition of "provocative" itself is conditional on the culture). In other cases, however, the properties of clothing were determined by the cultural background: Orthodox Jews wear hats for reasons that go back to antiquity. The ease with which Jews could be identified by

5. Some of the seminal papers are those of Campbell ([1960] 1987) and Toulmin (1967). For an application to science, see especially Hull (1988a, 1988b) and Ruse (1986, ch. 2).

6. The term "adaptive" will be used here in its biological meaning, which is a trait that confers upon its possessor a higher fitness, that is, a higher probability of survival and reproduction.

their hats was clearly a negatively adaptive trait in areas where Jews were discriminated against. Yet the custom has survived. In many other cases it is unknown whether a change is adaptive or not.[7] In examining technological change, however, when the "species" under investigation is the technique itself, such issues are relatively simple. The advantages of a new technique can be described by a few variables, most of which can be summarized by the economist's notion of social saving, that is, the difference in total social costs (including costs external to the firm, such as pollution) between producing the same goods with the old technique and producing them with the new. Traits such as lower production costs, higher durability, and safer production processes are examples of well-defined adaptive features. Other factors could matter as well. Some techniques might survive because they are attractive in terms of criteria outside the strict economic calculus, such as their aesthetic features. Still, given a relatively small amount of information, such as factor prices and the valuation that society places on matters such as noise, clean air, virgin forests, and so on, techniques can be ranked by their adaptiveness.

To repeat: in the paradigm I am proposing it is neither the firm nor societies nor the people that are the analogue to species, but the technique itself. Consider first how a static technology is similar to a hypothetical world of unchanging species. In the absence of progress, technological information is passed from generation to generation in unaltered form. The intergenerational transmission of "technological DNA" occurs through training, usually of apprentices or sons, by people possessing the information. Although some random noise may accompany this information, the mechanisms that eliminate these disturbances usually prevail, and any specimen displaying markedly different characteristics is maladjusted and dies out. The competition for resources among living creatures is analogous to competition in the marketplace, and natural selection ensures stability. Such periods of stasis are the rule rather than the exception in both biology and the history of technology.

Technological change occurs through the emergence of new ideas. They may occur perfectly randomly, or "blindly" (that is, unforeseen by the players themselves), or systematically with a large stochastic component. Like mutations, new ideas represent deviations from the displayed characteristics, and are subjected to a variety of tests of

7. It is, for example, unlikely that technological creativity itself is an adaptive feature for the populations that develop it and are enriched by it. In some cases technological progress has been accompanied by population growth, but modern technology has made war more destructive and led to such negatively adaptive traits as contraception and smoking.

their performance against the environment. Like mutations, most are stillborn or do not survive infancy. Of the few that do, a number of ideas are actually reproduced, that is, are transferred to other specimens. If the new technique is adaptive, it will generate new converts, or it will simply allow its bearers to reproduce faster so that eventually it can gradually replace the old technology. The analogy is not farfetched; like technological progress, evolution is widely understood to mean an increment in knowledge (Bartley, 1987, p. 23).

The analogy helps us distinguish between inventions and their diffusion and dissemination. During a period of change in a species, we observe at any moment both the old and the changed population. If the latter is better adapted to survival (e.g., a bacterium immune to antibiotics), it will eventually supplant the less adaptive population. Natural selection may well operate through changes in gene frequencies without mutation, simply by more adaptive species dominating the processes of Mendelian inheritance. Yet the stationary state in which a single superior species dominates the environment may not ever be observed. By the time the new species has replaced the old one, new mutations may have occurred creating an even more successful form. If they do not, evolution ceases, and thus "selection is like a fire that consumes its own fuel . . . unless variation is renewed periodically, evolution would have come to a stop almost at its inception" (Lewontin, 1982, p. 151). Technological change follows a similar dynamic. At any given time, we observe a best-practice (most up to date) technique, as well as an average-practice technique reflecting older practices still extant. For a one-shot innovation, the competitive process will, under certain circumstances, eventually eliminate the obsolete technologies, and produce uniformity in production methods. But if novel techniques are continuously "born," no single best-practice technique will ever dominate the industry.

The evolutionary model does not ensure that only the fittest will survive. Evolutionary biologists long ago rejected the Spencerian notion of the survival of the fittest as either incorrect or a tautology. Geographical isolation can create niches in which more primitive forms survive side by side with more efficient species. When isolation is then broken, the equilibrium is upset, and less efficient species may become extinct. Similarly, different techniques can coexist if in some way the less efficient producers control a specific resource over which they have property rights and that allows them to survive and reproduce without competing head-on with other producers. When circumstances change, competition for resources becomes tighter. Neither the dodo nor primitive manual production could survive the opening-up of the world.

The analogy between Darwinian evolution and technological change should not be pushed too far, because technological progress is basically non-Darwinian in nature.[8] Technology is information acquired by learning, not through genes. It is thus a culture, as customarily defined (Cavalli-Sforza and Feldman, 1981, p. 10). Cultural transmission of science, art, and technology involves the passing-on of acquired (that is, learned) traits to the next generation (Boyd and Richerson, 1985). In biological evolution, of course, the transmission of acquired traits would be Lamarckian and does not occur. The intergenerational transmission of information in technology occurs, after all, through the conscious training and teaching of apprentices, descendants, and students. Yet this training is not independent of what the parent has learned during his lifetime. Moreover, technology can change through the lateral transmission of information: unlike animals of the same species, firms and workers imitate and learn from each other.

The transmission of acquired characteristics is known as "biased transmission" and is, to a large extent, what the diffusion of innovations is all about. Such biases occur in a variety of ways: direct bias, in which each of the different options is tried and the most suitable is chosen; indirect bias, in which the option chosen follows some successful example; and frequency-dependent bias, in which a variant is chosen depending on the frequency with which the variant is used among the "parents." Each of these biases has an equivalent in the diffusion of new technology.[9] Cavalli-Sforza and Feldman (1981) show that the exact method of transmission of cultural information is crucial in determining how conservative or progressive a society is in terms of adopting an adaptive innovation. Good communications between teaching and learning individuals reduce the chances for a heterogeneity of techniques to survive.

The analogy is imprecise because epistemologies do not have a precise equivalent to DNA, and therefore do not change through mutations in the genes and chromosomes. Strictly speaking, muta-

8. Stephen Jay Gould remarks that "comparisons between biological evolution and cultural or technological change have done vastly more harm than good" and calls such analogies a common intellectual trap (Gould, 1987, p. 18). Given the past record of Social Darwinism and sociobiology, there is some justification for suspicion of facile comparisons. Yet, with a few exceptions, the analogy between technological change and biological evolution has not been pursued. It is my contention that it is one of those analogies that Gould has described elsewhere (1981, p. 328) as "useful but limited . . . [reflecting] common constraints but not common causes."

9. A detailed analysis of these models is presented by Boyd and Richerson (1985). Although their book is concerned with "culture," they define culture as information acquired by teaching or imitation, and an application to technological progress seems a promising area of research.

tions are nothing more than copying errors in the act of passing on genetic information from generation to generation. In cultural change, including technology, the changes occur in the living specimens. They are thus not copying errors in the standard sense of the word, but rather ideas and inspirations that arise in ways mostly unknown to us but not completely random.[10] Unlike science and technology, biological evolution involves no intentionality. No conscious entity operates in biology in a manner analogous to the individuals who are trying to make life better, reduce production costs, fight cancer, or achieve nuclear fusion.[11] New ideas tend to be affected by their environment through subtle and often subconscious mechanisms. Thus, cultural and technological innovations are directed toward problem solving and therefore will prove on average more adaptive than if innovation were purely random (Cavalli-Sforza and Feldman, 1981, p. 66). This is not necessarily a fatal flaw in the analogy. Although genetic mutations are copying errors, they are not truly random either. Mutations, too, are affected by their environment in that they are induced by mutagens, such as x-rays. Moreover, not all genes are equally sensitive to these mutagens, and some parts of the chromosomes, known as "hot spots," have very high mutation rates. Moreover, we now know that the probability of mutations is not symmetric: mutations in some direction are more likely than in the reverse, a phenomenon known as "mutation pressure" (Dawkins, 1987, p. 307).[12]

A further difference between the two involves cross-lineage borrowings. In evolutionary biology, different species do not often exchange genes, that is, information. In the organic world, hybridization can happen, but only between very closely related species. We share 99 percent of our genes with primates, yet hybridization does not occur. In the history of technology, such exchanges occur all the time. In biology the main effect of the evolution of one species on others is through its impact on the environment, whereas in tech-

10. It is interesting, however, that some notable technological ideas did occur during and as a result of the intergenerational transmission of information, that is, during university lectures. Among the inventions inspired in this fashion were Tesla's polyphase electrical motor, inspired by a professor at the Graz Polytechnic; Hall's aluminum smelting technique, which was begun after a professor at Oberlin College mentioned the possibility; and Diesel's engine, which was first clearly defined in his mind during a lecture at the Munich Polytechnic (T. P. Hughes, 1987, p. 60).

11. As Hull (1988b, pp. 468–74) argues convincingly, this difference in no way invalidates the analogy.

12. Even more encouraging for the analogy is recent research in evolutionary biology that seems to suggest that "directed mutation" may occur in nature, that is, some mutations may occur because the organism needs them. See "How Blind is the Watchmaker," *The Economist,* Sept. 24, 1988, p. 114.

nology changes in different "species" affect others, both directly, through cross-borrowing and indirectly, through their impact on the economic environment. The cross-fertilization of different techniques has not been widely appreciated in the history of technology. An exception is Sahal (1981, pp. 71–74; see also Basalla, 1988, pp. 137–38), who argues for the importance of what he calls "creative symbiosis" in which previously disjoint technologies are merged. Advances in metallurgy and boring technology made the high-pressure steam engine possible; radical changes in the design of clocks and ships suggested to others how to make better instruments and windmills; fuels and furnaces adapted to beer brewing and glassblowing turned out to be useful to the iron industry; technical ideas from organmaking were applied successfully to weaving. It is not clear, however, to what extent this difference damages the validity of the analogy. Hull (1988b, p. 451) concludes that "the merger of lineages is more common in biological evolution and much less common in sociocultural evolution than superficial appearances would lead one to expect."

Moreover, in zoology reproduction is usually biparental. Since different species are by definition reproductively isolated, the possibilities of sudden leaps in evolution are greatly limited. An individual mutant will find it difficult to find a mate with similar attributes, and the mutation will disappear. The transmission of cultural or technological information to future generations requires no mating, and thus the dynamics of invention are quite different from those of speciation. Yet again, the difference is not as profound as it might seem. Sexual reproduction requiring two matched separate parents is not universal in the biological world. As Hull (1988b, p. 444) points out, if biparental inheritance were necessary for selection mechanisms to operate, selection would have been inapplicable to biological evolution throughout most of its history. Moreover, wherever sexual reproduction exists, it implies a complementarity that constrains mutations. A similar complementarity is imposed by the laws of genetics. When a mutation results in a recessive allele, it does not manifest itself in the phenotype until it mates with another such mutation to form a homozygote, that is, an individual whose maternal and paternal alleles, carrying the genetic information, are the same. This history of technology is replete with similar complementarities. We have seen many examples of inventions that required accompanying complementary inventions without which the new species might have been as doomed as a mutant without a mate. I shall return to this issue below.

An important difference between biological and technological evolution is that a central element of biological evolution is generally believed to be adaptation to a changing environment, whereas technological change is above all a manipulation of the environment (though some nonhuman species such as beavers and ants do, of course, sometimes manipulate their environment). In evolutionary biology, adaptation to a changing environment more frequently involves what economists would regard as the equivalent of factor substitution. That is, it is more like a choice among existing techniques (substitution) than the creation of a new and superior alternative. The distinction between substitution and technological improvement is often difficult because it is hard to know in practice whether a technique is known and available but not used, or simply unknown, but the logical distinction is important.[13] Yet, as we shall see, this difference between biological evolution and technological change may not be decisive. Speciation is not adaptation, though it is often accompanied by it (Eldredge, 1985, p. 97), just as invention is not substitution, though it is often accompanied by it. Indeed, Darwin's original view that evolutionary changes are largely propelled by the species' need to adjust to environmental changes is no longer accepted. Successful species do not maintain a "tightly adjusted relationship" with their environment (Stanley, 1981, p. 11). The rate of evolutionary change is believed to be determined primarily by the rate of mutation and the characteristics of the population, such as its size. Even in a constant environment mutations occur, and therefore speciation can take place (although some environments are of course more conducive to speciation than others). Because each species regards all others as part of its environment, such changes mean that a stable environment may be upset by evolution itself. Each species has to change if only because others do. Such a model of the dynamics of change should come as no surprise to the student of the economic history of technological progress. What modern biologists have called "the red queen hypothesis"—the need of species to evolve in order to keep up with others (Maynard Smith, 1988, p. 183) seems an apt description of the waves of innovation that Schumpeter and

13. Some other differences that allegedly invalidate the analogy are that biological evolution is very slow and irreversible, whereas cultural and technological change are rapid and bidirectional (Gould, 1981, p. 325). The velocity issue strikes me as specious, since there is no way to put the two processes on the same scale. As far as reversibility is concerned, Gould is, of course, correct in claiming that cultural changes are not coded in our genes. But technological information *was* written down—in the medieval parchments of Friars Roger Bacon and Theophilus, in the great Renaissance engineering manuals of Agricola and Ramelli, in the blueprints of John Smeaton, and in the thousands of modern journals of applied science and engineering. Technological change can be reversed, and there are cases in which it has been; yet the preponderance of historical evidence is that it shares with biological evolution its irreversibility.

others have described. Species that fail to conform become extinct. Evolution, like technological progress, is creative destruction.

This brings us to the most confounding logical difficulty in the analogy: a new species is (or eventually becomes) reproductively isolated from other species. It is therefore usually simple to make the crucial distinction between adaptation and speciation. In technology we cannot readily define a precise line separating a new "species" (i.e., technique) from existing ones, and therefore it is inevitably arbitrary to speak of "improvement within an old" or the "emergence of a new" technique. Yet, as I shall argue, despite the inevitability of gray areas, such distinctions are not only possible but essential to the understanding of technological progress. It is perhaps of some consolation that evolutionary biologists, too, have been unable to agree on a suitable definition of a species (Bush, 1982; Maynard Smith, 1988, p. 127).

The analogy is thus imperfect, but we should not underrate the communalities. In the final analysis, both biological reproduction and economic activity are dynamics of nature constrained by the finiteness of resources. Darwin himself, after all, hit upon the idea of natural selection after reading Malthus's *Essay on Population,* the quintessential statement of scarcity in a dynamic context.[14] Population pressure provides natural selection with its main modus operandi (Ruse, 1986, p. 24), although other causes for differential survival and reproduction exist. By analogy, technological change will be hampered in an economic environment in which firms cannot freely enter or exit an activity, or where they are constrained by traditions and institutions to a fixed share of the market. Yet here, too, competition among producers is not the only reason for the selection of techniques in use. Even in an industry with no competition among techniques, a technique can disappear if it is used to produce something consumers no longer want. The theory of natural selection is, of course, closely related to Adam Smith's idea of an invisible hand that, through competition among individuals struggling for their own benefit, created a higher order.[15] Notwithstanding the enormous differences in time scales, there are identifiable secular trends in both the history of technology and evolutionary biology. Long-term trends toward increasing efficiency and complexity can be discerned, although such progress is more easily identified in technological history than in evolution. In both, the se-

14. This famous case of economics inspiring science is itself a classical case of cross-lineage borrowing of a kind that could not take place in biology, where all information is transmitted genetically.

15. Gould (1980b, p. 68), who points this out, adds caustically that "Darwin may have cribbed the idea of natural selection from economics, but it may still be right."

lection process is extremely wasteful. The vast majority of both new technological ideas and mutations have been useless, either inherently or relative to their environment. Yet in technological change too, it is probably true, as Hull (1988a) points out for the cases of scientific change and biological evolution, that the selection process cannot be made too efficient without neutralizing its effects. After all, technological change ventures into the unknown, not into the uncertain. The risks cannot be diversified away.

At the same time, the selection processes often seem oddly incomplete. Some seemingly "unfit" species cling tenaciously to life. In other cases, biologists have pointed out that random genetic drift can lead to uncertain outcomes. Mildly deleterious genes can and do become homozygous in populations by drift, despite the pressures of natural selection (Lewontin, 1982, p. 159). Population geneticists have long realized that selective processes responsible for adaptation of gene frequencies do not maximize fitness. Although economic historians have tried valiantly over the past several decades to show that some apparently inefficient techniques that survived despite the emergence of newer and better ones were in reality sophisticated adaptations to their environment, their efforts have not proved the strong functionalist position that long-term survival of a technique necessarily implies efficiency. The clever analogy between the Panda's thumb and the QWERTY typewriter keyboard both of which are patently inefficient outcomes of natural selection (Gould, 1987) illustrates the obvious suboptimality of evolutionary processes in biology and technology. Not all problems are solved, and not all possible solutions are achieved. As Campbell (1987, p. 105) remarks, "The knowledge we do encounter is achieved against terrific odds."

The analogy can be extended. Both technology and evolution develop via processes that are sequential in time, that is, we can identify generations, sometimes overlapping, sometimes not. As Hull (1988b, p. 441) puts it, "Each time an old man shows a young boy how to make a slingshot, that is a conceptual generation." As a result, in both biological evolution and technological change, the number of outcomes is constrained because selection occurs at every step. Variations on unsuccessful experiments in a previous generation are rarely tested. Thus, Arthur (1989) maintains that when steam cars were rejected, no further improvements on them were made, although it is possible that further learning by doing would have created a steam vehicle that was ultimately superior to the gasoline engine. It is this property—cumulating outcomes from a blind variation and then exploring future variations on those paths that passed earlier tests—that prevents a divergence of experimentation to infinity and, as Campbell puts it, makes the improbable inevitable. Selection operates on what exists, not on what could have been (Nelson and Winter, 1982, p. 142). In other words, evolution is path dependent, a property of technological change recently emphasized by David (1987, 1988) and Arthur (1989). As Schumpeter (1934, p. 6) noted, the net of economic and social connections bequeathed from the past holds us "with iron fetters fast in [our] tracks." In evolutionary biology, scholars have recently begun to recognize that when the entities upon which selection occurs are components in larger systems, the past imposes structural constraints on the selection process. Evolution cannot change too many things at once, and the results are often bizarre. Hull (1988b, p. 449) points to the human epiglottis as one such example.

To be sure, in a path-dependent world, outcomes are never inevitable, and worlds that could have been but never were might be fruitfully contemplated, much as they may be distasteful to orthodox historians.[16] Basalla (1988, p. 190) calls this property the "branched character" of technology and insists that "despite widespread belief that the world could not be otherwise than it is . . . different choices could have been made." Had things been different, we could all be driving steam-driven cars, running our factories on water-power, crossing the Atlantic on Zeppelins or subsisting primarily on potatoes. Nevertheless, the history of technology surveyed in Part I shows that in the majority of instances the system that prevailed did so because it was "adaptive," that is, it worked better. Natural selection may at times have thrown up an anomaly, or given way to a fad or superstition. Not everything that ever was, was good. But by and large there was order and logic in the evolvement of techniques, and when necessary the shackles of the past could be broken. Precisely for that reason, path dependency in biological evolution is much stronger than in technological progress. Conceivably, steam cars or windmills could be brought back under the right set of circumstances, and sudden leaps in which the economy hopped from one technique to another did occur when the advantages of a totally different technology became apparent.

16. An interesting feature of path dependency is inertia, or resistance to change, which is observed in both biological evolution and technological progress. Mayr (1988, p. 424) explains that variability is limited because "genes are tied together into balanced complexes that resist change." Successful speciation requires the breaking up of the previous cohesion and its replacement by new ones. Genetic cohesion is reminiscent of what is known as technological interrelatedness (Frankel, 1955; David, 1975, pp. 245–46). Attempts to introduce one piece of modern machinery in an otherwise obsolete plant or a modern locomotive on outdated tracks and switches are usually doomed for reasons of technological cohesiveness. The constraint this cohesion places on technological systems is less severe, because technological systems are far less complex than biological systems.

Living creatures belong to a species in the same sense that different producers "belong" to a given technique. This imposes constraints on what individuals can do, and introduces some arbitrariness in the determination of certain features. For instance, long-playing records turn at 33⅓ revolutions a minute and manual gear-shifts turn counterclockwise. In most countries, driving is on the right side of the road. There is no optimality per se in these standards, but given that they exist, they have to be accepted, and thus impose a constraint on the techniques that can be used. No language is "optimal," yet children learn the language of the society in which they happen to be born. Words by themselves are meaningless, but when they become part of a system, they become a common adaptive feature adopted generally through natural selection. Yet not all specific behavior has definite adaptive meaning, just as not all technical conventions are necessarily efficient (David, 1985).

Moreover, in both evolution and technological change, the ex post course is not the one taken ex ante. There is a large accidental component in long-term change, and the record suggests that serendipity, opportunism, and the "King Saul effect" (which occurs when the search for the solution of a specific problem inadvertently leads to a totally new opportunity set) play an important role.[17] The importance of this effect in biological evolution is still in dispute. Gould (1982a, p. 384) points out that such effects would be nonadaptive (or preadaptive) in nature, and emphasizes their crucial role in evolution. For example, the growth of the human brain was the result of a set of complex factors related to selection. But once the brain reached a certain size, it could perform in ways that bore no relation to the selective reasons for its initial growth. Such effects are common in technological history; many major inventions were preadaptive in the sense that they were designed to solve a small local problem and mushroomed into something entirely different.

Both biological species and production techniques in use are bounded in time. Many evolutionary biologists believe that species, like individuals, are born and die. So did undershot water mills, water-driven clocks, and Newcomen engines.[18] Similarly, species emerge out of something else. In technology, this occurs through the process

17. Young Saul, it will be recalled, went to search for asses and found a Kingdom (1 Sam., chs. 9–10). The classic examples in technological history of this effect are Perkin's discovery of aniline mauve while trying to make quinine and Edison's accidental invention of the phonograph while trying to perfect a device that recorded telegraphic impulses on paper discs.

18. Ruse (1986, p. 52) maintains that this part of the analogy fails as far as scientific progress is concerned because scientific theories rarely become extinct. Yet scientific theories are superseded. One could argue over whether Galenian medicine and Phlogiston physics did not "vanish without a trace." De Bresson (1987) argues that, unlike biological species, no invention is lost and that technical knowledge is cumulative. But

of invention; in evolution, primarily through a process known as "allopatric," or "geographic," speciation, that is, the creation of new species as a result of geographic isolation that prevents interbreeding (Mayr, 1970). Species die out either by extinction, often caused by some outside ecological shock, or by pseudoextinction, in which the species transforms itself into something else. Both phenomena have their analogues in technological history. Above all, both technological change and evolution contain elements that make prediction all but impossible. Historians can explain the past pattern of technology, but cannot predict what is to come any more than evolutionary biologists can predict the species of the future. Basalla (1988, p. 210), whose analysis is in some ways parallel to mine, laments that "we who postulate theories of technological evolution likewise have our Darwins but not our Mendels." This misses the essential point of the analogy. The study of genetics is the study of the causes of genetic variation in the population. Yet genetics has contributed little to our understanding of speciation and nothing to our understanding of extinction (Lewontin, 1974, p. 12). Economic analysis, which postulates that techniques will be chosen by profit-maximizing firms employing engineers in whose minds the genotypes of various techniques are lodged, plays a role analogous to genetics. It explains how demand and supply produce a variety of techniques, and points to the constraining influences of environment and competition as a limit to the degree of variety. Just as genetics by itself does not explain speciation, economic analysis has difficulty explaining macroinventions. Like evolution, technological progress was neither destiny nor fluke. Yet the power of Darwinian logic—natural selection imposed on blind variation—is that we need not choose between the two.

Whether there is a trend or direction in the evolutionary process is a highly controversial issue (Hull, 1988c). Laszlo (1987, p. 83) maintains that biological evolution led to higher and higher organizational levels, producing species that are more and more complex but also more and more specialized and hence vulnerable. In these views, he follows some of the most distinguished scientists who created modern evolutionary biology, from Charles Darwin himself to Ronald Fisher. Most contemporary evolutionary biologists, however, deny these claims. There is in fact, some dispute over whether evolution is progressive at all. The current state of biological science seems to gravitate toward the position that the record is consistent with local and reversible direction at best, but the issue remains in

techniques are often abandoned or else preserved only in musea or enclaves (such as the Amish communities, for example). Distaff-and-spindle spinning or the Roman neckstrap horse harness are arguably every bit as extinct as dinosaurs.

dispute. Biological evolution is not teleological; it does not lead to any specific goal, unless one takes the view that the human race is the final objective of biological evolution. Simpson (1967, pp. 239–262) discusses this issue and concludes that determining whether evolution is progressive depends on the choice of criterion.[19] This theme is elaborated upon in detail by Ayala (1988), who rejects claims to a general criterion of progress such as survival probability or amount of genetic information. Depending on the criterion chosen, *Homo sapiens* could indeed find itself as the most progressive species (for example, if the criterion is ability to gather and process information about the environment) or at the bottom (if the criterion chosen is the ability to synthesize its own biological material from inorganic sources). There is, of course, no reason to prefer one criterion over another.

Similarly, it is futile for historians to argue about whether there has been progress in the human condition over the centuries, unless the criterion is specified. The criterion economists would like to use is the capacity of the productive sector to satisfy human needs relative to resources. Technological progress, in that sense, is worthy of its name. It has led to something that we may call an "achievement," namely the liberation of a substantial portion of humanity from the shackles of subsistence living. But if that is the criterion, we run into a dilemma: the increase in living standards is at most a century and a half old. Does what came before 1850 not, then, deserve the term "progress"? In the 1930s, the eminent archaeologist V. Gordon Childe ([1936] 1965, p. 7) wrote that "the historian's progress may be the equivalent of the zoologist's evolution." What he meant by that was that economic progress could be termed a success in the limited sense that it allowed the species to propagate and multiply. If we use that criterion, the difficulty in our evolution of the historical effects of technological progress is eliminated. Technological change can increase numbers, or it can increase economic welfare. For most of history technological progress led mostly to population growth, as the classical economists, led by Malthus, observed. By the purely biological criterion proposed by Childe, the size of the species, it was indeed a success. Seen in this light, the history of technology as opportunities created by humans, selected for survival by the relentless economic mechanisms that make societies prefer more to less and

19. Simpson (1967, p. 251) suggests that it is possible to define the development of humans as progress because of their virtually exclusive ability to control their environment. ". . . it is a peculiarly *human* sort of progress, part of the larger wonder that man is a new sort of animal that has discovered new possibilities in ways of life—and this is progress, whether referred specifically to the human viewpoint or not" (emphasis in original).

cheaper to costlier, can dismiss charges of "Whiggishness." Even if what we today define as living standards did not rise all that much before the mid-nineteenth century, the larger families and reduced infant mortality made possible by better production techniques are in and of themselves a form of progress. It is clear that historians today reject the cruder ideas of progress that regard history as a stepwise progression toward some kind of copious utopia. Yet it is possible to go too far, and to deny that the mechanisms of natural selection impart any direction at all onto the economic history of technology. Direction is not purpose and discerning a trend is not teleology. To deny the idea of progress in its more limited sense altogether is to deny the very fact of economic growth, which this book set out to explain.

To sum up, then, if it is granted that a prima facie case can be established for looking at technology from an evolutionary point of view, it might seem as if the gradualist school in economic history—including those who deny outright the usefulness of the concept of an Industrial Revolution—would be able to find some indirect support for their position from a comparison with evolutionary biology. The neo-Darwinian synthesis, which until recently was the unchallenged paradigm among evolutionary biologists, held that evolution was slow, continuous, and gradual. Species evolved from each other through infinitely small steps, each one too insignificant to be noticed. Natural selection, however, guided what would otherwise have been a stationary process into directed adaptive genetic change. Clearly, a gradualist view of technological history, including an interpretation of the Industrial Revolution that denied its revolutionary character, would be perfectly consistent with that view of historical change. Every change in production methods is a minor improvement on the previous technique in use, and it is through the accretion of thousands of these small changes that technological progress is ultimately achieved. This extreme gradualism, associated with the work of the sociologist S. C. Gilfillan (1935), was at the time a much-needed antidote to the naive "great men" accounts of invention of the early twentieth century, and has steadily gained influence.

Many economic historians believe in an evolutionary model of change, where "evolutionary" is used in its sense of gradual. They have argued that technological change consists of small incremental steps, a "steady cumulation of innumerable minor improvements and modifications with only very infrequent major innovations" (Rosenberg, 1982, p. 63; see also Sahal, 1981, p. 37 and Basalla 1988, pp. 26–63). They have not fully realized, however, that biologists have been far from unanimous about the gradual nature of biological evolution itself. In a classic but controversial work, the geneticist

Richard Goldschmidt (1940) proposed a distinction between micro- and macromutations. The former accounted for changes within a species, and were more or less continuous and cumulative. The latter accounted for great leaps in biological evolution that created new species. Goldschmidt believed that at times evolution moved in leaps and bounds by means of those macromutations. He argued, probably too rigidly, that speciation occurred only by way of macromutations and not by the continuous accumulation of micromutations. After a new species emerges, its evolution follows the standard adaptive process of cumulative micromutations. Goldschmidt remained an isolated and maligned critic of standard evolutionary theory.[20] The conventional neo-Darwinian wisdom in evolutionary biology had no place for Goldschmidt's saltationism.

Recently, however, the modern Darwinism synthesis has come under attack from an unexpected corner. A number of prominent paleontologists and paleobiologists, especially Stephen Jay Gould and Niles Eldredge, have maintained that the emphasis on continuity and gradual change is exaggerated, if not completely misplaced. Instead of gradualism, they maintain, the fossil records show that extended periods of history are characterized by stasis, followed by periods of short, feverish evolutionary change. These periods of stasis are known as "punctuated equilibria." Because they believe that evolutionary change occurred in short and abrupt leaps, these scholars have found much inspiration in the saltationist views of Goldschmidt.

It is significant, perhaps, that the challenge to the gradualist synthesis has come from paleontologists studying fossils. The source of their skepticism is the lack of evidence for any change during very long periods, which supports the idea of stasis, and the lack of evidence for intermediary forms between different species, as gradualism would require (Eldredge 1985). Paleontologists are to the study of evolution what historians are to the study of technology. They observe the records of the past rather than the abstract models or experimental data of the present, and in so doing they present a unique, long-term point of view. What the new punctuated equilibrium theorists, following Goldschmidt, maintain is that there are two forms of speciation, one that is continuous, cumulative, and adaptive, and one that creates new species by discontinuous and often nonadaptive change. These macromutations are believed to be the result of chromosomal alterations, occurring early in the ontogeny and "leading to cascading effects" throughout embryology (Gould, 1980a, p. 127).

20. See Gould (1980a, p. 124). For a particularly scornful and misleading analysis of Goldschmidt's views, see Dawkins (1986, pp. 223–52).

Something similar holds for technological change. In the history of technology there are long periods of stasis as well as major discontinuous changes that parallel macromutations. To adapt Goldschmidt's terminology to the history of technology, I propose to refer to major inventions that constitute discontinuous changes as "macroinventions." Goldschmidt referred to macromutations as "hopeful monstrosities," a powerful metaphor that is remarkably apt for new inventions as well. Such macroinventions constitute a tiny minority of all inventions ever made. Yet their number is not crucial. Inventions do not obey the laws of arithmetic, and the importance of macroinventions was far greater than their small number would suggest. These macroinventions may have been few and far in between, but they were the stuff from which new "species," i.e., techniques, were made.

A macroinvention is an invention without clear-cut parentage, representing a clear break from previous technique. Morison (1966, pp. 8–9) has argued with some exaggeration that "almost any" study of a new invention indicates that long before the inventor did his work, others were doing much the same thing "unconsciously or by accident."[21] It is equally misleading to argue that, just because many macroinventions do have some precursors, all change is continuous (Basalla, 1988, pp. 26–63).[22] Precursors have to be evaluated in terms of their success and their influence on the macroinvention itself.

For a macroinvention to succeed, it must be able to compete and survive. The requirements for viability are first, that the new idea must be technically feasible, that is to say, within the ability of contemporaries to reproduce and utilize.[23] Second, the new idea must be economically feasible, that is, at least as efficient as existing technologies. Third—and here there is no direct analogue in evo-

21. This generalization was based on all of two observations, that Bessemer-type steel was made accidentally as a by-product of puddling furnaces, and that European mothers tended to heat cow milk before Pasteur's discovery of the sterilization process. Other examples may be found; yet as a rule the generalization is surely false.

22. Nobody would dispute the discontinuity that took place when Martin Luther nailed his Ninety-five Theses to the Wittenberg church door, despite the fact that he had been preceded by Wycliffe and Hus. Neither would anyone contest the breakthrough achieved by Joseph de Montgolfier when he first produced his hot-air balloon, even if a century earlier Lana Terzi, a Jesuit priest, had proposed a flying boat buoyed by thin copper globes evacuated of air. Similarly, Beau de Rochas's anticipation of Otto's four-stroke principle is an interesting curiosity; it does not detract from the discontinuity achieved by the inventor. Basalla himself (1988, p. 95) admits that in the case of Newcomen's steam engine, there was very little in Papin's apparatus that could have served as a guide to the atmospheric steam engine.

23. Thus, Leonardo Da Vinci's many radical technological ideas, many of which were potential macroinventions, failed because the materials and workmen of the time were incapable of carrying them out.

lutionary biology—the new invention has to be born into a socially sympathetic environment. As we have seen, technologically and economically feasible inventions were at times suppressed and delayed by reactionary governments or worried competitors.

Once a macroinvention occurs, and a "new species" emerges, it creates a fertile ground for further adaptive microinventions. The macroinvention itself need not be economically very important right away. It is defined as a new conceptual departure and it thus raises the marginal product of subsequent microinventions.[24] The new technology has to be debugged and adapted to local conditions. Additional improvements through learning processes continue after the new technology has emerged. In some instances, a subsequent improvement was so important that we may say that a macroinvention occurred in two or more discrete stages. Such macroinventions consisted of an original idea that had to await a critical revision to become workable, so that one of the subsequent improvements could be regarded as part of the macroinvention itself. Newcomen needed his Watt, Bouchon his Jacquard, Cartwright his Roberts, Pacinotti his Gramme, Lenoir his Otto. In these cases pinpointing the discontinuity with precision becomes difficult, but the principle is not invalidated as long as the number of critical revisions is small. The process of gradual improvement by a series of microinventions sets in after the critical revision has occurred. It is odd to argue, as Sahal (1981, p. 37) does, that major inventions are made possible by numerous minor inventions. The complementarity between the two is mutual. Without microinventions, most macroinventions would not be implemented and their economic rents not realized. But without macroinventions, what would there be to improve?

I argued at the beginning of this book that without invention, innovation will eventually exhaust itself. In symmetric fashion, without macroinventions, microinventions are likely eventually to run into diminishing returns, and technology will begin to look more and more like the "stasis" observed by Gould and his colleagues in the fossil record, unless and until a new major invention occurs. Without the macroinventions of the Industrial Revolution, we might have a world of almost perfectly designed stagecoaches and sailing ships. It is precisely the prediction of extended periods of stasis and stagnation punctuated by periods of rapid and intensive change that makes the Gould–Eldredge theory of evolutionary change such an attractive paradigm for technological history. As Spengler (1932, p. 37) put it, somewhat crudely, "World history strides on from catastrophe to ca-

24. As Usher (1920, p. 274) points out, it should not surprise us that the "subordinate" invention yields larger returns than the "principal" invention.

tastrophe, whether we can comprehend and prove it or not."[25] Technological progress was neither continuous nor persistent. Genuinely creative societies were rare, and their bursts of creativity usually short-lived.

The traditional neo-Darwinist view of evolution argues that natural selection operates at the level of the individual organism. The technological equivalent of that is the Nelson–Winter view that economic selection works primarily at the level of the firm. Against that, the Gould–Eldredge view of evolution is one of hierarchy; it does not deny that evolution works at the level of the individual, it merely seeks to extend selection to higher levels, such as species, as well. The idea of species selection has had a mixed reception among evolutionary biologists, and its empirical importance is still in dispute (Maynard Smith, 1988, pp. 138–142). In technological history, however, the idea of a hierarchy of selection is illuminating. The analogue here is that natural selection works at the level of the technique in use (the analogue of species or deme) as well as on the level of the firm (the analogue of individual organism) (Gould, 1982b).

The modern assessment of Goldschmidt's view of evolution is mixed, in part because his magnum opus is a difficult and contradictory book. Mayr (1988, p. 465) concedes that "hopeful monsters are at least in theory possible," but adds that recent authors are virtually unanimous that there is no evidence that saltations such as envisaged by Goldschmidt and others occur (ibid., p. 414). Yet some modified version of Goldschmidt's view of macromutations has survived in modern evolutionary theory. Even without "drastic mutations," it has become clear that Gould's (1982c) interpretation of Goldschmidt's definition of macroevolution as small genetic mutations early in the ontogeny with large phenotypic effects is not incompatible with Darwinian evolution and the evidence (Mayr, 1988, p. 413; Maynard Smith, 1988, p. 153). The parallel with technological change is immediate: many major technological breakthroughs occurred through small conceptual changes or simple ideas yet led to drastic differences in production methods.

Where gradualism is inevitable is in populational, rather than phenotypic, change. In other words, when a new species with new features appears, it does not immediately predominate. Hopeful monsters have to spread through populations by interbreeding with individuals with normal phenotypes, or by their superior survival capacity. The change in gene frequency in the population through selective mechanisms is by necessity slow, not unlike diffusion lags in

25. It is interesting to note that Spengler quoted the Dutch biologist Hugo de Vries, a convinced saltationist, as the analogue of his own theory of history.

technological change (though here the analogy becomes a bit forced). Contemporary criticism of Goldschmidt is most effective not in denying the existence of macromutations, but in refuting the exclusivity he claimed for macromutations in generating new species. It is possible that new species eventually evolve through lengthy cumulative series of micromutations. No evolutionary biologist would concur today that saltations are the only (or even the primary) source of new species. Similarly, not *all* new techniques require macroinventions.

Techniques may be subject to both micro- and macromutations. A hierarchy of change means that technological progress takes place through both the improvement of existing techniques and the emergence of new ones. Whether such distinctions are useful or not is a matter of judgment. As noted earlier, the demarcation line marking where an old technique ends and a new one begins is arbitrary, though I believe that the number of truly ambiguous cases is not large. There is not always a direct correlation between the novelty of an idea and its economic impact. Ballooning must be regarded as one of the most radical new ideas of all times, yet its direct impact on economic welfare was small. On the other hand, Neilson's hot blast and Richard Roberts's self-actor, of immense economic importance, must be regarded as improvements on existing ideas, and thus as microinventions.

New techniques can thus evolve in two ways. One is through a sudden macroinvention, followed by a series of microinventions that modify and improve it to make it functional without altering its concept. The other is through a sequence of microinventions that eventually lead to a technique sufficiently different from the original one as to classify it as a novel technique rather than an improved version of the original one. It is erroneous to argue, as Persson (1988) does, that all technological progress in the premodern era consisted of continuous sequences of microinventions. The great inventions of the later Middle Ages—the windmill, spectacles, the mechanical clock, moveable type, and the casting of iron—are classic examples of new techniques emerging from macroinventions, as are many of the great inventions of the halcyon days of the First and Second Industrial Revolutions. The five examples discussed in Mokyr (1991) of macroinventions during the late eighteenth century are gaslighting, the breastwheel, the Jacquard loom, chlorine bleaching, and ballooning. To be sure, examples of new techniques emerging through "drift" can be found in the historical record. In shipping, mining, construction, and especially agriculture, undramatic, cumulative, barely perceptible improvements led to increased productivity. The archgradualist Gilfillan chose as his case study the development of the ship, in which macroinventions were rare. The eighteenth-century ship was significantly different from the early fifteenth-century ship, yet with few exceptions these changes were the result of cumulative microinventions. In agriculture, gradualism was even more pronounced. It took centuries for the heavy plow to become universally accepted and almost a millenium for the three-field system to spread throughout Europe. What must be regarded as new techniques in seed selection, tools, and rotations led to rising productivity after 1500, but the trend was so slow that even some historians have missed it. These developments took centuries to complete; in farming, few inventions as dramatic as Gutenberg's or Montgolfier's can be identified as the beginning of the process.[26]

The distinction between micro- and macroinventions matters because they appear to be governed by different laws. Microinventions generally result from an intentional search for improvements, and are understandable—if not predictable—by economic forces. They are guided, at least to some extent, by the laws of supply and demand and by the intensity of search and the resources committed to them, and thus by the signals emitted by the price mechanism. Furthermore, insofar as microinventions are the by-products of experience through learning by doing or learning by using they are correlated with output or investment. Macroinventions are more difficult to understand, and seem to be governed by individual genius and luck as much as by economic forces. Often they are based on some fortunate event, in which an inventor stumbles on one thing while looking for another, arrives at the right conclusion for the wrong reason, or brings to bear a seemingly unrelated body of knowledge that just happens to hold the clue to the right solution. The timing of these inventions is consequently often hard to explain.[27] Much of the economic literature dealing with the generation of technological progress through market mechanisms and incentive devices thus explains only part of the story. This does not mean that we have to give up the attempt to try to understand macroinventions. We must, however, look for explanations largely outside the trusted and familiar market mechanisms relied upon by economists.

Is it possible to conjecture when technological progress will ad-

26. Even in agriculture, however, some discontinuous leaps did occur. There is no continuous way to turn oats into potatoes or oxen into horses, nor a smooth gradual path from organic to chemical fertilizer.

27. In discussing coke smelting, Flinn (1978) maintains that "the element of pure luck should not be underestimated . . . the actual lag in the case of coke smelting was about 125 years but it could easily have been 275 years and might have been a mere 75 years." A similar statement holds for Gutenberg's invention of moveable type, in which the type metal had to be made out of a soft special alloy that only a mechanically gifted person with a strong background in metallurgy could have developed.

vance by leaps and bounds, dominated by macroinventions and when it will progress in continuous and smooth sequences of small inventions alone? Some technological systems, such as ships, mines, and farms, are complex and interrelated. Dramatic sudden changes are not impossible in such systems, but are less likely because of the need to preserve compatibility with other components. Because of the resistance of other parts in the system, large changes were slow in the making. In complex systems, improvements can be evaluated only when a single component is altered while all others are held constant. In practice, this is often very difficult to carry out.[28] As we have seen, the transition from sailing ships to steamships in the nineteenth century gradually transformed a sailing ship with auxiliary engines into a steamship with auxiliary sails over a period of half a century. During those decades every part of the ship, from mast to rudder, was redesigned. Although some of the improvements in the ship between 1820 and 1880 were more radical than others (the change from wood to iron, for example), the process as a whole was clearly gradual.

A large and sudden change in technology was more likely when it was not location specific. A Newcomen machine ran in Cornwall in just the same way it did in Germany or Spain. Moveable type that printed Latin books worked just as well for the Cyrillic, Hebrew, or Arabic alphabets. But in mining and in agriculture, what worked in one place might not work elsewhere if the topographical, climatic, or soil conditions were different. The American reaper, for example, could not be applied to the British landscape (David, 1975, pp. 233–75). Fertilizing, drainage, irrigation, seed selection, animal breeding, the erection of fences and hedges—all were functions of local conditions and could not be made to work universally. By providing insight into the mechanism behind the invention, modern science has made inventions more universal. We now understand better why some things work in one place and not in another, and are able to adapt a technique to make it more universal. Chemical pesticides in agriculture, screw propellers in ships, and the application of compressed air as a source of power in mines are examples of macroinventions that transcended the specificity of local conditions.

Gradual change will be the rule when the complementary technical support system is inadequate to support macroinventions. Many of the most innovative ideas of Renaissance and baroque Europe came to naught because they could not be built at all or at a cost that

28. Cardwell (1968, p. 120) attributes this methodology to Smeaton, and argues that it is the basis for the piecemeal improvement of any system, but that it cannot yield a major leap forward.

was within reason. High-brow technology depended on advances in low-brow technology. The technological ideas of the Industrial Revolution became macroinventions because they could be built, reproduced, and they worked. Complementary microinventions were as much at the center of the Industrial Revolution as the great ideas themselves. Indeed, Britain's success after 1750 depended primarily on the skilled mechanics and engineers who refined and corrected the ideas of famous inventors. Most of the revolutionary technological ideas of the sixteenth- and seventeenth centuries—such as Jacques Besson's screw-cutting lathe (1569); Simon Stevin's sailing chariot (1600), which made one celebrated journey; Giovanni Branca's steam turbine (1629), which was never built; and Blaise Pascal's *sautoir* or adding machine (1642) of which 70 prototypes were made—could not be made practicable even when conceptually sound.

The main reason I have dwelled so long on the distinction between micro- and macroinventions is that *both* were very much part of the story. Here, then, is the most fundamental complementarity of the economic history of technological change. Without new big ideas, the drift of cumulative small inventions will start to run into diminishing returns. When exactly this will occur depends on the technique in question, but it seems clear that additional improvements in the sailing ship were becoming more difficult by the 1870s, that best-practice grain yields were approximating some kind of ceiling by the mid-nineteenth century, and that crucible steel had been taken a long way by 1856. Macroinventions such as the screw propeller, chemical fertilizers, and the Bessemer process revitalized a movement that was approaching something close to a technological ceiling. It is not necessary for the ideas to emerge in the receptive economy itself; some of the inventions Britain exploited during the Industrial Revolution came from France. After 1860 the inventions it generated were increasingly adopted in other countries more quickly than in Britain itself. Regardless of where they came from, genuinely important new ideas were neither cheap nor elastically supplied. Technology was, as I have argued repeatedly, constrained by supply. What made societies poor was not that they had too few resources, but that they did not know how to produce more wealth with the resources they had. Ideas alone may not have been enough either; all the same, they were indispensable.

But more is involved. The historical survey in Part II has indicated that macroinventions rarely occurred alone, but rather tended to appear in clusters. In other words, the occurrence and timing of macroinventions is partly explained by other macroinventions. The later Middle Ages and the Industrial Revolution, both rich in macroinventions, punctuated two-and-a-half centuries of gradual and continuous improvements in which most technologies, from ship design to ironworking to textiles, were improved and refined, but in which large macroinventions were rare.

Two factors help explain these clusters. One possibility, discussed in Chapter 7, is that macroinventions were not independent events, but rather influenced each other. Social scientists have long realized that when the behavior of economic agents depends on what other agents do, critical-mass types of models are relevant.[29] One or two lonely inventors may not be enough to start an industrial revolution, but with a few more, the mutual effects of imitation and learning may become strong enough to start something much larger. Thus, a wave of macroinventions may occur when inventive activity has reached a certain critical level. Despite their obvious relevance to social sciences, critical-mass models by themselves are not satisfying explanatory devices in that they tend to relate large, possibly momentous, outcomes to small and insignificant beginnings. In these models, luck and randomness alone determine where and when the sequence of innovations occurs (Crafts, 1985).

Second, exogenous changes in the institutional and social environment upon which the new ideas were imposed may have changed the receptiveness of the overall economy to macroinventions. Once more, the analogy with evolutionary biology turns out to be helpful. An exogenous change, such as an earthquake that separates a peninsula from the mainland, or a flood that changes the course of a river, creates the reproductive isolation necessary for allopatric speciation. We have to search for similar environmental events in technological history to explain the major upheavals that punctuated persistent stasis.

Some societies were less receptive to radical changes than others. With a few exceptions, Renaissance and baroque Europe were permeated by reactionary elements, heralded in by the Reformation and Counter-reformation, struggling to suppress intellectual novelties of any kind. The conservative forces failed to stop the great scientific achievements of the seventeenth century, but not for lack of trying. In technology, where the strength of the forces of reaction was greater than it was among the small cosmopolitan scientific elite of Huygens, Boyle, Leibniz, and Descartes, radical ideas were understandably unfashionable. Craft guilds, anxious to protect their turf, strengthened the conservative elements. Although these conservative forces were ultimately defeated here as well, their resistance may have had an effect on the rate of progress, not so much in slowing down, but in smoothing its path and restricting progress to local microinventions.

29. For an excellent exposition, see Schelling (1978, pp. 81–134).

The great engineering books of Zonca, Ramelli, Agricola, and their colleagues were mostly summaries of existing technology, not novel suggestions. Leonardo Da Vinci, the most original technological spirit of the age, never published a word of his thousands of pages of engineering notes. The adventuresome spirit returns in full swing only in the late seventeenth century. Nonetheless, there was progress in these years. Technological creativity was by this time so strongly entrenched in Europe that it manifested itself in the cumulative small changes of the sixteenth and seventeenth centuries. Yet the environmental changes in Britain after the Glorious Revolution of 1688 produced the background against which the discontinuous technological breakthroughs of the eighteenth century could unfold. Once they were established and viable, the new species that sprung up turned out to be irresistible.

To conclude, then, we have to face again the question of what causes great technological "events" such as the Industrial Revolution, during which we observe clusters of macroinventions in a variety of industries. Macroinventions are seeds sown by individual inventors in a social soil. There is no presumption that the flow of macroinventions is the same at all times and places. Some of the factors determining the supply of ideas—such as religion, education, willingness to bear risk, and the social status of physical production in society—were discussed in Chapter 7. But the environment into which these seeds are sown is, of course, the main determinant of whether they will sprout. There are no one-line explanations here, no simple theorems. It is hard to think of conditions that would be either necessary or sufficient for a high level of technological creativity. A variety of social, economic, and political factors enter into the equation to create a favorable climate for technological progress. At the same time, a favorable environment may itself be insufficient if the new technological ideas fail to arise. The dynamic of this evolutionary process generated a rich historical pattern of long periods of stasis or very gradual change, punctuated at times by clusters of feverish progress in which radical inventions created new techniques with an abruptness that does not square with the adage that Nature does not make leaps.

CHAPTER TWELVE

Epilogue

Professional historians, like evolutionary biologists, do not view the study of history as a tool with which to predict the future. Still, after reviewing the achievements of technology, it is tempting to wonder whether the cheap lunches will continue to be served, or whether the long-feared stationary state predicted by economists for centuries will eventually be reached. Schumpeter (1942) had no doubts on the matter. "Technological possibilities are an uncharted sea," he insisted, "there is no reason to expect slackening of the rate of output through exhaustion of technological possibilities" (ibid., p. 118). Ten years ago one of Schumpeter's disciples expressed his concern that the United States was approaching something like technological maturity, and that technological opportunities were reaching exhaustion (Scherer, 1984, pp. 261–69). For the global economy as a whole rather than the United States alone, I see no evidence of such an exhaustion. In the past ten years a veritable revolution has occurred in a wide variety of fields, from genetic engineering to consumer electronics. The essence of technological progress is its unpredictability. Nothing in the historical record seems to indicate that the creation of new technological opportunities—as opposed to their exploitation—is subject to diminishing returns, fatigue, old age, or exhaustion. I am not implying that such exhaustion could not ever happen, only that the historical record has little to suggest why or how it should occur.

Nonetheless, if there is one lesson to be drawn from this search for the causes of technological progress, it is that it should not be taken for granted. If Cardwell's Law can be extrapolated into the future, no single society should expect to be on the cutting edge of technology forever. Contrary perhaps to what economic logic suggests, most societies have not been particularly amenable to the application of new ideas to production. As Schumpeter stressed, the enemies of technological progress were not the lack of useful new ideas, but social forces that for one reason or another tried to preserve the status quo. These forces represented different interests and came in a variety of forms: environmental lobbies, labor unions, clayfooted giant corporations, professional associations, reactionary or incompetent bureaucracies may all, in one way or another, try to block the kind of relentless and aggressive thrusts forward that occurred in the nineteenth and twentieth centuries.

The West may run out of stamina long before it runs out of ideas. We are, after all, living in an exception. Technologically creative societies are unusual in history. Our age is unique: only in the last two centuries has Western society succeeded in raising the standard of living of the bulk of the population beyond the minimum of subsistence. What made the West successful was neither capitalism, nor science, nor an historical accident such as a favorable geography. Instead, political and mental diversity combined to create an ever-changing panorama of technologically creative societies. From its modest beginnings in the monasteries and rain-soaked fields and forests of western Europe, Western technological creativity rested on two foundations: a materialistic pragmatism based on the belief that the manipulation of nature in the service of economic welfare was acceptable, indeed, commendable behavior, and the continuous competition between political units for political and economic hegemony. Upon those foundations rested the institutions and incentive structures needed for sustained technological progress. But the structure is and will always be shaky. There are no sufficient conditions for technological creativity in any society. Precisely for that reason it is crucial that the world preserve a measure of diversity. If technological progress is ephemeral and rare, multiplying the number of societies in which the experiment is carried out and allowing some measure of competition between them improves the chances for continued progress. As long as *some* societies remain creative, others will eventually be dragged along.

It seems almost inevitable that twentieth-century historiography should have rebelled against the idea that technological progress was indeed "progress," that is, that it improved the lot of the people who lived in the societies in which it took place. There will always be what Pollard has called "doubters and pessimists."[30] There are two responses to these doubts. One is simply to argue that technological change made an increase in population possible without reducing long-term average living standards. As Childe ([1936] 1965, pp. 13–

1. An eminent anthropologist (Harris, 1977, pp. x, xi) wrote a decade ago that "the great industrial cornucopia has been spewing forth increasingly shoddy, costly, and defective goods and services . . . our culture is not the first that technology has failed." Similarly, Basalla (1988, p. 218) writes that "the historical record [does not] justify a return to the idea that a causal connection exists between advances in technology and the overall betterment of the human race."

14) wrote, "[after 1750] the people multiplied as never before since the arrival of the Saxons. Judged by the biological standard . . . the Industrial Revolution was a huge success." It has facilitated the survival and multiplication of the species. Others doubt that the post-1850 increase in per capita income was in fact an improvement. The most resounding and eloquent response that an economic historian can give to those was given half a century ago by Sir John Clapham, a man otherwise little given to eloquence. The pessimists were disposed to argue, wrote Clapham [1938 (1965, Vol. III, p. 507)], that with all their inventions people [in the modern industrial age] were no wiser, no happier, no clearer-eyed than those of other ages, and possibly less upright. "Of uprightness, wisdom, and the clearness of the eye, the economic historian as such may not profess to speak. He moves on the lower plane of commodities and comforts. Moving there, he does not hesitate to compare the [present] time to its advantage—not only with other times in the industrial age but with any time certainly known to him."

The "lower plane" of material goods includes more than just comfort. The riches of the post-industrial society have meant longer and healthier lives, liberation from the pains of hunger, from the fears of infant mortality, from the unrelenting deprivation that were the part of all but a very few in preindustrial society. The luxuries and extravagances of the very rich in medieval society pale compared to the diet, comforts, and entertainment available to the average person in Western economies today.

It is true, of course, that technological progress is not a universal panacea for human want. Some desires and needs cannot be satisfied by inventiveness. Social prestige, political influence, and personal services supplied by other human beings cannot be increased readily by technology. Hirsch (1976) refers to these goods as "positional" goods. Insofar as positional goods are based on a ranking of individuals, their supply is fixed, and their allocation constitutes the quintessential zero-sum game. No amount of technological progress can ever increase their supply. We should not, however, underrate the capacity of technology to satisfy some of the needs underlying the demand for positional goods. Valet services might be supplied by robots; the need to manipulate others could be satisfied for some through computer simulation. Still, as long as ambition and envy are part of human nature, the free lunches served by technological progress will never be quite enough to satiate our appetites.

To return to the question of technological progress as a tale of success. E. H. Carr has written that history is by and large an account of what people did, not what they failed to do. For technological history this holds a fortiori: what people failed to do is simply not recorded. By the very nature of technology, we rarely miss what was not invented. It seems oddly anachronistic and out of place to "blame" the Romans for not inventing the horse collar or the Chinese for not inventing the mechanical clock. Still, when medieval Europe produced these and similar ideas and other societies did not, the outlines of the current economic map of the world were drawn. It is because what happened in western Europe in the past thousand years that non-Western nations, from Japan to Mauritania, are trying to compete with the West by emulating its technology. So far, the most economically successful nonwestern nations are those that have been the best at imitating the West. Nations such as Korea and Japan clearly are capable of creating new technologies, and not just copying them. But the inventions that are being made today in research laboratories in the Far East are quintessentially Western in nature. Although some of these societies have preserved much of their own culture and traditions, in the field of technology the Western heritage is unchallenged.

The long-predicted decline of the West has thus far failed to materialize. Though it no longer sends gunboats to foreign ports, the West continues to prosper, thanks to its past technological creativity. If it is threatened at all, it is by non-Western nations that appear to be able to beat it at its own game. From a global point of view, this threat is desirable. If the West as a whole succumbs to Cardwell's Law, the torch of creativity will be carried on by others. As long as some segment of the world economy is creative, the human race will not sink into the technological stasis that could eventually put an end to economic growth. In the politically competitive world of today, nations will be forced not to fall behind. Psychological shocks such as the fear that engulfed the West after the launching of the Soviet Sputnik will prevent the smugness and self-righteousness that in the past permeated societies in which innovation was repressed.

If economists and policy makers are to understand this fundamental point, they must study the economic history of technological change. To paraphrase another statement of Carr's: a society that has ceased to concern itself with the progress of the past will soon lose belief in its capacity to progress in the future.

Bibliography

Agricola, Georgius. [1556;1912] 1950. *De Re Metallica.* Translated and annotated by Herbert Clark Hoover and Lou Henry Hoover. Reprint edition. New York: Dover Publications.

Aitken, Hugh G. J. 1976. *Syntony and Spark: The Origins of Radio.* New York: John Wiley and Sons.

Al-Hassan, Ahmad Y. and Hill, Donald R. 1986. *Islamic Technology.* Cambridge: Cambridge University Press.

Allen, Robert C. 1977. "The Peculiar Productivity History of American Blast Furnaces, 1840–1913." *Journal of Economic History* 37(September):605–33.

Allen, Robert C. 1981. "Entrepreneurship and Technical Progress in the Northeast Coast Pig Iron Industry: 1850–1913." *Research in Economic History* 6:35–71.

Allen, Robert C. 1983. "Collective Invention." *Journal of Economic Behavior and Organization* 4:1–24.

Arthur, Brian. 1989. "Competing Technologies, Increasing Returns, and Lock-in by Historical Events." *Economic Journal* 99 (March):116–131.

Ashby, Eric. 1958. "Education for an Age of Technology." In *A History of Technology.* Vol. 5, *The Late Nineteenth Century,* edited by Charles Singer et al., 776–98. New York and London: Oxford University Press.

Ashton, Thomas S. 1924. *Iron and Steel in the Industrial Revolution.* Manchester: Manchester University Press.

Ashton, Thomas S. 1948. *The Industrial Revolution, 1760–1830.* Oxford and New York: Oxford University Press.

Audin, Maurice. 1969. "Printing." In *A History of Technology and Invention.* Vol. 22, *The First Stages of Mechanization, 1450–1725,* edited by Maurice Daumas, 620–67. New York: Crown.

Audin, Maurice. 1979. "Printing." In *A History of Technology and Invention.* Vol. 3, *The Expansion of Mechanization, 1725–1860,* edited by Maurice Daumas, 656–705. New York: Crown.

Ayala, Francisco J. 1988. "Can 'Progress' be Defined as a Biological Concept?" In *Evolutionary Progress,* edited by Matthew Nitecki, 75–96. Chicago: University of Chicago Press.

Ayres, Clarence. [1944] 1962. *The Theory of Economic Progress.* 2d ed. New York: Schocken Books.

Babbage, Charles. 1864. *Passages from the Life of a Philosopher.* London: Longman Green.

Baines, Edward. 1853. *History of the Cotton Manufacture in Great Britain.* London: H. Fisher.

Baldwin, William L., and Scott, John T. 1987. *Market Structure and Technological Change.* Chur, Switzerland: Harwood Academic Publishers.

Ballot, Charles. [1923] 1978. *L'Introduction du Machinisme dans l'Industrie Française.* Edited by Claude Gével. Geneva: Slatkine.

Barclay, Harold B. 1980. *The Role of the Horse in Man's Culture.* London: J. A. Allen.

Bartley, W. W. 1987. "Philosophy of Biology Versus Philosophy of Physics." In *Evolutionary Epistemology, Rationality, and the Sociology of Knowledge,* edited by Gerard Radnitzky and W. W. Bartley III, 7–45. La Salle, IL: Open Court.

Barraclough, K. C. 1984. *Steelmaking Before Bessemer.* 2 vols. London: The Metals Society.

Basalla, George 1988. *The Evolution of Technology.* Cambridge: Cambridge University Press.

Baumol, William J. 1988. "Entrepreneurship: Productive, Unproductive and Imitative; or The Rule of the Rules of the Game." Unpublished manuscript, Princeton University.

Beaumont, Olga. 1958. "Agriculture: Farm Implements." In *A History of Technology.* Vol. 4: *The Industrial Revolution, 1750–1850,* edited by Charles Singer et al., 1–12. New York and London: Oxford University Press.

Beer, John Joseph. 1959. *The Emergence of the German Dye Industry.* Urbana, IL: University of Illinois Press.

Benz, Ernst. 1966. *Evolution and Christian Hope: Man's Concept of the Future from the Early Fathers to Teilhard de Chardin.* Garden City, NJ: Doubleday.

Berg, Maxine. 1985. *The Age of Manufactures.* London: Fontana Press.

Bernal, J. D. 1965. *Science in History.* Vol. 1, *The Emergence of Science.* Cambridge, MA: MIT Press.

Bijker, Wiebe E. 1987. "The Social Construction of Bakelite: Toward a Theory of Invention." In *The Social Construction of Technological Systems,* edited by Wiebe E. Bijker, Thomas P. Hughes, and Trevor J. Pinch, 159–87. Cambridge, MA: MIT Press.

Blaine, Bradford B. 1976. "The Enigmatic Water Mill." In *On Pre-Modern Technology and Science: Studies in Honor of Lynn White,* edited by Bert S. Hall and Delno C. West, 163–76. Malibu, CA: Undena Publications.

Bloch, Marc. 1966. *Land and Work in Medieval Europe.* New York: Harper & Row.

Boserup, Ester 1981. *Population and Technological Change.* Chicago: University of Chicago Press.

Boulding, Kenneth. 1981. *Evolutionary Economics.* Beverly Hills: Sage Publications.

Boulding, Kenneth. 1983. "Technology in the Evolutionary Process." In *The Trouble with Technology,* edited by S. McDonald, D. McL. Lamberton, and T. Mandeville, 4–10. New York: St. Martin's Press.

Boyd, Robert, and Richerson, Peter J. 1985. *Culture and the Evolutionary Process.* Chicago: University of Chicago Press.

Bray, Francesca. 1984. *Agriculture.* In *Science and Civilization in China,* Vol. 6, Part 2, edited by Joseph Needham. Cambridge: Cambridge University Press.

Bray, Francesca. 1986. *The Rice Economies.* Oxford: Basil Blackwell.

Brenner, Reuven. 1983. *History: The Human Gamble.* Chicago: University of Chicago Press.

Brenner, Reuven. 1987. *Rivalry.* New York: Cambridge University Press.

Bromehead, C. N. 1956. "Mining And Quarrying in the Seventeenth Century." In *A History of Technology.* Vol. 2, *The Mediterranean Civilizations and the Middle Ages, 700 B.C. to A.D. 1500,* edited by Charles Singer et al., 1–40. New York and Oxford: Oxford University Press.

Brown, Shannon R. 1979. "The Ewo Filature: A Study in the Transfer of Technology to China." *Technology and Culture* 20(July):550–68.

Brown, Shannon R., and Wright, Tim. 1981. "Technology, Economics, and Politics in the Modernization of China's Coal-mining Industry, 1850–1895." *Exploration in Economic History* 18(January):60–83.

Bruland, Tine. 1982. "Industrial Conflict as a Source of Technical Innovation: three Cases." *Economy and Society* 11(May):91–121.

Brumbaugh, Robert S. 1966. *Ancient Greek Gadgets and Machines.* New York: Thomas Y. Crowell.

Bryant, Lynwood. 1966. "The Silent Otto." *Technology and Culture* 7(Spring):184–200.

Bryant, Lynwood. 1967. "The Beginnings of the Internal Combustion Engine." In *Technology in Western Civilization,* Vol. 1, edited by Melvin Kranzberg and Carroll W. Pursell, Jr., 648–63. New York: Oxford University Press.

Bryant, Lynwood. 1969. "Rudolf Diesel and his Rational Engine." *Scientific American.* 221(August):108–17.

Bryant, Lynwood. 1973. "The Role of Thermodynamics: The Evolution of the Heat Engine." *Technology and Culture* 14:152–65.

Buchanan, R. A. 1986. "The Diaspora of British Engineering." *Technology and Culture* 27 (July):501–24.

Bulliet, Richard W. 1975. *The Camel and the Wheel.* Cambridge, MA: Harvard University Press.

Burford, A. 1960. "Heavy Transport in Classical Antiquity." *Economic History Review* 13:1–19.

Burstall, Aubrey F. 1965. *A History of Mechanical Engineering.* Cambridge, MA: MIT Press.

Bush, Guy L. 1982. "What Do We Really Know About Speciation?" In *Perspectives on Evolution,* edited by Roger Milkman, 119–28. Sunderland, MA: Sinauer Publishing Co.

Campbell, Donald T. [1960] 1987. "Blind Variation and Selective Retention in Creative Thought as in Other Knowledge Processes." In *Evolutionary Epistemology, Rationality, and the Sociology of Knowledge,* edited by Gerard Radnitzky and W. W. Bartley III 91–114. La Salle, IL: Open Court.

Cardwell, D. S. L. 1968. "The Academic Study of the History of Technology." In *History of Science, an Annual Review of Literature, Research and Teaching,* Vol. 7 edited by A. C. Crombie and M. A. Hoskin, 112–24.

Cardwell, D. S. L. 1971. *From Watt To Clausius: The Rise of Thermodynamics in the Early Industrial Age.* Ithaca, NY: Cornell University Press.

Cardwell, D. S. L. 1972. *Turning Points in Western Technology.* New York: Neale Watson Science History Publication.

Carr, Edward Hallett. 1961. *What is History?* New York: Vintage Books.

Casson, Lionel. 1971. *Ships and Seamanship in the Ancient World.* Princeton, NJ: Princeton University Press.

Cavalli-Sforza, L. L., and M. W. Feldman. 1981. *Cultural Transmission and Evolution: a Quantitative Approach.* Princeton, NJ: Princeton University Press.

Chao, Kang. 1977. *The Development of Cotton Textile Production in China.* Cambridge, MA: Harvard University Press.

Chao, Kang. 1986. *Man and Land in Chinese History: An Economic Analysis.* Stanford, CA: Stanford University Press.

Chapman, S. D. 1972. *The Cotton Industry in the Industrial Revolution.* London: Macmillan.

Chatterton, E. Keble. 1909. *Sailing Ships and Their Story.* London: Sidgwick and Jackson.

Childe, V. Gordon. [1936] 1965. *Man Makes Himself.* Fourth edition. London: Watts & Co.

Cipolla, Carlo M. 1965a. *The Economic History of World Population.* Harmondsworth, England: Pelican Books.

Cipolla, Carlo M. 1965b. *Guns and Sails in the Early Phase of European Expansion, 1400–1700.* London: Collins.

Cipolla, Carlo M. 1967. *Clocks and Culture, 1300–1700.* New York: Norton.

Cipolla, Carlo M. 1969. *Literacy and Development in the West.* Harmondsworth, England: Penguin Books.

Cipolla, Carlo M. 1972. "The Diffusion of Innovation in Early Modern Europe." *Comparative Studies in Society and History* 14:46–52.

Cipolla, Carlo M. 1980. *Before the Industrial Revolution.* 2d ed. New York: W. W. Norton.

Clapham, John. [1938] 1963. *An Economic History of Modern Britain.* 3 vols. Cambridge: Cambridge University Press.

Clapham, Michael. 1957. "Printing." In *A History of Technology.* Vol. 3, *From the Renaissance to the Industrial Revolution, 1500–1750,* edited by Charles Singer et al., 377–410. New York and London: Oxford University Press.

Clow, Archibald, and Clow, Nan L. 1952. *The Chemical Revolution.* London: The Batchworth Press.

Clow, Archibald, and Clow, Nan L. 1956. "The Timber Famine and the Development of Technology." *Annals of Science* 12(June):85–102.

Clow, Archibald, and Clow, Nan L. 1958a. "The Chemical Industry: Interaction with the Industrial Revolution." In *A History of Technology.* Vol. 4, *The Industrial Revolution, 1750–1850,* edited by Charles Singer et al., 230–57. New York and London: Oxford University Press.

Clow, Archibald, and Clow, Nan L. 1958b. "Ceramics from the Fifteenth Century to the Rise of the Staffordshire Potteries." In *A History of Technology,* Vol. 4, *The Industrial Revolution, 1750–1850,* edited by Charles Singer et al., 328–57. New York and London: Oxford University Press.

Coleman, D. C. 1958. *The British Paper Industry, 1495–1860.* Oxford: Clarendon Press.

Coombs, Rod, Saviotti, Paolo, and Walsh, Vivien. 1987. *Economics and Technological Change.* Totowa, NJ: Rowman and Littlefield.

Cooper, Carolyn. 1984. "The Portsmouth System of Manufacture." *Technology and Culture* 25(April)182–225.

Crafts, N. F. R. 1985. "Industrial Revolution in England and France: Some Thoughts on the Question 'Why was England First.'" In *The Economics of the Industrial Revolution,* edited by Joel Mokyr, 119–31. Totowa, NJ: Rowman and Allanheld.

Crosby, Alfred. 1972. *The Columbian Exchange: Biological and Cultural Consequences of 1492.* Westport, CT: Greenwood Publishing Co.

Crosby, Alfred. 1986. *Ecological Imperialism: The Biological Expansion of Europe, 900–1900.* Cambridge: Cambridge University Press.

Crouzet, François. 1985. *The First Industrialists.* Cambridge: Cambridge University Press.

Cunliffe, Barry. 1979. *The Celtic World.* New York: McGraw-Hill.

Cutting, C. L. 1958. "Fish Preservation." In *A History of Technology.* Vol. 4: *The Industrial Revolution, 1750–1850,* edited by Charles Singer et al., 44–55. New York and London: Oxford University Press.

Daumas, Maurice, 1964. "The Extraction of Chemical Products." In *A History of Technology and Invention.* Vol. 2, *The First Stages of Mechanization, 1450–1725,* edited by Maurice Daumas, 172–211. New York: Crown.

Daumas, Maurice, and Paul Gille. 1979a. "Methods of Producing Power." In *A History of Technology and Invention.* Vol. 3, *The Expansion of Mechanization, 1725–1860,* edited by Maurice Daumas, 17–80. New York: Crown.

Daumas, Maurice, and Paul Gille. 1979b. "Transportation and Communication." In *A History of Technology and Invention.* Vol. 3: *The Expansion of Mechanization, 1725–1860,* edited by Maurice Daumas, 233–390. New York: Crown.

David, Paul A. 1975. *Technical Choice, Innovation, and Economic Growth.* Cambridge: Cambridge University Press.

David, Paul A. 1985. "Clio and the Economics of QWERTY." *American Economic Review* 75(May):332–37.

David, Paul A. 1986. "Understanding the Economics of QWERTY: The Necessity of History." In *Economic History and the Modern Economist,* edited by William N. Parker. Oxford: Basil Blackwell.

David, Paul A. 1987. "The Hero and the Herd in Technological History: Reflections on Thomas Edison and the 'Battle of the Systems.'" Unpublished paper, Stanford University.

David, Paul A. 1988. "Path Dependence: Putting the Past into the Future of Economics." Unpublished paper, Stanford University.

Davis, Lance E., Robert G. Gallman, and T. D. Hutchins. 1991. "Call Me Ishmael Not Domingo Floresta: The Rise and Fall of the American Whaling Industry." In *Research In Economic History, Supplement IV*, Joel Mokyr: *The Vital One: Essays Presented to Jonathan R. T. Hughes*, 191–233. Greenwich, CT: JAI Press.

Dawkins, Richard. 1987. *The Blind Watchmaker*. New York: W. W. Norton.

De Bresson, Chris. 1987. "The Evolutionary Paradigm and the Economics of Technological Change." *Journal of Economic Issues* 21(June):751–61.

De Camp, L. Sprague, 1960. *The Ancient Engineers*. New York: Ballantine Books.

Derry, T. K., and T. I. Williams. 1960. *A Short History of Technology*. Oxford: Oxford University Press.

Deshayes, Jean. 1969. "Greek Technology." *In A History of Technology and Invention*. Vol. 1, *The Origins of Technological Civilization*, edited by Maurice Daumas, 181–214. New York: Crown.

Dickinson, H. W. 1958. "The Steam Engine to 1830." In *A History of Technology*. Vol. 4, *The Industrial Revolution, 1750–1850*, edited by Charles Singer et al., 168–98. New York and London: Oxford University Press.

Dosi, Giovanni. 1984. *Technical Change and Industrial Transformation*. New York: St. Martin's Press.

Dresbeck, LeRoy. 1976. "Winter Climate and Society in the Northern Middle Ages: The Technological Impact." In *On Pre-Modern Technology and Science: Studies in Honor of Lynn White*, edited by Bert S. Hall and Delno C. West, 177–99. Malibu, CA: Undena Publications.

Dutton, H. I. 1984. *The Patent System and Inventive Activity during the Industrial Revolution*. Manchester: Manchester University Press.

Duval, Paul-Marie. 1962. "The Roman Contribution to Technology." In *A History of Technology and Invention*. Vol. 1, *The Origins of Technological Civilization*, edited by Maurice Daumas, 216–59. New York: Crown.

Eldredge, Niles. 1985. *Time Frames: The Rethinking of Darwinian Evolution and the Theory of Punctuated Equilibria*. New York: Simon and Schuster.

Elster, Jon. 1983. *Explaining Technical Change*. Cambridge: Cambridge University Press.

Elton, Sir Arthur. 1958. "Gas for Light and Heat." In *A History of Technology*. Vol. 4, *The Industrial Revolution, 1750–1850*, edited by Charles Singer et al., 258–76. New York and London: Oxford University Press.

Elvin, Mark 1973. *The Pattern of the Chinese Past*. Stanford, CA: Stanford University Press.

Elvin, Mark. 1984. "Why China Failed to Create an Endogenous Industrial Capitalism." *Theory and Society* 13:379–92.

English, W. 1958. "The Textile Industry: Silk Production and Manufacture, 1750–1900." In *A History of Technology*. Vol. 4: *The Industrial Revolution, 1750–1850*, edited by Charles Singer et al., 308–27. New York and London: Oxford University Press.

Ercker, Lazarus. [1580] 1951. *Treatise on Ores and Assaying*. Translated by Anneliese G. Sisco and Cyril Stanley Smith. Chicago: University of Chicago Press.

Eudier, Walter, and Jacques Payen. 1979. "The Spinning of Textile Fibers."

In *A History of Technology and Invention.* Vol. 3, *The Expansion of Mechanization, 1725–1860,* edited by Maurice Daumas, 583–98. New York: Crown.

Evans, Francis T. 1981. "Roads, Railways, and Canals: Technical Choices in Nineteenth Century Britain." *Technology and Culture* 22(January):1–34.

Fei, Hsia-tung, 1953. *China's Gentry.* Chicago: University of Chicago Press.

Fenoaltea, Stefano. 1984. "Slavery and Supervision in Comparative Perspective: A Model." *Journal of Economic History* 44(September):635–68.

Ferguson, Eugene S. 1981. "History and Historiography." In *Yankee Enterprise: The Rise of the American System of Manufactures,* edited by Otto Mayr and Robert C. Post, 1–23. Washington, DC: The Smithsonian Institution.

Feuerwerker, A. 1984. "The State and the Economy in Late Imperial China." *Theory and Society* 13:297–326.

Finley, M. I. 1965. "Technical Innovation and Economic Progress in the Ancient World." *Economic History Review* 18(August):29–45.

Finley, M. I. 1973. *The Ancient Economy.* Berkeley, CA: University of California Press.

Fletcher, R. A. 1910 *Steam-Ships: The Story of Their Development to the Present Day.* Philadelphia: J. B. Lippincott.

Flinn, Michael W. 1959. "Timber and the Advance of Technology: a Reconsideration." *Annals of Science* 15(June):109–20.

Flinn, Michael W. 1978. "Technical Change as an Escape from Resource Scarcity: England in the Seventeenth and Eighteenth Centuries." In *Natural Resources in European History,* edited by William N. Parker and Antoni Maczak, 139–59. Washington, DC: Resources for the Future.

Foley, Vernard. 1983. "Leonardo, the Wheel Lock, and the Milling Process." *Technology and Culture* 24(July):399–427.

Forbes, R. J. 1956a. "Metallurgy." In *A History of Technology.* Vol. 2, *The Mediterranean Civilizations and the Middle Ages, 700 B.C. to A.D. 1500,* edited by Charles Singer et al., 41–80. New York and London: Oxford University Press.

Forbes, R. J. 1956b. "Food and Drink." In *A History of Technology.* Vol. 2, *The Mediterranean Civilizations and the Middle Ages, 700 B.C. to A.D. 1500,* edited by Charles Singer et al., 103–146. New York and London: Oxford University Press.

Forbes, R. J. 1956c. "Power." In *A History of Technology.* Vol. 2, *The Mediterranean Civilizations and the Middle Ages, 700 B.C. to A.D. 1500,* edited by Charles Singer et al., 589–628. New York and London: Oxford University Press.

Forbes, R. J. 1956d. "Hydraulic Engineering and Sanitation." In *A History of Technology.* Vol. 2, *The Mediterranean Civilizations and the Middle Ages, 700 B.C. to A.D. 1500,* edited by Charles Singer et al., 688–94. New York and London: Oxford University Press.

Forbes, R. J. 1957. "Food and Drink." In *A History of Technology.* Vol. 3, *From the Renaissance to the Industrial Revolution, 1500–1750,* edited by Charles Singer et al., 1–26. New York and London: Oxford University Press.

Forbes, R. J. 1958a. "Power to 1850." In *A History of Technology*. Vol. 4, *The Industrial Revolution, 1750–1850*, edited by Charles Singer et al., 148–67. New York and London: Oxford University Press.

Forbes, R. J. 1958b. *Man the Maker*. London: Abelard-Shuman.

Fossier, Robert. 1982. *Enfance de l'Europe*. 2 vols. Paris: Presses Universitaires de France.

Foster, John. 1987. *Evolutionary Macroeconomics*. London: Allen and Unwin.

Frankel, Marvin. 1955. "Obsolescence and Technological Change." *American Economic Review* 45(June):298–319.

Friedel, Robert. 1979. "Parkesine and Celluloid: The Failure and Success of the First Modern Plastic." *History of Technology* 4:45–62.

Fussell, G. E. 1958. "Agriculture: Techniques of Farming." In *A History of Technology*. Vol. 4, *The Industrial Revolution, 1750–1850*, edited by Charles Singer et al., 13–43. New York and London: Oxford University Press.

Futia, C. A. 1980. "Schumpeterian Competition." *Quarterly Journal of Economics* 94:675–95.

Gernet, Jacques. 1982. *A History of Chinese Civilization*. Cambridge: Cambridge University Press.

Gernsheim, Helmut, and Gernsheim, Alison. 1958. "The Photographic Arts: Photography." In *A History of Technology*. Vol. 5, *The Late Nineteenth Century*, edited by Charles Singer et al., 716–33. New York and London: Oxford University Press.

Gerschenkron, Alexander. 1966. "The Discipline and I." *Journal of Economic History* 27(December):443–59.

Gibbs, F. W. 1957. "Invention in Chemical Industries." In *A History of Technology*. Vol. 3, *From the Renaissance to the Industrial Revolution, 1500–1750*, edited by Charles Singer et al., 676–708. New York and London: Oxford University Press.

Gibbs, F. W. 1958. "Extraction and Production of Metals." In *A History of Technology*, Vol. 4, *The Industrial Revolution, 1750–1850*, edited by Charles Singer et al., 118–47. New York and London: Oxford University Press.

Giedion, Siegfried. 1948. *Mechanization Takes Command*. New York: W. W. Norton.

Gilbert K. R. 1958. "Machine-Tools." In *A History of Technology*, Vol. 4, *The Industrial Revolution, 1750–1850*, edited by Charles Singer et al., 417–41. New York and London: Oxford University Press.

Gilfillan, S. C. 1935. *The Sociology of Invention*. Cambridge, MA: MIT Press.

Gille, Bertrand. 1956. "Machines." In *A History of Technology*. Vol. 2, *The Mediterranean Civilizations and the Middle Ages, 700 B.C. to A.D. 1500*, edited by Charles Singer et al., 629–57. New York and London: Oxford University Press.

Gille, Bertrand. 1966. *Engineers of the Renaissance*. Cambridge, MA: MIT Press.

Gille, Bertrand. 1969a. "The Medieval Age of the West, Fifth Century to 1350." In *A History of Technology and Invention*. Vol. 1, *The Origins of Technological Civilization*, edited by Maurice Daumas, 422–572. New York: Crown.

Gille, Bertrand. 1969b. "The Fifteenth and Sixteenth Centuries in the Western World." In *A History of Technology and Invention.* Vol. 2, *The First Stages of Mechanization, 1450–1725,* edited by Maurice Daumas, 16–148. New York: Crown.

Gille, Bertrand. 1978. *Histoires des Techniques: Technique et Civilisations, Technique et Sciences.* Paris: Editions Gallimard.

Gille, Bertrand. 1979. "The Evolution of Metallurgy." In *A History of Technology and Invention.* Vol. 3, *The Expansion of Mechanization, 1725–1860,* edited by Maurice Daumas, 527–53. New York: Crown.

Gille, Paul. 1969a. "Sea and River Transportation." In *A History of Technology and Invention.* Vol. 2: *The First Stages of Mechanization, 1450–1725,* edited by Maurice Daumas, 361–436. New York: Crown.

Gille, Paul. 1969b. "From the Traditional Methods to Steam." In *A History of Technology and Invention.* Vol. 2, *The First Stages of Mechanization, 1450–1725,* edited by Maurice Daumas, 437–63. New York: Crown.

Gillispie, Charles. 1983. *The Montgolfier Brothers and the Invention of Aviation.* Princeton, NJ: Princeton University Press.

Gimpel, J. 1976. *The Medieval Machine: The Industrial Revolution of the Middle Ages.* Harmondsworth, England: Penguin Books.

Golas, P. J. 1982. "Chinese Mining: Where Was the Gunpowder?" In *Explorations in the History of Science and Technology in China,* edited by Li Guohao, Zhang Mengwen, and Cao Tianqin, 453–58. Shanghai: Chinese Classics Publishing House.

Goldschmidt, Richard. 1940. *The Material Basis of Evolution.* New Haven, CT: Yale University Press.

Goldstone, Jack A. 1987. "Geopolitics, Cultural Orthodoxy, and Innovation." *Sociological Theory* 5(Fall):119–35.

Gordon, Robert B. 1985. "Hydrological Science and the Development of Waterpower for Manufacturing." *Technology and Culture* 26(April):204–35.

Gould, Stephen Jay. 1980a. "Is a New and General Theory of Evolution Emerging?" *Paleobiology* 6(1):119–30.

Gould, Stephen Jay. 1980b. *Ever Since Darwin.* New York: W. W. Norton.

Gould, Stephen Jay. 1981. *The Mismeasurement of Man.* New York: W. W. Norton.

Gould, Stephen Jay. 1982a. "Darwinism and the Expansion of Evolutionary Theory." *Science* 216(April):380–87.

Gould, Stephen Jay. 1982b. "The Meaning of Punctuated Equilibrium and its Role in Validating a Hierarchical Approach to Macroevolution." In *Perspectives on Evolution,* edited by Roger Milkman, 83–104. Sunderland, MA: Sinauer Publishing Co.

Gould, Stephen Jay. 1982c. "The Uses of Heresy: An Introduction to the Reprint of Goldschmidt's *The Material Basis of Evolution.*" New Haven, CT: Yale University Press.

Gould, Stephen Jay. 1987. "The Panda's Thumb of Technology." *Natural History* 1:14–23.

Grantham, George. 1984. "The Shifting Locus of Agricultural Innovation in Nineteenth Century Europe." In *Technique, Spirit and Form in the*

Making of the Modern Economies: Essays in Honor of William N. Parker, edited by Gary Saxonhouse and Gavin Wright, 191–214. Greenwich, CT: JAI Press.

Greene, Kevin. 1986. *The Archaeology of the Roman Economy.* Berkeley and Los Angeles: University of California Press.

Griffiths, Dot. 1985. "The Exclusion of Women from Technology." In *Smothered by Invention: Technology in Women's Lives* edited by Wendy Faulkner and Erik Arnold. London and Sydney: Pluto Press.

Guha, Ashok. 1981. *An Evolutionary View of Economic Growth.* Oxford: The Clarendon Press.

Guilmartin, John F., Jr. 1988. "The Second Horseman: War, the Primary Causal Agent in History." Paper presented to the Texas A&M University conference on "What is the Engine of History."

Guohao, L., Mengwen, Z., and Tianqin, C. 1982. *Explorations in the History of Science and Technology in China.* Shanghai: Chinese Classics Publishing House.

Gutmann, Myron P. 1988. *Toward the Modern Economy: Early Industry in Europe, 1500–1800.* New York: Alfred A. Knopf.

Habakkuk, H. J. 1962. *American and British Technology in the Nineteenth Century.* Cambridge: Cambridge University Press.

Haber, L. F. 1958. *The Chemical Industry during the Nineteenth Century.* Oxford: The Clarendon Press.

Hacker, Barton. 1968. "Greek Catapults and Catapult Technology: Science, Technology and War in the Ancient World." *Technology and Culture* 9(January):34–50.

Hacker, Barton. 1977. "The Weapons of the West: Military Technology and Modernization in Nineteenth Century China and Japan." *Technology and Culture* 18(January):43–55.

Hall, A. Rupert. 1967. "Early Modern Technology to 1600." In *Technology in Western Civilization.* Vol. 1. Edited by Melvin Kranzberg and Carroll W. Pursell, Jr., 79–106. New York: Oxford University Press.

Hall, A. Rupert. 1978. "On Knowing and Knowing how to. . . ." *History of Technology* 3:91–104.

Hall, Marie Boas. 1976. "The Strange Case of Aluminum." *History of Technology* 1:143–57.

Harley, C. Knick. 1971. "The Shift from Sailing Ships to Steamships, 1850–1890: A Study in Technological Change and its Diffusion." In *Essays on a Mature Economy: Britain After 1840,* edited by Donald N. McCloskey, 215–23. London: Methuen.

Harley, C. Knick. 1973. "On the Persistence of Old Techniques: The Case of North American Wooden Shipbuilding." *Journal of Economic History* 33(June):372–89.

Harley, C. Knick. 1988. "Ocean Freight Rates and Productivity 1740–1913: The Primacy of Mechanical Invention Reaffirmed." *Journal of Economic History* 48(December):851–76.

Harms, Robert. 1978. *Games Against Nature.* Cambridge: Cambridge University Press.

Harris, Marvin. 1977. *Cannibals and Kings: The Origins of our Cultures.* New York: Vintage.

Harrison, J. A. 1972. *The Chinese Empire.* New York: Harcourt Brace.

Hartwell, Robert M. 1971. "Historical Analogism, Public Policy, and Social Science in Eleventh- and Twelfth Century China." *American Historical Review* 76(June):690–727.

Headrick, Daniel R. 1981. *The Tools of Empire.* New York: Oxford University Press.

Headrick, Daniel R. 1988. *The Tentacles of Progress.* New York: Oxford University Press.

Headrick, Daniel R. 1990. *The Invisible Weapon: Telecommunications and International Politics, 1851–1945.* New York: Oxford University Press.

Heertje, Arnold. 1983. "Can We Explain Technical Change?" In *The Trouble with Technology,* edited by S. McDonald, D. McL. Lamberton, and T. Mandeville, 37–49. New York: St. Martin's Press.

Hicks, J. R. 1969. *A Theory of Economic History.* Oxford: Oxford University Press.

Hill, Donald R. 1977. "The Banu Musà and their 'Book of Ingenious Devices.' " *History of Technology* 2:39–76.

Hill, Donald R. 1984a. "Information on Engineering in the Works of Muslim Geographers." *History of Technology* 9:127–42.

Hill, Donald R. 1984b. *A History of Engineering in Classical and Medieval Times.* London: Croom Helm.

Hills, Richard L. 1970. *Power in the Industrial Revolution.* Manchester: Manchester University Press.

Hills, Richard L. 1979. "Hargreaves, Arkwright, and Crompton: Why Three Separate Inventors?" *Textile History* 10:114–26.

Hindle, Brooke. 1981. *Emulation and Invention.* New York: New York University Press.

Hirsch, Fred. 1976. *Social Limits to Growth.* Cambridge, MA: Harvard University Press.

Hobsbawm, Eric J. 1968. *Industry and Empire.* Harmondsworth, England: Penguin Books.

Hodges, H. 1970. *Technology in the Ancient World.* London: Allen Lane, Penguin Press.

Hohenberg, Paul. 1967. *Chemicals in Western Europe, 1850–1914.* Chicago: Rand-McNally.

Hollister-Short, G. J. 1976. "Leads and Lags in Late Seventeenth Century English Technology." *History of Technology* 1:159–83.

Hollister-Short, G. J. 1985. "Gunpowder and Mining in Sixteenth- and Seventeenth Century Europe." *History of Technology* 10:31–66.

Holmyard, E. J. 1958a. "The Chemical Industry: Developments in Chemical Theory and Practice." In *A History of Technology.* Vol. 4: *The Industrial Revolution, 1750–1850,* edited by Charles Singer et al. 214–29. New York and London: Oxford University Press.

Holmyard, E. J. 1958b. "Dyestuffs in the Nineteenth Century." In *A History of Technology* Vol. 5, *The Late Nineteenth Century,* edited by Charles Singer et al., 257–83. New York and London: Oxford University Press.

Hounshell, David A. 1975. "Elisha Gray and the Telephone." *Technology and Culture* (April):133–61.

Hounshell, David A. 1984. *From the American System to Mass Production, 1800–1932.* Baltimore: The Johns Hopkins Press.

Howard, Robert. 1978. "Interchangeable Parts Reexamined: The Private Sector of the American Arms Industry on the Eve of the Civil War." *Technology and Culture* 19(October):633–49.

Huang, Philip C. C. 1985. *The Peasant Economy and Social Change in North China.* Stanford, CA: Stanford University Press.

Huard, Pierre et al. 1969. "The Techniques of the Ancient Far East." In *A History of Technology and Invention.* Vol. 1, *The Origins of Technological Civilization,* edited by Maurice Daumas, 216–59. New York: Crown.

Hucker, Charles O. 1975. *China's Imperial Past.* Stanford, CA: Stanford University Press.

Hughes, Jonathan R. T . 1986. *The Vital Few.* 2d ed. New York: Oxford University Press.

Hughes, J. Donald. 1975. *Ecology in Ancient Civilizations.* Albuquerque: University of New Mexico Press.

Hughes, Thomas P. 1983. *Networks of Power Electrification in Western Society, 1880–1930.* Baltimore: Johns Hopkins University Press.

Hughes, Thomas P. 1987. "The Evolution of Large Technological Systems." In *The Social Construction of Technological Systems,* edited by Wiebe E. Bijker, Thomas P. Hughes, and Trevor J. Pinch, 51–82. Cambridge, MA: MIT Press.

Hull, David L. 1988a. "A Mechanism and its Metaphysics: An Evolutionary Account of the Social and Conceptual Development of Science." *Biology and Philosophy* 3:123–55.

Hull, David L. 1988b. *Science as a Process.* Chicago: University of Chicago Press.

Hull, David L. 1988c. "Progress in Ideas of Progress." In *Evolutionary Progress,* edited by Matthew Nitecki, 27–48. Chicago: University of Chicago Press.

Hunter, Dard. 1930. *Papermaking through Eighteen Centuries.* New York: William Edwin Rudge.

Hyde, Charles. 1977. *Technological Change and the British Iron Industry.* Princeton, NJ: Princeton University Press.

Jacob, Margaret C. 1988. *The Cultural Meaning of the Scientific Revolution.* New York: Alfred A. Knopf.

Jewkes, J., Sawers, D., and R. Stillerman. 1969. *The Sources of Invention.* 2d ed. New York: Norton.

Jones, Eric L. 1981. *The European Miracle.* Cambridge: Cambridge University Press.

Jones, Eric L. 1988. *Growth Recurring: Economic Change in World History.* Oxford: The Clarendon Press.

Jones, Eric L. 1989. "The Real Question about China: Why Was the Song Economic Achievement Not Repeated." unpublished working paper, Latrobe University.

Jones, Stephen R. G. 1984. *The Economics of Conformism.* Oxford: Basil Blackwell.

Jope, E. M. 1956a. "Vehicles and Harness." In *A History of Technology.* Vol. 2, *The Mediterranean Civilizations and the Middle Ages, 700 B.C. to A.D. 1500,* edited by Charles Singer et al., 537–62. New York and London: Oxford University Press.

Jope, E. M. 1956b. "Agricultural Implements." In *A History of Technology.* Vol. 2, *The Mediterranean Civilizations and the Middle Ages, 700 B.C. to A.D. 1500,* edited by Charles Singer et al., 81–102. New York and London: Oxford University Press.

Kahn, Arthur D. 1970. "Greek Tragedians and Science and Technology." *Technology and Culture* 11(April):133–62.

Kamien, Morton I., and Schwartz, Nancy L. 1982. *Market Structure and Innovation.* Cambridge: Cambridge University Press.

Kanefsky, J., and J. Robey. 1980. "Steam Engines in Eighteenth Century Britain: A Quantitative Assessment." *Technology and Culture* 21(April):161–86.

Kaufer, Erich. 1989. *The Economics of the Patent System.* Chur, Switzerland: Harwood Academic Publishers.

Klemm, Friedrich 1964. *A History of Western Technology.* Cambridge, MA: MIT Press.

Krantz, John C. 1974. *Historical Medical Classics Involving New Drugs.* Baltimore: Williams and Wilkins.

Kreutz, Barbara M. 1973. "Mediterranean Contributions to the Medieval Mariner's Compass." *Technology and Culture* 14(July):367–83.

Kuhn, Dieter. 1988. *Textile Technology: Spinning and Reeling.* In *Science and Civilization in China,* Vol. 6, part 2, edited by Joseph Needham. Cambridge: Cambridge University Press.

Kuhn, Thomas S. 1969. "Comment on Folke Dovring." *Comparative Studies in Society and History* 11(October):426–30.

Kuhn, Thomas S. 1977. *The Essential Tension: Selected Studies in Scientific Tradition and Change.* Chicago: University of Chicago Press.

Kuran, Timur. 1988. "The Tenacious Past: Theories of Personal and Collective Conservatism." *Journal of Economic Behavior and Organization* 10:143–71.

Lach, Donald F. 1977. *Asia in the Making of Europe.* Vol. 2, *A Century of Wonder.* Chicago: University of Chicago Press.

Landels, J. G. 1978. *Engineering in the Ancient World.* Berkeley, CA: University of California Press.

Landes, David. 1969. *The Unbound Prometheus.* Cambridge: Cambridge University Press.

Landes, David. 1983. *Revolution in Time.* Cambridge, MA: Harvard University Press.

Landes, David. 1986. "What Do Bosses Really Do?" *Journal of Economic History* 46(September):585–623.

Landström, Björn. 1961. *The Ship.* Garden City, NY:: Doubleday.

Lane, Frederic C. 1963. "The Economic Meaning of the Compass." *American Historical Review* 68(April):605–17.

Langdon, John. 1986. *Horses, Oxen, and Technological Innovation.* Cambridge: Cambridge University Press.

Langrish, J. et al. 1972. *Wealth from Knowledge: Studies of Innovation in Industry.* London: Macmillan.

Laszlo, Ervin. *Evolution: The Grand Synthesis.* Boston: New Science Library.

Lazonick, William. 1979. "Industrial Relations and Technical Change: the Case of the Self-acting Mule." *Cambridge Journal of Economics* 3:231–62.

Lazonick, William. 1981. "Production Relations, Labor Productivity, and Choice of Technique: British and U.S. Spinning." *Journal of Economic History* 41(September):491–516.

Lazonick, William. 1986. "The Cotton Industry." In *The Decline of the British Economy,* edited by Bernard Elbaum and William Lazonick, 18–50. Oxford: The Clarendon Press.

Lazonick, William. 1987. "Theory and History in Marxian Economics." In *The Future of Economic History,* edited by Alexander J. Field, 255–312. Boston: Kluwer-Nijhoff.

Lee, Desmond. 1973. "Science, Philosophy, and Technology in the Greco-Roman World." *Greece and Rome* 2d ser., 20(April):65–78 and (October):180–93.

LeGoff, Jacques. 1980. *Time, Work, and Culture in the Middle Ages.* Chicago: University of Chicago Press.

Leighton, Albert C. 1972. *Transport and Communication in Early Medieval Europe.* New York: Barnes and Noble.

Lethbridge, T. C. 1956. "Shipbuilding." In *A History of Technology.* Vol. 2, *The Mediterranean Civilizations and the Middle Ages, 700 B.C. to A.D. 1500,* edited by Charles Singer et al., 563–88. New York and London: Oxford University Press.

Lewontin, Richard C. 1974. *The Genetic Basis of Evolutionary Change.* New York: Columbia University Press.

Lewontin, Richard C. 1982. *Human Diversity.* New York: Scientific American Books.

Lewis, Bernard. 1982. *The Muslim Discovery of Europe.* New York: W. W. Norton.

Lilley, S. 1965. *Men, Machines, and History.* New York: International Publishers.

Lindert, Peter H., and Trace, Keith. 1971. "Yardsticks for Victorian Entrepreneurs." In *Essays on a Mature Economy: Britain After 1840,* edited by Donald N. McCloskey, 239–74. London: Methuen.

Machabey, Armand. 1969. "Techniques of Measurement." In *A History of Technology and Invention.* Vol. 2, *The First Stages of Mechanization, 1450–1725,* edited by Maurice Daumas, 306–343. New York: Crown.

MacLeod, Christine. 1988. *Inventing the Industrial Revolution.* Cambridge: Cambridge University Press.

MacMullen, Ramsay. 1988. *Corruption and the Decline of Rome.* New Haven, CT: Yale University Press.

Mann, Julia De L. 1958. "The Textile Industry: Machinery for Cotton, Flax,

Wool, 1760–1900." In *A History of Technology,* Vol. 4, *The Industrial Revolution, 1750–1850,* edited by Charles Singer et al., 277–307. New York and London: Oxford University Press.

Mantoux, Paul. [1905] 1961. *The Industrial Revolution in the Eighteenth Century.* New York: Harper and Row.

Manuel F. 1938. "The Luddite Movement in France." *Journal of Modern History* 10:180–211.

Marshall, Alfred. 1919. *Industry and Trade.* London: Macmillan.

Marshall, Alfred. [1890] 1930. *Principles of Economics.* London: Macmillan.

Matthews, Derek. 1987. "The Technical Transformation of the Late Nineteenth Century Gas Industry." *Journal of Economic History* 48(December):967–80.

Maynard Smith, John. 1972. *On Evolution.* Edinburgh: Edinburgh University Press.

Maynard Smith, John. 1988. *Did Darwin Get it Right?* New York: Chapman and Hall.

Mayr, Ernst. 1970. *Population, Species, and Evolution.* Cambridge, MA: Belknap.

Mayr, Ernst. 1988. *Toward a New Philosophy of Biology.* Cambridge, MA: Belknap.

Mayr, Otto. 1976. "The Science-Technology Relationship as a Historiographical Problem." *Technology and Culture* 17(October):663–72.

McCloskey, Donald N. 1981. "The Industrial Revolution 1780–1860: A Survey." In *The Economic History of Britain Since 1700,* Vol. 1, edited by Roderick Floud and Donald N. McCloskey, 103–27.

McCloy, Shelby T. 1952. *French Inventions of the Eighteenth Century.* Lexington, KY: University of Kentucky Press.

McLaughlin, Charles C. 1967. "The Stanley Steamer: a Study in Unsuccessful Innovation." In *Explorations in Enterprise,* edited by Hugh G. J. Aitken, 259–72. Cambridge, MA: Harvard University Press.

McNeill, W. H. 1982. *The Pursuit of Power.* Chicago; University of Chicago Press.

Merton, Robert K. [1938] 1970. *Science, Technology and Society in Seventeenth Century England.* Reprint ed. New York: Fertig.

Merton, Robert K. 1973. *The Sociology of Science.* Chicago: University of Chicago Press.

Metzger, Thomas. 1979. "On the Historical Roots of Economic Modernization in China: The Increasing Differentiation of the Economy from the Polity During Late Ming and Early C'hing Times." In *Modern Chinese Economic History,* edited by C. Hou and T. Yu, 3–14. Taipei: Institute of Economics, Academia Sinica.

Miller, Harry. 1980. "Potash from Wood Ashes: Frontier Technology in Canada and the U.S." *Technology and Culture* 21(April):187–208.

Minchinton, W. E. 1979. "Early Tide Mills: Some Problems." *Technology and Culture* 20(October):777–86.

Mokyr, Joel, ed. 1985. *The Economics of the Industrial Revolution.* Totowa, NJ: Rowman and Allanheld.

Mokyr, Joel. 1991. "Was There a British Industrial Evolution?" in *Research*

In Economic History, Supplement IV, Joel Mokyr: *The Vital One: Essays Presented to Jonathan R. T. Hughes,* 253–286. Greenwich, CT: JAI Press.

Mokyr, Joel. 1990b. *Twenty Five Centuries of Technological Change: An Historical Survey.* Chur, Switzerland: Harwood Academic Publishers.

Molenda, Danuta. 1988. "Technological Innovation in Central Europe between the XIVth and the XVIIth Centuries." *Journal of European Economic History* 17:63–84.

Morison, Elting E. 1966. *Men, Machines, and Modern Times,* Cambridge, MA: MIT Press.

Morris, T. N. 1958. "Management and Preservation of Food." In *A History of Technology.* Vol. 5, *The Late Nineteenth Century,* edited by Charles Singer et al., 26–52. New York and London: Oxford University Press.

Mowery, David. 1986. "Industrial Research, 1900–1950." In *The Decline of the British Economy,* edited by Bernard Elbaum and William Lazonick, 189–222. Oxford: The Clarendon Press.

Multhauf, Robert P. 1967. "Industrial Chemistry in the Nineteenth Century." In *Technology in Western Civilization,* Vol. 1, edited by Melvin Kranzberg and Carroll W. Pursell, Jr., 468–88. New York: Oxford University Press.

Multhauf, Robert P. 1971. "The French Crash Program for Saltpeter Production, 1776–94." *Technology and Culture* 12(April):163–81.

Mumford, Lewis. [1934] 1963. *Technics and Civilization.* New York: Harcourt, Brace & World.

Munro, John. 1988. "Textile Technology." In *Dictionary of the Middle Ages,* Vol. 11, edited by Joseph R. Strayer, 693–711. New York: Scribner.

Musson, A. E. 1972. *Science, Technology and Economic Growth in the Eighteenth Century.* London: Methuen.

Musson, A. E. 1975a. "Joseph Whitworth and the Growth of Mass-Production Engineering." *Business History* 17(July):109–49.

Musson, A. E. 1975b. "Continental Influences on the Industrial Revolution in Great Britain." In *Great Britain and Her World: Essays in Honor of W. O. Henderson,* edited by Barrie M. Ratcliffe, 71–85. Manchester: Manchester University Press.

Musson, A. E. 1981. "British Origins." In *Yankee Enterprise: The Rise of the American System of Manufactures,* edited by Otto Mayr and Robert C. Post, 25–48. Washington, DC: The Smithsonian Institution.

Musson, A. E., and Robinson, Eric. 1969. *Science and Technology in the Industrial Revolution.* Manchester: Manchester University Press.

Naish, George. 1958. "Ship-Building." In *A History of Technology.* Vol. 4, *The Industrial Revolution, 1750–1850,* edited by Charles Singer et al., 574–95. New York and London: Oxford University Press.

Needham, Joseph. 1959. *Mathematics and the Sciences of Heaven.* In *Science and Civilization in China,* Vol. 3, edited by Joseph Needham. Cambridge: Cambridge University Press.

Needham, Joseph. 1964. *The Development of Iron and Steel Technology in China.* Cambridge: W. Heffer.

Needham, Joseph. 1965. *Physical and Physical Technology: Mechanical Engi-*

neering. In *Science and Civilization in China,* Vol. 4, part 2, edited by Joseph Needham. Cambridge: Cambridge University Press.

Needham, Joseph. 1969. *The Grand Titration.* Toronto: University of Toronto Press.

Needham, Joseph. 1970. *Clerks and Craftsmen in China and the West.* Cambridge: Cambridge University Press.

Needham, Joseph. 1975. "History and Human Values: A Chinese Perspective for World Science and Technology." Occasional Papers, Centre for East Asian Studies, McGill University, Montreal.

Needham, Joseph. 1981. *Science in Traditional China.* Cambridge, MA: Harvard University Press.

Nef, John U. 1950. *War and Human Progress.* Cambridge, MA: Harvard University Press.

Nef, John U. 1957. "Coal Mining and Utilization." In *A History of Technology.* Vol. 3: *From the Renaissance to the Industrial Revolution, 1500–1750,* edited by Charles Singer et al., 72–88. New York and London: Oxford University Press.

Nef, John U. 1964. *The Conquest of the Material World.* Chicago: University of Chicago Press.

Nelson, Richard R. 1987. *Understanding Technical Change as an Evolutionary Process.* Amsterdam: North Holland.

Nelson, Richard R., and Winter, Sidney. 1982. *An Evolutionary Theory of Economic Change.* Cambridge, MA: Belknap.

North, Douglass C. 1981. *Structure and Change in Economic History.* New York: W. W. Norton.

North, Douglass C. 1984. "Government and the Cost of Exchange in History." *Journal of Economic History* 44(June):255–64.

North, Douglass C., and Thomas, Robert P. 1973. *The Rise of the Western World.* Cambridge: Cambridge University Press.

Nye, John V. (1991). "Lucky Fools and Cautious Businessmen: Entrepreneurship and the Measurement of Entrepreneurial Failure." In *Research In Economic History, Supplement IV,* Joel Mokyr: *The Vital One: Essays Presented to Jonathan R. T. Hughes,* 131–152. Greenwich, CT: JAI Press.

Oleson, John Peter. 1984. *Greek and Roman Mechanical Water-Lifting Devices: The History of a Technology.* Toronto: University of Toronto Press.

Olmstead, Alan. 1975. "The Mechanization of Reaping and Mowing in American Agriculture, 1833–1870." *Journal of Economic History* 30(June):327–52.

Olson, Mancur. 1982. *The Rise and Decline of Nations.* New Haven, CT: Yale University Press.

Ovitt, George, Jr. 1986. "The Cultural Context of Western Technology: Early Christian Attitudes Toward Manual Labor." *Technology and Culture* 27(July):477–500.

Ovitt, George, Jr. 1987. *The Restoration of Perfection: Labor and Technology in Medieval Culture.* New Brunswick, NJ: Rutgers University Press.

Pacey, Arnold. 1975. *The Maze of Ingenuity: Ideas and Idealism in the Development of Technology.* New York: Holmes and Meier.

Pacey, Arnold. 1986. *The Culture of Technology.* Cambridge, MA: MIT Press.

Parker, William. 1984. *Europe, America, and the Wider World.* Cambridge: Cambridge University Press.

Parry, J. H. 1974. *The Discovery of the Sea.* Berkeley, CA: University of California Press.

Passer, Harold. C. 1953. *The Electrical Manufacturers, 1875–1900.* Cambridge, MA: Harvard University Press.

Patterson, R. 1956. "Spinning and Weaving." In *A History of Technology.* Vol. 2, *The Mediterranean Civilizations and the Middle Ages, 700 B.C. to A.D. 1500,* edited by Charles Singer et al., 191–220. New York and London: Oxford University Press.

Patterson, R. 1957. "Spinning and Weaving." In *A History of Technology.* Vol. 3, *From the Renaissance to the Industrial Revolution, 1500–1750,* edited by Charles Singer et al., 151–180. New York and London: Oxford University Press.

Paulinyi, Akos. 1986. "Revolution and Technology." In *Revolution in History,* edited by Roy Porter and Mikulas Teich. Cambridge: Cambridge University Press.

Payen, Jacques, and Jean Pilisi. 1979. "Weaving and Mechanical Finishing." In *A History of Technology and Invention.* Vol. 3, *The Expansion of Mechanization, 1725–1860,* edited by Maurice Dumas, 599–618. New York: Crown.

Perkin, H. J. 1969. *The Origins of Modern English Society, 1780–1880.* London: Routledge and Kegan Paul.

Perkins, Dwight H. 1967. "Government as an Obstacle to Industrialization: The Case of Nineteenth Century China." *Journal of Economic History* 27(December):478–92.

Perkins, Dwight H. 1969. *Agricultural Development in China, 1368–1968.* Chicago: Aldine.

Persson, Karl Gunnar. 1988. *Pre-Industrial Growth: Social Organization and Technological Progress in Europe.* Oxford: Basil Blackwell.

Pinch, Trevor J. and Bijker, Wiebe, E. "The Social Construction of Facts and Artifacts, or How the Sociology of Science and the Sociology of Technology Might Benefit Each Other." In *The Social Construction of Technological Systems,* edited by Wiebe E. Bijker, Thomas P. Hughes, and Trevor J. Pinch, 17–50. Cambridge, MA: MIT Press.

Pleket, H. W. 1967. "Technology and Society in the Graeco-Roman World." *Acta Historiae Neerlandica* 2:1–25.

Pounds, Norman J. G., and Parker, William. 1957. *Coal and Steel in Western Europe.* London: Faber and Faber.

Price, Derek J. de Solla. 1957. "Precision Instruments to 1500." In *A History of Technology.* Vol. 3: *From the Renaissance to the Industrial Revolution, 1500–1750,* edited by Charles Singer et al., 582–619. New York and London: Oxford University Press.

Price, Derek J. de Solla. 1975. *Science Since Babylon.* 2d ed., rev. and enl. New Haven, CT: Yale University Press.

Rae, John. 1967a. "The Invention of Invention." In *Technology in Western Civilization.* Vol. 1, edited by Melvin Kranzberg and Carroll W. Pursell, Jr., 325–336. New York: Oxford University Press.

Rae, John. 1967b. "Energy Conversion." In *Technology in Western Civilization,* Vol. 1, edited by Melvin Kranzberg and Carroll W. Pursell, Jr., 336–49. New York: Oxford University Press.

Rae, John. 1967c. "The Internal Combustion Engine on Wheels" in *Technology in Western Civilization,* Vol. 2, edited by Melvin Kranzberg and Carroll W. Pursell, Jr., 119–37. New York: Oxford University Press.

Ramelli, Agostino. [1588] 1976. *The Various and Ingenious Machines of Agostino Ramelli.* Translated by Martha Teach Gnudi, annotated by Eugene S. Ferguson, New York: Dover Publications.

Randall, Adrian J. 1986. "The Philosophy of Luddism: The Case of the West of England Workers, ca. 1790–1809." *Technology and Culture* 27(January):1–18.

Reti, Ladislao. 1970. "The Double-Acting Principle in East and West." *Technology and Culture* 11(April)178–200.

Reynolds, Terry S. 1979. "Scientific Influences on Technology: The Case of the Overshot Waterwheel, 1752–54." *Technology and Culture* 20(April):270–95.

Reynolds, Terry S. 1983. *Stronger than a Hundred Men: A History of the Vertical Water Wheel.* Baltimore: Johns Hopkins University Press.

Robinson, Eric. 1972. "James Watt and the Law of Patents." *Technology and Culture* 13(April)115–39.

Ronan, Colin A., and Needham, Joseph. 1986. *The Shorter Science and Civilization in China.* Vol. 3, Cambridge: Cambridge University Press.

Rosenberg, Nathan, ed. 1969. *The American System of Manufactures: The Report of the Committee on the Machinery of the United States, 1855.* Edinburgh: Edinburgh University Press.

Rosenberg, Nathan 1976. *Perspectives on Technology.* Cambridge: Cambridge University Press.

Rosenberg, Nathan. 1982. *Inside the Black Box: Technology and Economics.* Cambridge: Cambridge University Press.

Rosenberg, Nathan, and Birdzell, L. E. 1986. *How the West Grew Rich: The Economic Transformation of the Industrial World.* New York: Basic Books.

Rosenberg, Nathan, and Vincenti, Walter G. 1978. *The Britannia Bridge: The Generation and Diffusion of Technological Knowledge.* Cambridge, MA: MIT Press.

Rostoker, William, Bronson, Bennett, and Dvorak, James. "The Cast-Iron Bells of China." *Technology and Culture* 25(October):750–67.

Rostow, W. W. 1975. *How it All Began.* New York: McGraw-Hill.

Rothstein, Natalie. 1977. "The Introduction of the Jacquard Loom to Great Britain." In *Studies in Textile History,* edited by Veronika Gervers, 281–90. Toronto: Royal Ontario Museum.

Ruse, Michael. 1986. *Taking Darwin Seriously.* Oxford: Basil Blackwell.

Ruse, Michael. 1988. "Molecules to Men: Evolutionary Biology and Thoughts of Progress." In *Evolutionary Progress,* edited by Matthew Nitecki, 97–126. Chicago: University of Chicago Press.

Ruttan, Vernon. 1971. "Usher and Schumpeter on Invention, Innovation and Technological Change." In *The Economics of Technological Change,* edited by Nathan Rosenberg, 73–85. Harmondsworth, England: Penguin Books.

Sabel, Charles, and Zeitlin, Jonathan. 1985. "Historical Alternatives to Mass Production: Politics, Markets, and Technology in Nineteenth-Century Industrialization." *Past and Present* 108(August):133–76.

Sahal, Devendra. 1981. *Patterns of Technological Innovation.* Reading, MA: Addison Wesley.

Samuel, Raphael. 1977. "Workshop of the World: Steampower and Hand Technology in mid-Victorian Britain." *History Workshop* 3(Spring):6–72.

Sandberg, L. G. 1974. *Lancashire in Decline.* Columbus, OH: Ohio State University Press.

Schelling, Thomas. 1978. *Micromotives and Macrobehavior.* New York: Norton.

Scherer, F. Michael. 1980. *Industrial Market Structure and Economic Performance.* Chicago: Rand-McNally.

Scherer, F. Michael. 1984. *Innovation and Growth.* Cambridge, MA: MIT Press.

Schiff, Eric. 1971. *Industrialization without National Patent.* Princeton, NJ: Princeton University Press.

Schivelbusch, Wolfgang. 1988. *Disenchanted Night: The Industrialization of Light in the Nineteenth Century.* Translated by Angela Davies. Berkeley, CA: University of California Press.

Schmiechen, James A. 1984. *Sweated Industries and Sweated Labor.* Urbana, IL: University of Illinois Press.

Schubert, H. R. 1958. "Extraction and Production of Metals: Iron and Steel." In *A History of Technology.* Vol. 4, *The Industrial Revolution, 1750–1850,* edited by Charles Singer et al., 99–117. New York and London: Oxford University Press.

Schumpeter, Joseph A. 1934. *The Theory of Capitalist Development.* Cambridge, MA: Harvard University Press.

Schumpeter, Joseph A. 1939. *Business Cycles.* New York: McGraw-Hill.

Schumpeter, Joseph A. 1950. *Capitalism, Socialism and Democracy.* 3d edition New York: Harper & Row.

Scott, Joan W. 1974. *The Glassworkers of Carmaux.* Cambridge, MA: Harvard University Press.

Scoville, Warren C. 1950. *Capitalism and French Glassmaking.* Berkeley, CA: University of California Press.

Sewell, William H. 1980. *Work and Revolution in France.* Cambridge: Cambridge University Press.

Shang, Hung-k'uei. 1981. "The Process of Economic Recovery, Stabilization, and its Accomplishments in the Early Ch'ing, 1681–1735." *Chinese Studies in History* 14:20–61.

Sharlin, Harold I. 1967a. "Applications of Electricity." In *Technology in Western Civilization.* Vol. 1, edited by Melvin Kranzberg and Carroll W. Pursell, Jr., 563–78. New York: Oxford University Press.

Sharlin, Harold I. 1967b. "Electrical Generation and Transmission." In *Technology in Western Civilization.* Vol. 1, edited by Melvin Kranzberg and Carroll W. Pursell, Jr., 578–91. New York: Oxford University Press.

Shorter, Edward, and Tilly, Charles. 1974. *Strikes in France, 1830–1968*. Cambridge: Cambridge University Press.

Simon, Julian. 1977. *The Economics of Population Growth.* Princeton, NJ: Princeton University Press.

Simon, J. 1983. "The Effects of Population on Nutrition and Economic Well-Being." In *Hunger and History: The Impact of Changing Food Production and Consumption Patterns on Society,* edited by R. I. Rotberg and T. K. Rabb, 215–40. Cambridge: Cambridge University Press.

Simpson, George Gaylord. 1967. *The Meaning of Evolution.* 2nd revised edition. New Haven, CT: Yale University Press.

Singer, Charles. 1957. "Epilogue: East and West in Retrospect." In *A History of Technology.* Vol. 2, *The Mediterranean Civilizations and the Middle Ages, 700 B.C. to A.D. 1500,* edited by Charles Singer et al., 753–776. New York and London: Oxford University Press.

Smith, Adam. [1776] 1976. *The Wealth of Nations.* Cannan ed. Chicago: University of Chicago Press.

Smith, Cyril Stanley. 1967a. "Metallurgy in the Seventeenth and Eighteenth Century." In *Technology in Western Civilization.* Vol. 1, edited by Melvin Kranzberg and Carroll W. Pursell, Jr., 142–67. New York: Oxford University Press.

Smith, Cyril Stanley. 1967b. "Mining and Metallurgical Production." In *Technology in Western Civilization,* Vol. 1, edited by Melvin Kranzberg and Carroll W. Pursell, Jr., 349–66. New York: Oxford University Press.

Smith, Cyril Stanley, 1967c. "Metallurgy: Science and Practice before 1900." In *Technology in Western Civilization.* Vol. 1, edited by Melvin Kranzberg and Carroll W. Pursell, Jr., pp. 592–602. New York: Oxford University Press.

Smith, Cyril Stanley. 1981. *A Search for Structure.* Cambridge, MA: MIT Press.

Smith, Cyril Stanley, and Forbes, R. J. 1957. "Metallurgy and Assaying." In *A History of Technology,* Vol. 3, *From the Renaissance to the Industrial Revolution, 1500–1750,* edited by Charles Singer et al., 27–71. New York and London: Oxford University Press.

Smith, Merritt Roe. 1977. *Harpers Ferry Armory and the New Technology.* Ithaca, NY: Cornell.

Smith, Norman A. F. 1977. "The Origins of the Water Turbine and the Invention of its Name." *History of Technology* 2:215–59.

Sokoloff, Kenneth, 1988. "Inventive Activity in Early Industrial America: Evidence from Patent Records, 1790–1846." *Journal of Economic History* 48(December):813–50.

Spengler, Oswald. 1932. *Man and Technics.* London: Allen and Unwin.

Spruytte, J. 1977. *Etudes Experimentales sur l'Attelage.* Paris: Crépin—Leblond.

Stanley, Steven M. 1981. *The New Evolutionary Timetable.* New York: Basic Books.

Stern, Bernard J. 1937. "Resistances to the Adoption of Technological Innovations." In *Technological Trends and National Policy.* Washington, D.C.: United States Government Printing Office.

Stevenson, John. 1979. *Popular Disturbances in England, 1700–1870.* New York: Longman.

Strayer, Joseph R. 1980. "Review of Lynn White, Medieval Religion and Technology." *Technology and Culture* 21(January):82–85.

Sullivan, Richard. 1989. "England's 'Age of Invention': The Acceleration of Patents and Patentable Invention during the Industrial Revolution." *Explorations in Economic History* 26(October):424–52.

Swetz, Frank J. 1987. *Capitalism and Arithmetic: The New Math of the 15th Century.* La Salle, IL: Open Court.

Szostak, Richard. 1986. "The Role of Transportation in the English Industrial Revolution." Unpublished manuscript, University of Alberta, Edmonton, Canada.

Tang, Anthony M. 1979. "China's Agricultural Legacy." *Economic Development and Cultural Change* 28(October):1–22.

Taylor, E. G. R. 1957. *The Haven-Finding Art: A History of Navigation from Odysseus to Captain Cook.* New York: Aberlard-Schuman.

Taylor, F. Sherwood, and Singer, Charles. 1956. "Pre-Scientific Industrial Chemistry." In *A History of Technology.* Vol. 2, *The Mediterranean Civilizations and the Middle Ages, 700 B.C. to A.D. 1500,* edited by Charles Singer et al., 347–73. New York and London: Oxford University Press.

Temin, Peter. 1964. *Iron and Steel in Nineteenth Century America.* Cambridge, MA: MIT Press.

Thirtle, Colin G., and Ruttan, Vernon W. 1987. *The Role of Demand and Supply in the Generation and Diffusion of Technical Change.* Chur, Switzerland: Harwood Academic Publishers.

Thomson, Ross. 1984. "The Eco-technic Process and the Development of the Sewing Machine." In *Technique, Spirit and Form in the Making of the Modern Economies: Essays in Honor of William N. Parker,* edited by Gary Saxonhouse and Gavin Wright, 243–69. Greenwich, CT: JAI Press.

Toulmin, Stephen E. 1967. "The Evolutionary Development of Natural Science." *American Scientist* 55(December):456–71.

Trescott, Martha M. 1979. "Julia B. Hall and Aluminum." In *Dynamos and Virgins Revisited: Women and Technological Change in History,* edited by Martha M. Trescott. 149–79. Metuchen, NJ: Scarecrow Press.

Tsuen-Hsuin, Tsien. 1982. "Why Paper and Printing were Invented First in China and Used Later in Europe." In *Explorations in the History of Science and Technology in China,* edited by Li Guohao, Zhang Mengwen, and Cao Tianqin, 459–70. Shanghai: Chinese Classics Publishing House.

Tsuen-Hsuin, Tsien. 1985. *Paper and Printing.* In *Science and Civilization in China.* Vol. 5, part 1, edited by Joseph Needham. Cambridge: Cambridge University Press.

Tweedale, Geoffrey. 1986. "Metallurgy and Technological Change: A Case Study of Sheffield Specialty Steel." *Technology and Culture* 27(April):189–226.

Tylecote, R. F. 1976. *A History of Metallurgy.* London: The Metals Society.

Unger, Richard. 1978. *Dutch Shipbuilding Before 1800.* Assen, The Netherlands: Van Gorcum.

Unger, Richard. 1980. *The Ship in the Medieval Economy, 600–1600.* London: Croom Helm.

Unger, Richard. 1981. "Warships and Cargo Ships in Medieval Europe." *Technology and Culture* 22(April):233–52.

Usher, Abbott P. 1920. *An Introduction to the Industrial History of England.* Boston: Houghton Mifflin.

Usher, Abbott P. 1954. *A History of Mechanical Inventions.* Cambridge, MA: Harvard University Press.

Usher, Abbott P. 1955. "Technical Change and Capital Formation." In *Capital Formation and Economic Growth.* National Bureau of Economic Research. Reprinted in *The Economics of Technological Change,* edited by Nathan Rosenberg. Harmondsworth, England: Penguin Books.

Usher, Abbott P. 1957. "Machines and Mechanisms." In *A History of Technology.* Vol. 3, *From the Renaissance to the Industrial Revolution, 1500–1750,* edited by Charles Singer et al., 324–46. New York and London: Oxford University Press.

Van Crefeld, Martin. 1989. *Technology and War.* New York: Free Press.

Veblen, Thorstein. 1914. *The Instinct of Workmanship.* New York: Macmillan.

Vitruvius, Marcus. 1962. *On Architecture.* Edited and translated by Frank Granger. Cambridge, MA: Harvard University Press.

Volti, Rudi. 1988. *Society and Technological Change.* New York: St. Martin's Press.

Von Tunzelmann, G. N. 1978. *Steam Power and British Industrialization to 1860.* Oxford: Oxford University Press.

Von Tunzelmann, G. N. 1981. "Technical Progress during the Industrial Revolution." In *The Economic History of Britain Since 1700.* Vol. 1, edited by Roderick Floud and Donald N. McCloskey, 143–63.

Wadsworth, Alfred P., and Mann, Julia De Lacy. *The Cotton Trade and Industrial Lancashire.* Manchester: Manchester University Press.

Wailes, Bernard. 1972. "Plow and Population in Temperate Europe." In *Population Growth: Anthropological Implications,* edited by Brian Spooner, 154–79. Cambridge, MA: MIT Press.

Wailes, Rex. 1957. "Windmills." In *A History of Technology.* Vol. 3, *From the Renaissance to the Industrial Revolution, 1500–1750,* edited by Charles Singer et al., 89–109. New York and London: Oxford University Press.

Watson, Andrew M. 1983. *Agricultural Innovation in the Early Islamic World.* Cambridge: Cambridge University Press.

Wells, F. A. 1958. "Hosiery and Lace." In *A History of Technology.* Vol. 5, *The Late Nineteenth Century,* edited by Charles Singer et al., 595–604. New York and London: Oxford University Press.

White, K. D. 1967. "Gallo-Roman Harvesting Machines." *Latomus: Revue d'études Latines* 16:634–47.

White, K. D. 1969. "The Economics of the Gallo-Roman Harvesting Machines." In *Hommages à Marcel Renard,* Brussels: Collection Latomus. 102:804–809.

White, K. D. 1975. *Farm Equipment of the Roman World.* Cambridge: Cambridge University Press.
White, K. D., ed. 1977. *Country Life in Classical Times.* Ithaca, NY: Cornell University Press.
White, K. D. 1984. *Greek and Roman Technology.* Ithaca, NY: Cornell University Press.
White, Lynn. 1962. *Medieval Technology and Social Change.* Oxford: Oxford University Press.
White, Lynn. 1968. *Dynamo and Virgin Reconsidered.* Cambridge, MA: MIT Press.
White, Lynn. 1972. "The Expansion of Technology, 500–1500." In *The Fontana Economic History of Europe.* Vol. 1, *The Middle Ages,* edited by Carlo M. Cipolla, 143–74. London: Collins.
White, Lynn. 1978. *Medieval Religion and Technology.* Berkeley, CA: University of California Press.
Wiet, G. 1969. "The Moslem World, Seventh to Thirteenth Century." In *A History of Technology and Invention.* Vol. 1, *The Origins of Technological Civilization,* edited by Maurice Daumas, 336–72. New York: Crown.
Wikander, Orjan. 1985. "Archaeological Evidence for Early Water-Mills: an Interim Report." *History of Technology* 10:151–79.
Wilkinson, Norman B. 1963. "Brandywine Borrowings From European Technology." *Technology and Culture* 4(Winter):1–13.
Wilkinson, Richard G. 1973. *Poverty and Progress: An Ecological Perspective on Economic Development.* New York: Praeger.
Williams, Martha W. 1988. "Infant Nutrition and Economic Growth in Western Europe from the Middle Ages to the Modern Period." Unpublished Ph.D. dissertation, Northwestern University.
Williams, Trevor I. 1958. "Heavy Chemicals." In *A History of Technology.* Vol. 5, *The Late Nineteenth Century,* edited by Charles Singer et al., 235–56. New York and London: Oxford University Press.
Wittfogel, Karl A. 1957. *Oriental Despotism: A Comparative Study of Total Power.* New Haven, CT: Yale University Press.
Woodbury, Robert S. 1972. *Studies in the History of Machine Tools.* Cambridge, MA: MIT Press.
Woodforde, John. 1970. *The Story of the Bicycle.* New York: Universe Books.
Woronoff, Denis. 1984. *L'industrie Sidérurgique en France Pendant la Révolution et l'Empire.* Paris: Ecole de Hautes Etudes en Sciences Sociales.
Wright, Gavin. 1987. "Labor History and Labor Economics." In *The Future of Economic History,* edited by Alexander J. Field, 313–48. Boston: Kluwer-Nijhoff.
Wrigley, E. A. 1987. *People, Cities and Wealth.* Oxford: Basil Blackwell.
Wrigley, Julia. 1986. "Technical Education and Industry in the Nineteenth Century." In *The Decline of the British Economy,* edited by Bernard Elbaum and William Lazonick, 162–88. Oxford: The Clarendon Press.
Wyatt, Geoffrey. 1986. *The Economics of Invention: A Study of the Determinants of Inventive Activity.* New York: St. Martin's Press.
Young, Allyn, 1928. "Increasing Returns and Economic Progress." *Economic Journal* 38(December):527–42.

Index